**The Forensic Analysis,
Comparison and Evaluation of
Friction Ridge Skin Impressions**

The Forensic Analysis, Comparison and Evaluation of Friction Ridge Skin Impressions

Dan Perkins
Metropolitan Police Service,
London, UK

Registered Offices

John Wiley & Sons, Inc., 111 River Street, Hoboken, NJ 07030, USA

John Wiley & Sons Ltd, The Atrium, Southern Gate, Chichester, West Sussex, PO19 8SQ, UK

Editorial Office

The Atrium, Southern Gate, Chichester, West Sussex, PO19 8SQ, UK

For details of our global editorial offices, customer services, and more information about Wiley products visit us at www.wiley.com.

Wiley also publishes its books in a variety of electronic formats and by print-on-demand. Some content that appears in standard print versions of this book may not be available in other formats.

Library of Congress Cataloging-in-Publication Data applied for

Hardback ISBN:9781119230892

Cover Design: Wiley

Cover Image: Courtesy of Dan Perkins

Set in 9.5/12.5pt STIXTwoText by SPi Global, Pondicherry, India

Printed and bound by CPI Group (UK) Ltd, Croydon, CR0 4YY

C9781119230892_130722

Contents

Wherever he steps, whatever he touches, whatever he leaves – even unconsciously – will serve as silent evidence against him. Not only his fingerprints and his shoeprints, but also his hair, the fibres from his clothes, the glass he breaks, the tool mark he leaves, the paint he scratches, the blood or semen that he deposits or collects – all these and more bear mute witness against him. This is evidence that does not forget. It is not confused by the excitement of the moment. It is not absent because human witnesses are. It is *factual evidence*. Physical evidence cannot be wrong; it cannot perjure itself; it cannot be wholly absent. Only in its interpretation can there be error. Only human failure to find, study, and understand it can diminish its value.

Paul L. Kirk
Crime Investigation (1985)

Acknowledgements

I am grateful to Henry Swofford, Chris Kadis and Glenn Langenburg for the time they committed to reading this work and the insights and feedback they provided.

Introduction

The skin on the undersides of the hands and feet is known as friction ridge skin. The ridges covering it are not continuous and, in places, end or split into two forming 'characteristics'. There is a wealth of evidence that the configurations of these characteristics are extremely variable and generally remain unchanged throughout life (Abraham et al., 2013; Cummins & Midlo, 1961; Faulds, 1912; Galton, 1892; Hale, 1952; Herschel, 1916; Jennings in Cummins & Midlo, 1961; Langenburg, 2011; Langenburg, 2012; Lin et al., 1982; Maceo, 2011; Monson et al., 2019; Neumann, 2013; Neumann, Evett & Skerrett, 2012; Okajima, 1967; Srihari et al., 2008; Stoney, 2001; Welcker, in Monson et al., 2019; Wertheim, 2011; Wertheim & Maceo, 2002; Wilder & Wentworth, 1918). These two factors make friction ridge skin suitable for personal identification.

When friction ridge skin contacts a surface an impression of it may be left behind, often as a result of the transfer of sweat or other material from the ridges to the surface. Impressions that are left inadvertently, for example, at a scene where a crime is alleged to have been committed, are known as 'marks'. Impressions taken intentionally from a person, for example, as a result of their arrest, are known as 'prints'.

The task of the Fingerprint Examiner is to compare a mark from a crime scene with a print from a person to provide an opinion as to how likely or unlikely it is both impressions were made by the same area of skin. The task is complicated by the fact that due to a range of factors, impressions made by the same area of skin can look very different (Figure 1).

  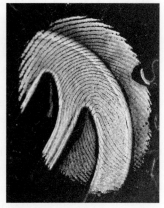

Figure 1 Though these marks look different they were all made by the same finger.

The Forensic Analysis, Comparison and Evaluation of Friction Ridge Skin Impressions, First Edition. Dan Perkins.
© 2022 John Wiley & Sons Ltd. Published 2022 by John Wiley & Sons Ltd.

The generally accepted framework for comparing impressions is known by the acronym 'ACE' as it comprises three stages: Analysis, Comparison and Evaluation (Ashbaugh, 1999). The use of ACE in forensic work was first articulated by Huber in 1959 and though the three stages are clear, what the examiner should do within each stage is not as well defined (Champod et al., 2016). This book introduces ten questions that can be used as steps to ensure all relevant aspects of each stage are considered. In addition, the appendices provide other information related to ACE and the wider aspects of the work of the fingerprint examiner.

References

Abraham, J, Champod, C, Lennard, C & Roux, C 2013, 'Modern Statistical Models for Forensic Fingerprint Examinations: A Critical Review', *Forensic Science International*, 232, pp. 131–150.

Ashbaugh, D, R 1999, *Quantitative-Qualitative Friction Ridge Analysis*, CRC Press LLC, Boca Raton, Florida.

Champod, C, Lennard, C, Margot, P & Stoilovic, M 2016, *Fingerprints and Other Ridge Skin Impressions*, Second Edition, CRC Press, Boca Raton.

Cummins, H & Midlo, C 1961, *Finger Prints, Palms and Soles*, Dover Publications, INC. New York, USA.

Faulds, H 1912, *Dactylography, or the Study of Finger-Prints*, Milner & Company, Halifax.

Galton, F 1892, *Fingerprints*, MacMillan and Co., London.

Hale, A, R 1952, 'Morphogenesis of Volar Skin in the Human Fetus', *The American Journal of Anatomy*, 91 (1), pp. 147–181.

Herschel, W, J 1916, *The Origin of Fingerprinting*, Oxford University Press, London.

Langenburg, G 2011, 'Scientific Research Supporting the Foundations of Friction Ridge Examinations', in National Institute of Justice, *The Fingerprint Sourcebook*, www.nij.gov.

Langenburg, G 2012, 'A critical analysis and study of the ACE-V process', PhD thesis, University of Lausanne.

Lin, C, H, Liu, J, H, Osterburg, J, W & Nicol, J, D 1982, 'Fingerprint Comparison. 1: Similarity of Fingerprints', *Journal of Forensic Sciences*, 27 (2), pp. 290–304.

Maceo, A, V 2011, 'Anatomy and Physiology of Adult Friction Ridge Skin', in National Institute of Justice, *The Fingerprint Sourcebook*, www.nij.gov.

Monson, K, L, Roberts, M, A, Knorr, K, B, Ali, S, Meagher, S, B, Biggs, K, Blume, P, Brandelli, D, Marzioli, A, Reneau, R & Tarasi, F 2019, 'The Permanence of Friction Ridge Skin and Persistence of Friction Ridge Skin and Impressions: A Comprehensive Review and New Results', *Forensic Science International*, 297, pp. 111–131.

Neumann, C 2013, 'Statistics and Probabilities as a Means to Support Fingerprint Identification', in Ramotowski, R, S (ed.), *Lee and Gaensslen's Advances in Fingerprint Technology*, Third Edition, CRC Press, Boca Raton, Florida.

Neumann, C, Evett, I, W & Skerrett, J 2012, 'Quantifying the Weight of Evidence from a Forensic Fingerprint Comparison: A New Paradigm', *Journal of the Royal Statistical Society, A*, 175 (Part 2), pp. 371–415.

Okajima, M 1967, 'Frequency of Epidermal-Ridge Minutiae in the Calcar Area of Japanese Twins', *American Journal of Human Genetics*, 19 (5), pp. 660–673.

Srihari, S, N, Srinivasan, H & Fang, G 2008, 'Discriminability of Fingerprints of Twins', *Journal of Forensic Identification*, 58 (1), pp. 109–127.

Stoney, D, A 2001, 'Measurement of Fingerprint Individuality', in Lee, H, C & Gaensslen, R, E (eds.), *Advances in Fingerprint Technology*, Second Edition, CRC Press, Boca Raton, Florida.

Wertheim, K, V 2011, 'Embryology and morphology of friction ridge skin', Chapter 3, *The Fingerprint Sourcebook*, National Institute of Justice, www.nij.gov.

Wertheim, K & Maceo, A 2002, 'The Critical Stage of Friction Ridge and Pattern Formation', *Journal of Forensic Identification*, 52 (1), pp. 35–85.

Wilder, H, H & Wentworth, B 1918, *Personal Identification*, The Gorham Press, Boston.

Part 1

Analysis Stage

The purpose of the analysis is to gather information – firstly about the mark and then the print. Most of the information will come from the impressions themselves, but crucial information may also be found in documentation produced by those who found the mark or recorded the print.

The Forensic Analysis, Comparison and Evaluation of Friction Ridge Skin Impressions, First Edition. Dan Perkins.
© 2022 John Wiley & Sons Ltd. Published 2022 by John Wiley & Sons Ltd.

1

Does the Mark Have Evidential Value?

There are three types of marks that may be found at a crime scene: latent, visible and indented.[1]

Latent marks are the most commonly encountered and generally require development to make them visible. Latent marks on non-porous surfaces are often developed with powder applied by a brush. The powder will contrast in colour with the surface the mark is on and make the mark visible by adhering to the material it is made in. Once the mark has been developed, it will either be photographed, 'lifted' or both. Lifting involves removing the mark from the surface with adhesive tape. Once the mark is on the tape, the tape is mounted on a backing material to preserve the mark (forming what is known as a 'lift'). Latent marks on porous surfaces such as paper will usually require chemical treatment to make them visible. These treatments are not generally carried out at a crime scene, and so suitable porous surfaces that may bear latent marks will likely be taken to a laboratory. As a laboratory offers a wider range of development techniques than is available at a crime scene, other items or surfaces including non-porous ones may also be taken there to maximise the likelihood of marks being developed.

Visible (or 'patent') marks do not require development because they will usually have been made in a material (such as blood) that contrasts in colour with the surface they are on. Visible marks will usually be recorded by photography.

Indented (or 'plastic') marks are made by impressing the ridges on the skin into a soft material such as putty or clay. Like visible marks, indented marks will usually be recorded by photography.

Whether found at a crime scene or in a laboratory, each mark will be assigned a unique reference. The reference, along with a description of where the mark was located, will be recorded in a report written at the time it was found. Once the scene or laboratory examination is complete, the marks and their accompanying documentation will be sent to the fingerprint laboratory/bureau.

1.1 Continuity and Integrity

Whether a mark has value as evidence in the investigation of an offence and how much value is dependent on several factors.

First, a mark will have no evidential value whatsoever if it is not possible to demonstrate its 'continuity'. From the moment it was found to the moment it was entered as evidence in court, an unbroken chain of custody should exist to demonstrate who has had possession of the mark and when they took possession of it. The organisation that employs the examiner will have procedures

---

1 Marks are also known as 'latent prints'.

The Forensic Analysis, Comparison and Evaluation of Friction Ridge Skin Impressions, First Edition. Dan Perkins.
© 2022 John Wiley & Sons Ltd. Published 2022 by John Wiley & Sons Ltd.

that dictate how the mark gets securely from location to location and will have a system to record its movements. The chain will begin with the person who found the mark recording information about the scene the mark was found at and where it was found at that scene. This documentation is crucial in demonstrating that, for example, at some point the mark has not been accidentally or deliberately mixed up with a mark from a different scene or a mark from a different surface at the same scene. How the continuity of a mark can be demonstrated will depend on the procedures of the organisation, but part of it will likely involve making sure that the unique reference assigned to the mark is the same reference that is recorded in the report produced by the person who found it. Additionally, the image of the mark itself may reveal indicators of the surface it was found on, which can be compared to the surface it was reported to have come from. If there is a photograph of the mark *in situ* on the surface that may conclusively establish where the mark was found but if not, there may also be indicators in a lift that show whether its appearance is consistent with the surface it is reported to have come from. For instance, in addition to revealing the mark, the application of powder to a surface may result in the edges of the surface or indentations and protrusions on the surface being recorded in the lift. The dimensions, shapes and textures of these features can therefore be compared with those of the surface the lift is reported to have come from (Figures 1.1 and 1.2).

Occasionally, as in Figure 1.3, such features may indicate whether the surface they originated from was flat or curved, as may creases in the tape used to lift the mark (Figure 1.4).

Powder may also adhere to contaminants on a surface whose presence may or not be consistent with the surface the mark is reported to have come from (Figures 1.5 and 1.6).

When considering whether the appearance of the lift is consistent with the surface, it must be recognised that some surfaces may appear differently than expected (Figure 1.5). For instance, the surface may be described as wood but depending on its condition, and whether it is painted or varnished, wood grain may or may not appear in a lift from a wooden surface. Also, whilst the use of one type of powder may result in a mark and features of the surface being recorded in the lift, the use of a different type of powder on the same surface may only result in the mark being

Figure 1.1 This mark was lifted from the reflective surface of the rear-view mirror of a car. The dimensions of the shape around the mark are consistent with those of a rear-view mirror, as is the appearance of the texture around the edge of the reflective surface. The 'clean' border around the inner side of the textured edge was caused by the lifting tape being unable to contact the point where the plastic frame meets the glass, due to the difference in the height of those surfaces (this has also created unnatural straight edges along the bottom right of the mark). Additionally, the size, ridge flow and orientation of the mark are consistent with it having been made by a left thumb – a digit that could commonly be expected to contact this surface (these indicators are covered in Chapters 6 and 7).

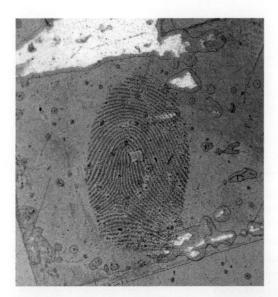

Figure 1.2 This mark was lifted from the painted black surface of an 'L' shaped piece of metal. The mark was developed with silver-coloured powder, and so the image has been inverted to make the ridges appear dark against a lighter background, as they would do in print on a fingerprint form. The edges of the metal can be seen in the lift forming the 'L' shape, and some of the black paint has been lifted off by the tape (the fragments of paint are the irregular shaped white areas). If required, this unique damage to the surface may allow the mark to be married up with a precise point on the surface it came from in the future (assuming the metal is in the same condition it was when the mark was lifted). The dark lines that outline the paint fragments mark the point where the tape rises slightly to accommodate the thickness of the paint. The lines can also be seen around tiny dots throughout the image that are likely small particles of dirt or other debris that were on the surface. The rectangular void near the centre of the mark was caused by recessed damage to the surface.

Figure 1.3 These marks were lifted from a mug. The powder used to develop the marks has also adhered to raised graphics and text on the mug. The marks are adjacent to the top edge of the mug, which appears curved in the lift, as do the horizontal lines that form part of the graphics. This effect is commonly seen when the lifting tape has been wrapped around a curved surface. Additionally, the area of the digits that has contacted the surface and the orientation of the marks is also consistent with the appearance of marks made on this type of surface.

Figure 1.4   This palm mark was lifted from a light bulb. The irregular shape of the mark and the voids running into it are consistent with it coming from this type of surface. The voids are caused by the rectangular tape creasing and folding over on itself as it is applied to a spherical surface.

Figure 1.5   These marks were lifted from the tiled wall of a bathroom. The examiner may expect the appearance of a lift from a smooth and likely clean surface such as this to be relatively featureless. However, in this case, the tiles had been wiped with a wet cloth, and the powder used to develop the ridge detail has also adhered to the dried watermarks left behind by the cloth. The application of powder has also resulted in cracks in the tiles being recorded in the lift as diagonal white lines.

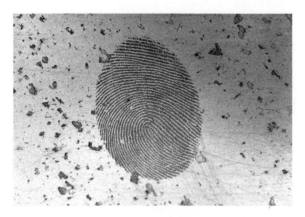

Figure 1.6   This mark was lifted from the outside glass surface of a window. The powder used to develop the ridge detail has also developed some marks left by rain.

recorded in the lift. However, the absence of features that may be expected or the presence of those that are not expected may require further investigation as it could indicate the donor of the mark has been wrongly implicated in the offence by a third party (see Appendix 1 on Fabrication, Transplantation and Forgery). This could include examining the original surface or a similar one and lifting marks from it to compare the appearance of those lifts with the original lift.

The mark may also have no evidential value if it cannot be demonstrated that it is in the same condition it was in when it was found, i.e. that it has not been altered either accidentally or deliberately since that time. This is known as the 'integrity' of the mark, and how it is demonstrated will depend on organisational procedures and the method used to retrieve or record the mark. For example, after a mark has been lifted from the surface and secured to backing material, the person who found it may cut away the excess tape on either side of the mark and sign across the cut edges. Therefore, by checking that the signatures are in place and that they still line up across the tape and backing material, it can be established that the tape has not been removed (which would be necessary to interfere with the mark). Alternatively, where a photograph of the mark *in situ* on the surface exists, that image can be compared with the image the examiner used.

If the continuity and integrity of the mark cannot be established, there may be little point in continuing with the analysis.

1.2 The Most Evidentially Valuable Mark

A scene or laboratory examination may result in the discovery of numerous marks, some of which may have more evidential value than others because they are more likely to have been made by someone involved in the offence. For example, if someone is stabbed in the street, a mark found on the knife used in the assault will likely be seen as having more evidential value than a mark found on a car parked where the attack took place. However, this is not to say that the mark on the car is not valuable. It could be the only mark that was made by the perpetrator as the mark on the knife may turn out to have been made by the victim, someone who rendered aid to the victim or someone who touched the knife before the offence. However, at the time, the mark on the knife would likely be considered the most evidentially valuable, and this is important as where possible the examiner should begin their work with the most evidentially valuable mark. It may be necessary to compare every mark that has been found, but starting with the one most likely to have been made by a person involved in the offence will increase the likelihood of identifying someone involved as quickly as possible, particularly in cases involving tens or hundreds of marks.

It will often be possible to establish which mark or marks are likely to be the most evidentially valuable using the documentation produced by those who found the marks as it may describe the offence type, how that offence was believed to have been perpetrated as well as revealing the surface the mark was found on and its position on that surface. For example, some marks may have been found on fixed surfaces and some on portable surfaces. Marks on fixed surfaces can provide evidence that the donor was likely to have been in a particular location (for example, inside a house that was burgled), whereas marks on portable surfaces usually cannot. However, marks found on portable surfaces can also be evidentially valuable if, for example, they are found on items used to commit an offence (like a knife), items that were not present at the scene prior to an offence or items that have been moved during an offence.

As well as the type of surface, the location of that surface can also indicate the evidential value of any marks on it. For instance, marks found around the point of entry or exit to premises that have been burgled may be in an area that the burglar is likely to have touched. Similarly, marks found inside a vehicle on the driver's side may indicate who the likely driver of the vehicle was.

If multiple marks are found on the same surface, their location on that surface can be significant. For example, the cowling that surrounds the steering column of a car may be removed to facilitate the theft of the car, and marks may be found on both the inner and outer surfaces of the cowling. Marks on the outer surface could have been left by anyone who had access to the car, but marks on the inner surface (because it is not usually accessible) may be more evidentially valuable.

A mark's orientation can also indicate its evidential value (see Chapter 6 for information on determining the orientation of a mark). For example, a mark found on the counter of a shop that has been robbed may have little evidential value as many people could have legitimately touched that surface. However, if the orientation of the mark indicates it was most likely made by a donor who was behind the counter, then its evidential value may be significant. Similarly, at the scene of a burglary in a shop, the orientation of marks found high up on a wall below a window may be consistent with being made by a person climbing through the window and reaching down the wall, but not consistent with that person reaching up to touch the wall whilst at ground level as a customer.

The material the marks are made in can also indicate their evidential value. For example, if blood is spilt as a result of an offence, then marks made in blood may indicate the presence of the donor around the time the offence took place.

Whilst these factors can be used by the examiner to indicate which mark was most likely to have been made by a person involved in the offence, they cannot be used to draw the conclusion that the donor of the mark was involved. For instance, in the previous scenario of the mark being found on the inner surface of the cowling, whilst the mark could have been made by someone who stole the car, it also could have been made by someone who fitted the cowling when it was installed, someone who repaired the car or who had access to the cowling after the offence took place.

If there are several marks with the same evidential value, the examiner may choose to begin their work with the clearest one or one that reveals a particularly distinctive or unusual detail as such marks may present the best opportunity for the examiner to complete their work in a timely fashion. If the appearance of some of the marks suggests they may have been made by the same area of skin, the examiner may choose to compare those marks with each other. This can be an efficient way to work as if some less clear marks can be identified as having been made by the same area of skin that made a clearer one, and the examiner may then only need to compare the clearer one.

Whilst gathering information during the analysis is crucial in ensuring the conclusion the examiner ultimately reaches is valid, some information has the potential to bias the examiner towards a particular conclusion – particularly with marks that reveal few or unclear details (Dror & Charlton, 2006). As a result, organisations will likely restrict the amount of information the examiner is exposed to. For example, there is no reason for the examiner to know the criminal history of a suspect in an offence as that information could influence their conclusion with some marks (see Appendix 3 on Bias). Most organisations do not routinely provide this information to examiners, and in general, examiners should not seek out any information that is not necessary to ensure the validity of their conclusion (Gardner et al., 2019).

References

Dror, I, E & Charlton, D 2006, 'Why Experts Make Errors', *Journal of Forensic Identification*, 56 (4), pp. 600–616.

Gardner, B, O, Kelley, S, Murrie, D, C & Blaisdell, K, N 2019, 'Do Evidence Submission Forms Expose Latent Print Examiners to Task-Irrelevant Information?', *Forensic Science International*, 297, pp. 236–242.

2

Does the Mark Require Enhancement?

The examiner's conclusions will be based on the detail they can see in the mark (and print). The more detail they can see and the clearer that detail is, the easier it will be to reach the correct conclusion.

The image of the mark from the crime scene or laboratory may require little or no enhancement if its quality was optimised by those who produced it. Where this is not the case, for example if a lift of a mark has been provided rather than a photograph of one, the examiner will have access to digital tools and filters that can significantly enhance the detail in the mark.

Any enhancements the examiner makes must be made to a 'working copy' of the image of the mark so that the original image is available for reference if required. Examiners may also be required to document any modifications they make or to use software that does this automatically so that it can be demonstrated later how the enhanced image was produced from the original.

One of the simplest ways to enhance the image of the mark is by modifying its histogram. Digital images are composed of millions of picture elements known as pixels. The examiner will usually be working with a greyscale image of the mark in which each pixel will have a value between 0 and 255 to indicate its tone (0 is black and 255 is white). The number of pixels at each tone in the mark will be represented in a graph called a histogram, and by modifying the histogram, the examiner can alter the brightness and contrast of the image to make the details easier to see.

The effects of any modification must be evaluated on the entirety of the details in the mark, as a modification that makes some details easier to see may make others more difficult to see. For example, stretching the histogram to increase the contrast will make darker pixels darker and lighter pixels lighter, so if the histogram is stretched too far, some details could become too dark or light to be seen. Modifications may also enhance other features in the image, such as dirt or damage to the surface. Where these features intersect with the mark, it can be difficult to differentiate them from actual details in the mark, so it is good practice to compare any modified image with the original before beginning a comparison to ensure that the modified image is a true representation of the original.

Most marks the examiner encounters will not require anything other than a basic modification to their brightness and contrast, but where this is not the case, there are a variety of other enhancement tools or filters available. These include ones that can sharpen the detail or remove or lessen the prominence of repeating patterns from the surface the mark is on to make the paths of the ridges easier to follow. As each image is different, the use of such tools is often conducted on 'trial and error' basis with experimentation used to establish the most effective approach for each image.

The Forensic Analysis, Comparison and Evaluation of Friction Ridge Skin Impressions, First Edition. Dan Perkins.
© 2022 John Wiley & Sons Ltd. Published 2022 by John Wiley & Sons Ltd.

On occasion, where enhancement of the image does not produce the desired result, it may be possible for a new image of the mark to be produced. For instance, if the mark still exists on the surface it was found on, it may be possible to re-photograph it to produce a new image that reveals additional or clearer detail. Alternatively, if the mark was developed with a chemical treatment, it may be possible to use an additional treatment to produce a mark that reveals more detail.

3

What Details Does the Mark Reveal?

Details in the mark can be categorised as first, second or third-level details (Ashbaugh, 1999). In general terms, first-level details describe overall ridge flow, second-level details describe the paths of specific ridges, and third-level details describe the shapes of individual ridges.

3.1 Persistence and Uniqueness

Traditionally the basis for the use of fingerprints as a means of personal identification has rested on two premises: that fingerprints are persistent and that they are unique.

The first premise supposes persistence rather than permanence because friction ridge skin does change over time. For instance, digits increase in size from childhood to adulthood, and as the body ages, ridges become flatter and new creases develop (Ashbaugh, 1999; Champod et al., 2016; Maceo, 2011; Monson et al., 2019; Schaumann & Alter, 1976). However, research examining the way skin renews itself indicates that despite those changes, the friction ridges should be persistent, and studies involving the comparison of impressions taken from the same areas of skin over long periods of time have validated this research (Faulds, 1912; Galton, 1892; Herschel, 1916; Jennings in Cummins & Midlo, 1961; Monson et al., 2019; Welcker, in Monson et al., 2019; Wilder & Wentworth, 1918). The varying support for the persistence of each level of detail is discussed in the relevant sections of this chapter.

In terms of the second premise, research carried out to study the way friction ridge skin forms has found that it is influenced by a nearly infinite number of factors, many of which are random and entirely independent from each other (Cummins & Midlo, 1961; Hale, 1952; Herschel, 1916; Lin et al., 1982; Maceo, 2011; Srihari et al., 2008; Wertheim, 2011; Wertheim & Maceo, 2002; Wilder & Wentworth, 1918). Wertheim & Maceo (2002) consider it therefore, 'completely inconceivable' that all these influences could be exactly duplicated in the skin of different people. The authors equated the likelihood of that occurring to 'an identical dump truck load of sticks being scattered twice along the same stretch of road [and] each stick [landing] in the exact same position both times'. Whilst the research that supports persistence can easily be validated by examining impressions made by the same area of skin at different times; it is not possible to demonstrate that a person's friction ridge skin is unique in the same way as this would involve comparing the impressions of everyone who is alive now with everyone who has lived before and who will live in the future. Nevertheless, it is generally accepted that because of the random and complex process involved in its formation, friction ridge skin is unique. However, it does not necessarily follow that because the skin is unique, an impression made by one person's skin will always look different from that made

The Forensic Analysis, Comparison and Evaluation of Friction Ridge Skin Impressions, First Edition. Dan Perkins.
© 2022 John Wiley & Sons Ltd. Published 2022 by John Wiley & Sons Ltd.

by another person. This is because each impression will not reveal all the detail on the skin, and different people can share similar configurations of details in small areas. As a result, different people can leave impressions revealing small areas of detail with configurations that appear very similar – as demonstrated by the 'close non-matches' found by examiners. Therefore, the basis for identification is not the uniqueness of the donor's skin but the extent to which the details in a mark allow the examiner to differentiate between people – known as the 'specificity' of the details (Langenburg, 2012). A mark revealing details with high specificity would be one in which the examiner considers it extremely unlikely those details would be found in a print made by anyone other than the donor of the mark, whereas a mark with low specificity would be one in which the examiner may expect to see similar details in prints made by different donors. The specificity of details is dependent on their variability; the greater their capacity to vary is, the less likely they are to appear similarly in two impressions unless both were made by the same donor. The same research that has been used to support the uniqueness of friction ridge skin does support its potential for extreme variability, and there have also been studies that validated this variability using statistical modelling (reviews of these models can be found in Abraham et al., 2013; Langenburg, 2011; Neumann, 2013 & Stoney, 2001). Therefore, rather than discussing the support for the uniqueness of details, this chapter considers the variability of each level of detail and the support that exists for it.

3.2 Documentation

Examiner's failure to adequately document the basis for their conclusions has been widely criticised (Campbell, 2011; Champod et al., 2016; National Research Council (NRC), 2009; The State of New Hampshire v. Richard Langill, 2007, No. 2007-300; Office of the Inspector General (OIG), 2006; President's Council of Advisors on Science and Technology (PCAST) 2016; Regina v Peter Kenneth Smith, 2011, EWCA Crim 1296). Historically, whilst the examiner has always been required to document what their conclusion was, little if any documentation has been required during the analysis and comparison stages to show what that conclusion was based on. The absence of such documentation produced at the time of the examination can be problematic for several reasons.

First, without it, the conclusion may lack transparency. For example, whilst the nature of the details in some marks may be obvious to anyone, the details in others may not. With such marks, without a contemporaneous record indicating how the examiner interpreted the details, a judge, jury or another examiner may be left with just the conclusion, without any means to understand, scrutinise and challenge the 'working out' behind it.

Second, with some marks, if the examiner does not record the details, they see prior to seeing the print, there is an increased chance that the details in the print can improperly influence how the examiner sees the details in the mark. Confirmation bias describes a tendency in people to look for, interpret or favor information that supports an existing expectation or motivation, and for the examiner, there may be various potential sources of such expectation present in their working environment (see Appendix 3 on Bias). For example, the knowledge that another examiner has reached a conclusion of identification with the mark in question can lead to the examiner seeing that conclusion as being expected or more likely than another before they begin their work. This means that with some marks, particularly those revealing ambiguous details, the examiner could be influenced to interpret those details as agreeing with those in a print even if the impressions were not made by the same person. For instance, the investigation into the erroneous identification of Brandon Mayfield found that: 'Having found as many as 10 points of unusual similarity, the FBI examiners began to 'find' additional features in [the mark] that were not really there, but rather were suggested to the examiners by features in the Mayfield prints (OIG, 2006)'. Whilst recording the details in the mark prior to seeing the print cannot prevent the examiner from being influenced by it, where it has occurred, it can make that influence more obvious and allow the examiner to consider it before reaching a conclusion. Being able to compare how they saw the details in the analysis with how they saw them in the comparison should draw the examiner's attention to any details they did not origi-nally see or any they have changed their interpretation of and allow them to reflect on whether those differences can be justified. Such documentation would also make it transparent to others that the examiners' perception of the details had changed after they saw the print, which may affect the weight they attach to the examiner's conclusion. The investigation into the Brandon Mayfield erroneous identification concluded that there is a strong possibility that had a requirement for examiners to document their analysis and comparison been in place, the error may not have occurred (OIG, 2006). There is also a study that found that examiners who documented and annotated their work made significantly fewer errors than those who did not (Langenburg, Champod & Genessay, 2012).

3.3 Complex Marks

Whilst increased documentation can lead to greater transparency and reduce the risk of a conclusion being improperly influenced by the print, most organisations will need to balance it with the corresponding increase in time and resources it requires.

Requirements for documentation generally stipulate that it must be clear and in enough detail to allow another examiner to verify the validity of the original examiner's observations (Forensic Science Regulator (FSR), 2020; Scientific Working Group on Friction Ridge Analysis, Study and Technology (SWGFAST), 2012). Studies have demonstrated that with some marks, for example those revealing ambiguous details, there can be considerable variation in how details are perceived by examiners, so one examiner cannot assume that another would see the same details in the same way (Dror et al., 2011; Evett & Williams, 1995; Langenburg, 2004; Langenburg, Champod & Wertheim, 2009; Ulery et al., 2014). Whilst increased documentation would be warranted in those cases, with marks revealing clear and abundant detail, it may not be necessary as there is less scope for interpretation and, irrespective of which details are relied upon, the conclusion may be self-evident to another examiner. Similarly, there is some research to support the idea that examin-ers are fairly resistant to the influence of bias, with marks revealing clear and abundant details (Dror et al., 2005; Dror & Charlton, 2006). As a result, organisations may target requirements for more extensive documentation at marks that are considered more 'complex'.

Definitions of complexity vary (FSR, 2013; OSAC, 2020; SWGFAST, 2013b; Triplett, 2020), but in general, a mark may be considered complex as a result of some or all of the following factors:

- The details it reveals being unclear
  Clarity is a measure of how well the details on the skin of the donor have been reproduced in the mark (Ashbaugh, 1999). Some areas of a mark may only reveal first-level details, slightly clearer areas may also reveal second-level details and the clearest areas may allow the examiner to see third-level details such as the precise shapes of the edges of the ridges. The more clearly the details have been recorded, the more confidence the examiner can have that those details will appear similarly in a print made by the same area of skin (allowing for the fact that some types of distortion can significantly alter the appearance of details without affecting their clarity – for example, see 'Direction Reversal' in Chapter 4). If details have not been recorded clearly the

examiner may have to allow for them to appear differently in a print made by the same area of skin, which will make it more difficult to decide whether details agree or not.

- Only a small amount of detail being revealed

  If a mark reveals a very large amount of detail and corresponding detail is also found in a print, it will mean that it is extremely unlikely that the impressions were not made by the same area of skin. However, as different people can have small areas of ridge detail with similar configurations, where only a small amount of detail is revealed in the mark, and corresponding details are found in a print, the examiner may need to much more carefully consider whether that agreement is coincidental rather than a result of the impressions being made by the same area of skin.

- The details it reveals being distorted

  Normally, if details in a mark and print appear differently, then those differences would support the conclusion that the impressions were not made by the same area of skin. However, as distortion can alter the appearance of details, if indicators of distortion are observed (see Chapter 4), the examiner may need to consider whether any differences could also be due to the distortion. Eldridge et al. (2020) found that examiners' perception of distortion was a significant factor in deciding whether a mark was complex or not. The study reported that if there was a lot of distortion, even if there was also a large amount of detail revealed, examiners were likely to consider the mark as being complex.

Designating a mark as complex indicates the examiner anticipates the comparison and evaluation of that mark will not be simple; however, this may not always be the case. For example, a mark may only reveal a small area of unclear ridge flow, but if that flow includes part of a distinctive pattern that is different to that revealed in a print, then that comparison and evaluation may be simple. Likewise, a non-complex mark may result in a complex comparison if, for example, the print reveals few or unclear details.

There is no objective method for assessing what is or is not a complex mark. The decision is a subjective one that is based on the details in the mark but is also influenced by factors including the knowledge and experience of the examiner. Most marks the examiner encounters will not be complex, and the complexity of those that are will usually be obvious - some research has demonstrated that examiners were often able to identify which comparisons are difficult or likely to be error-prone (Mnookin et al., 2016). However, there will be some marks with which the decision is more difficult. With such marks, the examiner may find the graph in Figure 3.1 useful.

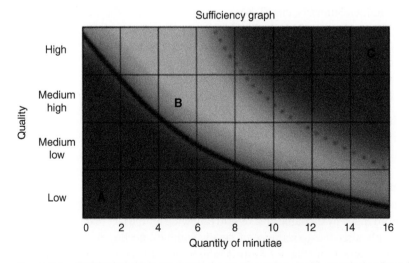

Figure 3.1 SWGFAST Sufficiency Graph. *Source:* Reproduced with permission from SWGFAST (2013a).

The graph was produced by SWGFAST based on the consensus of experienced examiners (SWGFAST, 2013a). It shows how determinations of complexity may relate to the number of characteristics ('minutiae') revealed in the mark and their clarity ('quality' – definitions for the quality levels are shown in Figure 3.2). The dotted line indicates the complexity boundary, a judgement that the details in the mark are above the line may indicate the mark is less likely to be complex than if they are judged to be below it.

The most recent criteria for estimating complexity utilises two categories (low and high) and was produced by OSAC (2020). OSAC recommends documentation conform to the NIST Markup Instructions for Extended Friction Ridge Features (NIST, 2013), which uses five levels to describe the confidence an examiner has in details in the mark. Level one is used for the least clear details, and level five for the clearest. Using the OSAC criteria, for a mark to be considered as low complexity, the examiner would need to either observe between 8 and 15 characteristics categorised as level 3 or higher (level 3 indicates the examiner has no doubt about the presence and location of a characteristic) or observe between 5 and 12 level 4 or higher characteristics (indicating the examiner is sure they can clearly determine not only the presence and location of a characteristic but also the shapes and contours along the edges of the ridge(s) that form it). Irrespective of the amount and clarity of the details it reveals, OSAC also allows the examiner to consider a mark as low complexity if the details revealed do not provide a strong indication of the area of skin that made the mark or its orientation. For a mark to be considered as high complexity, the examiner would need to observe fewer than 8 characteristics categorised as level 3 or higher or fewer than 5 categorised as level 4 or higher.

Alternatively, some examiners may use software such as that provided in the Universal Latent Workstation (ULW) to assist in their determination of complexity.[1] The ULW is a tool developed for examiners that, among other features, allows the clarity of a mark to be automatically assessed. The ULW includes the FBI's Latent Quality Metrics (LQMetrics) software that uses algorithms to assess the clarity of the mark based on data gathered from a study of examiners clarity assessments

Quality	
High	Level 1 is distinct; Level 2 details are distinct; There are abundant distinct Level 3 details.
Medium high	Level 1 is distinct; Most of the Level 2 details are distinct; There are minimal distinct Level 3 details.
Medium low	Level 1 is distinct; Few of the Level 2 details are distinct; There are minimal distinct Level 3 details.
Low	Level 1 may not be distinct; Most of the Level 2 details are indistinct; There are no distinct Level 3 details.

Figure 3.2   SWGFAST Quality Table. *Source:* Reproduced with permission from SWGFAST (2013a).

---

1 The ULW is available at no cost to authorised criminal justice agencies from https://www.fbibiospecs.cjis.gov/Latent/PrintServices.

by Hicklin et al. (2010). The software automatically generates a 'clarity map' and then calculates an 'overall clarity score' between 0 and 100 (Hicklin, Buscaglia & Roberts, 2013). The Defense Fingerprint Image Quality Index (DFIQI) application also provides a method of assessing the clarity of marks and estimating complexity (Swofford et al. 2021).[2]

Whilst complex marks require more documentation than non-complex marks, exactly what needs to be documented for each type of mark will likely vary from organisation to organisation. For non-complex marks, as a minimum, SWGFAST (2012) requires that the examiner record a marks orientation, the most likely area of skin to have made it (e.g. finger or palm), and whether there are level one or level two details present. SWGFAST also requires the examiner to record the surface the mark came from, the medium used to develop it, and whether it has been lifted or photographed (if that information is not already part of the documentation produced by those who found the mark). SWGFAST permits much of this information to be recorded in the form of symbols. For example, a semicircle symbol above a mark can be used to indicate that a mark is considered to have been made by a finger (a different symbol would be used for palm), what the marks orientation is (the semicircle is always put over the top of the mark) and that the mark reveals level 1 and 2 detail.

In addition to the above, with marks that are designated complex, the examiner may also record; the locations, orientations and type of all the details observed, the confidence they have that each of those details will appear similarly in a print made by the same area of skin, the clarity of the mark as a whole (and/or variances of clarity within it) and whether any details appear to have been affected by distortion. The documentation may include outlining or shading areas of the mark (for example, to indicate clear or unclear areas or areas affected by distortion), using symbols or different colours to indicate different types of details, or tracing over ridges to show their ridge flow. Various systems have been proposed to document the details in a mark. SWGFAST (2012) suggests four different symbols the examiner can use to quickly indicate how confident they are in the type and position of a characteristic. Alternatively, Langenburg & Champod (2010) and Laird & Lindgren (2011) both describe similar systems in which the examiner colour codes details in the mark to indicate their clarity or the confidence level they have that those details would appear similarly in a print made by the same area of skin. In Langenburg & Champod's 'GYRO' (Green, Yellow, Orange and Red) system, an examiner would mark a characteristic green on an image of the mark if they are very confident that it exists and will appear similarly in a print. Whereas they would mark it yellow if they have a medium level of confidence that the characteristic will appear similarly and red if they have a very low level of confidence[3]. Langenburg & Champod explain that the colours also determine how much weight each detail will contribute to a conclusion and how much tolerance the examiner will have for allowing the detail to appear differently as a result of distortion. For example, a characteristic that was very clear in the mark would likely be marked green and would contribute the most weight to a conclusion. The examiner would have a low tolerance for that characteristic to appear differently in a print made by the same area of skin – so if it did, it may be an indication that the impressions were not made by the same area of skin. GYRO can be implemented using hard copies of the impressions or digitally using software such as Adobe Photoshop, but there are also some programs specifically designed for examiners. For instance, as well as providing an automatic way of assessing the clarity of a mark, the ULW can be

---

2 The DFIQI can be accessed at: https://zenodo.org/record/4426344.

3 There have been few studies which evaluate how accurate examiners are with their perceived confidence levels, but John & Swofford (2020) found when examiners reported they were 'highly confident' in the existence of a characteristic they were correct approximately 96% of the time and when they had a medium level of confidence, they were correct approximately 82% of the time.

(a)　　　　　　　　　　　　(b)

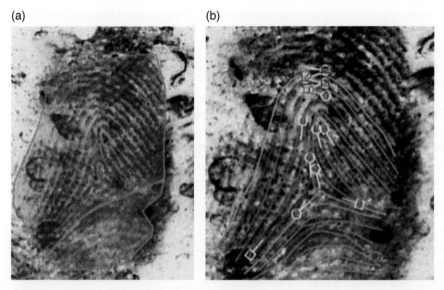

Figure 3.3 This mark has been annotated using the Picture Annotation Software (PiAnoS4), which uses a slightly different colour coding scheme than GYRO. The annotations in (a) indicate areas of differing clarity; green for high and orange for medium (the circles with tails indicate the positions and orientations of the characteristics). The annotations in (b) show the ridge flow and highlight the positions of the characteristics. *Source:* Reprinted with permission Neumann et al. (2013).

used by the examiner to produce manual quality maps by shading specific areas in different colours to denote clarity as well as annotating details within them. PiAnoS (Picture Annotation Software) also allows the examiner to outline or shade specific areas of the mark in different colours according to their clarity, as well as tracing the ridges and using symbols to record the confidence they have in the type and position of characteristics (Figure 3.3).

3.4 First-Level Detail

First-level details describe the overall flow of the ridges and any patterns or distinctive formations they assume.

Support for the persistence of first-level details comes from research into the way skin renews itself and the comparison of impressions taken from the same area of skin over long periods of time.

Skin is composed of three main layers: the epidermis, the dermis and the hypodermis. Friction ridges are present in their final configurations on the outer layer, the epidermis, prior to birth. The epidermis itself is made up of several different layers, the lowest of which is known as the basal layer. The skin cells that form the ridges are not permanent and are regularly shed from the surface of the skin and replaced by new cells generated in the basal layer. The process of cell generation and migration to the surface takes approximately 30 days, so over the course of a person's life, the cells forming the ridges will have been completely replaced hundreds of times (Wertheim & Maceo, 2002). The reason the configuration of the ridges does not change despite the constant replacement of the cells comprising them is that the basal layer acts as a template or blueprint for the layout of the ridges on the surface of the skin (Maceo, 2011). Therefore, when new cells are generated in the basal layer, they replicate those that have been shed, ensuring a persistent layout of the ridges on the surface.

This mechanism explains why those who have compared impressions taken from the same area of skin over intervals of many years have all observed that first-level details persist over time (Figure 3.4).[4] The longest intervals between impressions for each study were; 10 years (Wilder & Wentworth, 1918); 30 years (Faulds, 1912); 31 years (Galton, 1892); 41 years (Welcker, in Monson et al., 2019); 50 years (Jennings, in Cummins & Midlo, 1961); 53 years (Monson et al., 2019) and 57 years (Herschel, 1916).

In terms of the capacity for first-level details to be used to differentiate between different donors, whilst some are very unusual, most configurations of ridge flow are common. For example, marks made by the tips of the fingers typically reveal a slight curvature, so the presence of that flow in such a mark has very little specificity because it would likely appear similarly irrespective of who made the mark. If more of the ridge flow is revealed, the details may allow the examiner to categorise the ridge flow as conforming to a pattern. Whilst there are only a limited number of patterns, and so billions of people may share them, some are less common than others, and so an understanding of pattern frequency distribution (statistics are provided in Chapter 7) can allow the examiner to determine where the specificity of first-level details is higher. Occasionally the mark may reveal a very rare pattern or particularly unusual ridge flow, which may significantly increase its capacity to be used to differentiate between different people.

Whilst first-level details in most circumstances lack the specificity to be used on their own as the justification for a conclusion of identification, they can be the only detail used to support a conclusion of exclusion. For example, if the mark reveals a pattern that is not found in a person's prints,

(a) (b)

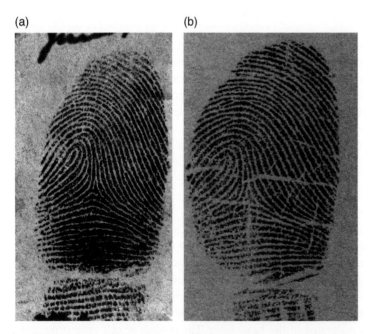

Figure 3.4 These prints were both made by the same area of skin. Print (b) was made 54 years after print (a) and demonstrates that the first (and second) level detail is persistent over time. *Source:* Reprinted with permission from Cherrill (1954).

---

4 Monson et al. (2019) did report some configurations of ridge detail on the skin becoming more elongated over time. The effect is unlikely to have significant implications for the examiner as a slight elongation or compression is routinely encountered in impressions made by the same area of skin due to distortion.

then the first-level details may allow the examiner to conclude that person was unlikely to have been the donor of the mark.

3.4.1 Ridge Flow

If enough of it is revealed, the ridge flow in a mark made by the ends of one of the digits may form a recognisable pattern that will make it easy to compare with a pattern revealed in a print (see 3.1.2). However, even if the mark does not reveal a pattern, the flow of the ridges, particularly their straightness, curvature and any areas where they diverge or converge, can also be compared with the ridge flow in a print (Figure 3.5).

3.4.2 Pattern Family and Pattern Type

Each finger is made up of three bones called phalanges. The proximal phalanges are closest to the palm; the medial phalanges are in the middle and the distal phalanges at the ends (the thumbs only have proximal and distal phalanges). Marks made by the skin of the distal phalanges will usually reveal ridge flow that conforms to one of three pattern families: Loops, Whorls or Arches.

Which of these patterns form on a digit is determined by the interaction of a range of influences on the skin prior to birth. Ridges begin to form in the womb just after a foetus reaches the 10th week of estimated gestational age and tend to develop with a flow that is perpendicular to the direction of growth (Wertheim, 2011). The predominant direction of growth at this time is along the length of the hand; therefore, the ridges are influenced to flow across the width of the hand. This general flow can be seen on the skin of the proximal and medial phalanges; however, the flow on the distal phalanges is much more variable. This is because at the time the ridges are forming, the distal phalanges are covered by transient swellings known as volar pads, and the shape and size of these pads also exerts an influence on the flow of the ridges. For example, if the volar pad is symmetrical, then a symmetrical pattern like a whorl or arch will likely result - whereas a loop pattern will result from an asymmetric pad (Wertheim, 2011). Similarly, if the pad is high and pronounced, a whorl pattern will typically result, whereas an arch pattern is more likely to form on a low pad. The pattern that forms on the pad is also influenced by the time at which formation begins. At the same time the ridges are developing, the growth rate of the fingers is making the volar pads less distinct and

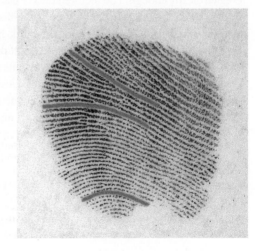

Figure 3.5 This mark was made by the tip of a right thumb, and though it does not reveal a pattern, its ridge flow is distinctive. Most of it is diagonal though some ridges begin to curve downwards towards the left side, with the curve becoming sharper towards the bottom of the mark. In the upper left area of the mark, there are also ridges that begin to diverge from each other.

ultimately results in them merging with the contours of the fingers (Wertheim, 2011). The precise time at which the pads begin to merge into the fingers varies from person to person, as does the time that ridges start to develop. Therefore, if the pad begins to merge with the contours of the finger at an early stage, then by the time ridges start to develop, it may be relatively low, and so an arch pattern may result. Conversely, if the pad only begins to merge with the finger at a late stage, then a whorl pattern may result (as the pad will still be relatively high and pronounced). Similarly, if the merging of the pad with the finger begins at the 'average' time, but the ridges begin to form early, the pad may be high, whereas if the ridges began to form late, then the pad may be low. As well as being influenced by other factors, the precise time the ridges begin to form and the time the volar pad begins to be absorbed into the contours of the digit is directed by genetics, so there is a hereditary influence on the pattern type that forms on the digit of the foetus (Wertheim, 2011).

Depending on how much detail they reveal, with some marks an examiner may only be able to determine the family the pattern comes from (loop, whorl or arch), whereas with others, they may be able to recognise one of the different types of patterns that exist within each family. Whilst there is agreement that almost all patterns can be categorised as loops, whorls or arches, there are different views on how many different types there are within each family, what those types are and what they are called. For example, most sources, including SWGFAST (2013b) recognise eight different pattern types, whereas the FSR (2017) recognises thirteen. The FSR classification agrees with six of the eight SWGFAST patterns though there are differences in the names for some. The remaining seven patterns are either rarely encountered types or slight divergences from one of the six agreed on patterns that are recognised as a separate type by one authority but not the other. Those described in this chapter are the FSR patterns though the SWGFAST terminology is also provided, and the key differences pointed out.

Historically, precisely classifying the pattern type of the mark was a critical aspect of the examiner's role. This is because sets of prints were traditionally filed in collections according to their pattern types. So, to be able to compare the mark with a print revealing a similar pattern type, the examiner had to be able to accurately classify the pattern in the mark and then go to that section of the collection to retrieve sets with similar patterns. Whilst it is usually simple to classify the pattern type, there are patterns that look very similar, so to differentiate between them, detailed rules were devised that specified the key requirements of each pattern. Whether such a pattern is classified as one type, or another may come down to minute details such as whether a single ridge that ends very close to another ridge is considered to be touching it or not. In those cases, even if the rules are considered it sometimes is still only possible to say that the mark could be one of two types because the relevant detail is not revealed clearly enough. The requirements of each pattern type are described in detail in sources such as FBI (2006); however, since fingerprint collections were computerised, the need to be able to precisely classify the pattern type of the mark has become less important. Automated Fingerprint Identification Systems (AFIS) use second-level details to search databases of prints rather than patterns, so the examiner can retrieve similar prints for comparison with the mark just by indicating the positions and orientations of the details it reveals on the AFIS. Recognising a pattern family and type is still a crucial part of the examiner's role, but now where the pattern could be one of two alternatives, it is more efficient to recognise what those are and ensure that the mark is compared with any prints that reveal similar patterns, rather than to attempt to classify the pattern precisely.

3.4.2.1 Loop Family Patterns

Loop patterns all feature some ridges that enter on one side of the pattern and loop back on themselves to exit on the same side they entered. All patterns in the loop family also feature a formation known as a delta (Figure 3.6). A delta is a triangular formation of ridge flow in which ridges flowing from three directions meet in the same place. Whilst whorl patterns have multiple deltas, loops only have one.

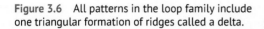
Figure 3.6 All patterns in the loop family include one triangular formation of ridges called a delta.

Deltas are thought to occur as a result of ridges forming on the skin of the foetus in three separate places simultaneously. The theory is that a delta is created where fields of ridge from the centre of the volar pad, the tip of the distal phalange and the base of the distal phalange meet (Wertheim, 2011).

Deltas are almost always located in the bottom half of a pattern, and in loops, they will be on the opposite side to the side the ridges that loop back on themselves enter and exit the pattern. The point at which those ridges begin to loop back on themselves or 'recurve', occurs at a point around the centre of the pattern known as the 'core'.

Within the loop family, there are three distinct types of loop: the plain loop, the converging loop and the nutant loop (SWGFAST (2013b) only recognise the plain loop).

3.4.2.2 Plain Loop
A plain loop is a pattern in which some ridges enter on one side of the pattern before looping back on themselves in the core and exiting the pattern on the same side they entered (Figure 3.7).

3.4.2.3 Converging Loop
A converging loop is like a plain loop, except it includes ridges which, after looping back on themselves around the core, converge towards the ridges forming the other side of the loop (Figure 3.8). This convergence often results in a series of ridges that end or join with other ridges, as in Figure 3.8.

3.4.2.4 Nutant Loop
A nutant loop is a rarely seen pattern that is like a plain loop except that at the point at which they loop back on themselves, some of the ridges turn or droop towards the delta (Figure 3.9).

3.4.2.5 Subdividing Loop Patterns
Many patterns slope one way or the other, but it is usually most noticeable in loops. If the ridges that loop around the core enter and exit on the right, the pattern is known as a right loop (as in Figure 3.7), and if they enter and exit on the left it is called a left loop (as in Figure 3.9). SWGFAST (2013b) recognise left and right sloping loops as two different patterns, whereas the FSR (2017) considers them the same pattern but with a different slope.

In addition to noting the slope of the pattern, the examiner may also count the number of ridges between the core and the delta to establish the 'ridge count' of the pattern (Figure 3.10).

Figure 3.7   This print reveals a plain loop pattern.

Figure 3.8   This print reveals a converging loop pattern. *Source:* Andrey Burmakin/Dreamstime LLC.

Figure 3.9   This print reveals a nutant loop pattern. *Source:* Printed with permission.

(a) (b)

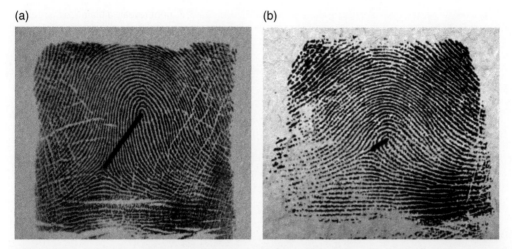

Figure 3.10 The ridge count is the number of ridges bisected by an imaginary line drawn from the delta to the core. Both these prints reveal patterns from the same family (loop) with the same pattern type (plain) and the same slope (right), but they can be easily differentiated from each other by their ridge count. When counting the ridges, the starting and ending ridges are not included, so the ridge count of print 'a' is 15 and that of print 'b' is 3. Print (b) licensed for use by the Mayor's Office for Policing and Crime under an Agreement between John Wiley & Sons Ltd and MOPAC dated 28th April 2016. *Source:* Courtesy of Mayor's Office for Policing and Crime, London.

Many of the same factors that influence the type of pattern that forms on a digit also affect the ridge count. For example, patterns that develop on a high, pronounced volar pad will have a high ridge count, whereas those that develop on low pads will have a low count. However, like pattern development, ridge count is also affected by the time the ridges start to form in relation to when the volar pad starts to merge with the digit. So, if this begins early or the ridges begin to form late, a low count pattern may result and vice versa.

The examiner determines the ridge count by counting all the ridges that are bisected by an imaginary line drawn between the delta and the core. Usually, this is done approximately so that prints with a very different ridge count can be quickly excluded in the comparison stage and those with a similar count can be compared further using other details such as characteristics. Often it may not be possible to obtain even an approximate count as many marks will not clearly reveal both the core and the delta and all the ridges in-between. With such marks, the examiner may only be able to say that the ridge count is greater than 'X' number of ridges (Figure 3.11).

On the rare occasions when an exact count is needed, and the mark reveals all the necessary details, there are rules for determining exactly where the count should begin and end depending on the types of ridge formations in the core and delta revealed in the mark as described in FBI (2006).

3.4.2.6 Whorl Family Patterns

The most commonly seen patterns in the whorl family feature circular or oval formations of ridge. The formation can be comprised of a single unbroken circuit of a ridge, a spiral, or can also be made up of multiple sections of ridge that collectively form a 360-degree circuit.

All whorls have at least two deltas that, in most cases, can be found on opposite sides, in the bottom half of the pattern. Whorl patterns may be symmetrical or asymmetrical, and one delta may be much closer to the core and/or higher or lower than the other.

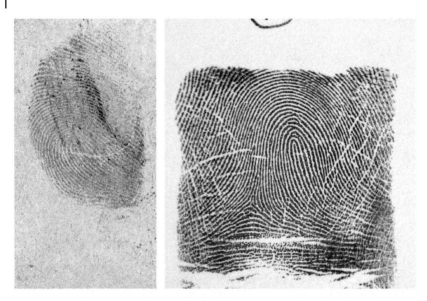

Figure 3.11 The mark on the left reveals ridges that enter from the right and recurve in the core before exiting on the right – indicating the pattern is likely to be a right loop. Therefore, though it is not revealed in the mark, there should be a delta on the lower left side of the pattern. By counting the ridges from the core to the lower-left edge of the mark, it can be inferred that the ridge count of the pattern is likely to be greater than 13. The print on the right was made by the same area of skin as the mark and revealed the ridge count is 15.

Figure 3.12 This print reveals a plain whorl pattern. *Source:* Arkadiusz Fajer/Dreamstime LLC.

Within the whorl family, there are seven distinct types of whorl: the plain whorl, the elongated whorl, the central pocket, the twinned loop, the lateral pocket, the composite and the accidental (SWGFAST (2013b) does not recognise the composite, elongated or lateral pocket patterns).

3.4.2.7 Plain Whorl

Plain whorl patterns are comprised of circular formations of ridge and always have two deltas (Figure 3.12).

3.4.2.8 Elongated Whorl

An elongated whorl has the same attributes as a plain whorl but features a more elliptical ridge flow around the core (Figure 3.13).

3.4.2.9   Central Pocket

Central pocket patterns (also known as 'Pocket Loops') are whorls in which one of the deltas is significantly closer to the core than the other (Figure 3.14).

3.4.2.10   Twinned Loop

Twinned loops (also known as 'Double Loops') feature two loop formations but are considered part of the whorl family because they have two deltas (Figure 3.15). Usually, one loop will be ascending and the other descending though examples in which the loops are almost horizontal can also be found.

3.4.2.11   Lateral Pocket

A rarely seen pattern that is like a nutant loop but includes another formation underneath the drooping loop (Figure 3.16). This formation will either be a tented arch (see Arch Family Patterns) or a second loop with the same slope. The lateral pocket pattern is part of the whorl family as it has two deltas though unusually both are located on the same side of the pattern.

Figure 3.13   This print reveals an elongated whorl pattern. *Source:* aluxum/Getty Images.

3.4.2.12   Composite

A very rarely seen pattern that is a combination of two or more basic patterns and normally has at least three deltas (Figure 3.17).

3.4.2.13   Accidental

A very rarely seen pattern that appears like an arch pattern but includes a small whorl-like formation at the core (Figure 3.18).

Figure 3.14   This print reveals a central pocket pattern in which the delta on the right is much closer to the core than the delta on the left. *Source:* ra-photos/Getty Images.

Figure 3.15 This print reveals a twinned loop pattern. *Source:* VCNW/Getty Images.

3.4.2.14 Subdividing Whorl Patterns

Like loops, some whorls may have a slope. If a whorl is elliptical and its major axis (the longer of the two lines of symmetry) rises from left to right, the whorl is said to be left sloping and vice versa (Figure 3.19).

As with loops, if a delta can be seen, then a ridge count can be used to differentiate between whorls of the same type. If both deltas can be seen, then the 'tracing' of the whorl can also be established and used in the same way (Figure 3.20).

Tracing involves considering the relative positions of the deltas and is determined by following the ridge that forms the lowest arm of the left delta towards the right delta. If that ridge ends, then the ridge immediately below it is followed. Similarly, if the ridge being followed splits into two, then the lower arm is followed. If the ridge being followed crosses an imaginary line between the right delta and the core with three or more ridges between it and the lower arm of the right delta, the whorl has an 'inner' tracing. If the ridge passes underneath the right delta with three or more ridges between it and the

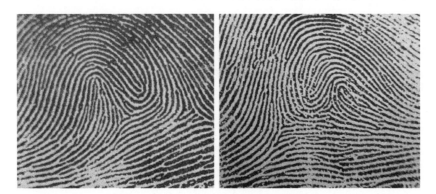

Figure 3.16 These prints reveal lateral pocket patterns. *Source:* Reprinted with permission from Cherrill (1954).

Figure 3.17 Though these patterns would be deemed composites using the classification scheme described by Cherrill (1954), other schemes such as that used by the FBI (2006) that do not recognise that pattern would consider them accidentals. *Source:* Reprinted with permission from Cherrill (1954).

Figure 3.18 These prints reveal accidental patterns. Some definitions of this pattern, such as the FBI (2006) one are broader than the one described here, which is based on the description in Cherrill (1954). *Source:* Reprinted with permission from Cherrill (1954).

(a)   (b)

Figure 3.19 Left (a) and right (b) sloping whorl patterns. (a) ar-chi/Getty Images, (b) Juri Bizgajmer/Dreamstime LLC.

(a)   (b)   (c)

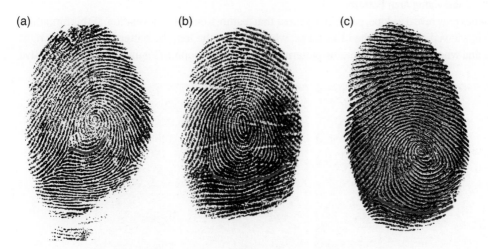

Figure 3.20 Print 'a' reveals an inner tracing, print 'b' reveals a meeting tracing and print 'c' reveals an outer tracing. (a) 4x6/Getty Images, (b) inkit/Getty Images, (c) kristo74/Getty Images.

Figure 3.21 This print reveals a plain arch pattern. *Source:* labsas/Getty Images.

lower arm, the whorl has an 'outer' tracing. If the ridge flows into the lower arm of the right delta or passes it on either side within one or two ridges, the tracing is described as 'meeting'.

Often both deltas will not be revealed in the mark, but if one can be seen, the examiner may be able to infer the position of the other and narrow the options for what the tracing could be.

3.4.2.15   Arch Family Patterns

The ridges in an arch pattern flow smoothly from side to side across the impression. Arches are distinct from loop and whorl patterns in that they do not feature formations that meet the precise requirements to be considered deltas.

Within the arch family, there are three distinct types of arch: the plain arch, the tented arch and the approximating arch (SWGFAST (2013b) does not recognise the approximating arch).

3.4.2.16   Plain Arch

The ridges in a plain arch flow smoothly across the pattern with only a slight hump or rise around the centre (Figure 3.21).

3.4.2.17   Tented Arch

Instead of a hump, the ridges in a tented arch have a prominent upward thrust around the centre of the pattern (Figure 3.22). A tented arch pattern features a formation that looks like a delta but is not considered one as its 'arms' do not enclose or attempt to enclose a formation of ridge in a core area – such as a recurve or circuit of ridge.

3.4.2.18   Approximating Arch

The ridge flow of an approximating arch is like that of the plain arch but includes a formation that resembles a delta that is not considered one as its 'arms' do not enclose or attempt to enclose a recurve or circuit of ridge in a core area (Figure 3.23).

3.4.2.19   Subdividing Arch Patterns

Some arches may reveal a slope, particularly those that feature a delta-like formation. If this formation were to be viewed as a delta, then like loops, where it is on the left, the pattern is said to be right sloping, and where it is on the right, the pattern is said to be left sloping (Figure 3.24). An alternative

Figure 3.22   These prints reveal tented arch patterns. *Source:* Reprinted with permission from Cherrill (1954).

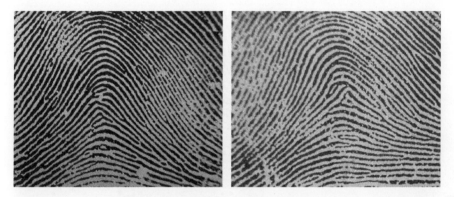

Figure 3.23 These prints reveal approximating arch patterns. The patterns appear similar to loops, but as the arms of the formation that resembles a delta only enclose a single ridge (rather than a recurve) the patterns are approximating arches. *Source:* Reprinted with permission from Cherrill (1954).

(a)   (b)

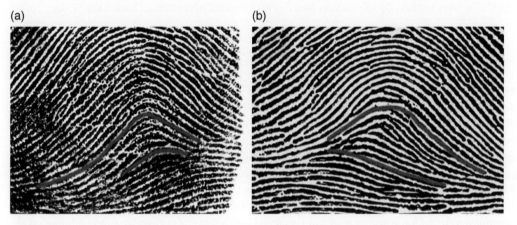

Figure 3.24 Print 'a' is left sloping and print 'b' is right sloping. (a) Iabsas/Getty Images, (b) CSA-Printstock/ Getty Images.

method of determining the slope is to follow the ridges around the delta-like formation. If they reveal a wedge shape, the shorter, steeper side will be on the right in left sloping patterns and on the left in right sloping patterns as in Figure 3.24 (New South Wales Police Force, 2014).

3.4.2.20 Summary

In many cases, the pattern type of the mark will be obvious, and where it is not, the examiner may be able to use the detail that is revealed to rule out some patterns, leaving one or two possibilities (Figure 3.25). For example, a mark that reveals a delta on the right must have been made by either a left sloping loop or a type of whorl - rather than an arch.

The observations the examiner may record of first-level detail include noting the pattern family, type, slope, ridge count or tracing. They may also describe or trace the ridge flow, note the appearance of any cores or deltas, and record the overall clarity of the mark or designate areas of differing clarity. Additionally, they may document the confidence level they have (for example, high, medium or low) that the details are present and will appear similarly in a print made by the same area of skin and the tolerance they have for allowing those details to appear differently as a result of distortion.

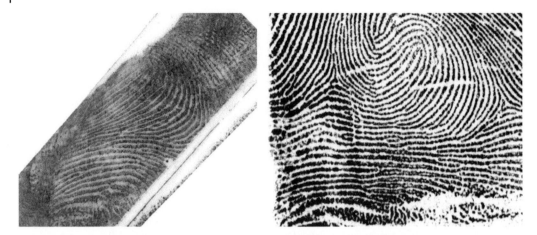

Figure 3.25 The mark on the left appears to be a left loop. However, the print on the right, which was identified as having been made by the same area of skin, shows that when more of the pattern is revealed, it is a twinned loop. *Source:* Reprinted from Complete Latent Print Examination, http://clpex.com, C.Parker, Texas, FIG 91.

3.5 Second-Level Detail

Second-level details are the distinct deviations in the natural flow of the ridges known as characteristics.

In addition to the research indicating how skin renews itself, the persistence of second-level details is also supported by studies that compared impressions made by the same area of skin at different times. The same studies that found that first-level details persist over time also found that second-level details were persistent over the same intervals, with the longest interval studied being 57 years (Faulds, 1912; Galton, 1892; Herschel, 1916; Jennings in Cummins & Midlo, 1961; Monson et al., 2019; Welcker, in Monson et al., 2019; Wilder & Wentworth, 1918). One study did report a difference in one characteristic in an impression taken from a person when they were two years old with one taken from that person at the age of fifteen (Galton, 1892). This difference could be due to damage to the skin, but it has also been suggested that it could have been due to changes in the underlying structure of the skin related to the aging process (Misumi & Akiyoshi, 1984). However, in repeated comparisons of the prints of a person at similar ages in a separate study, no differences were observed (Wilder & Wentworth, 1918). Whilst the most recent study did not find any new or missing characteristics in later impressions, and the authors did report encountering subtle differences in the appearance of characteristics over a long period of time (Monson et al., 2019). The authors attribute most of these differences to the condition of the skin or to distortion caused by the method used to record the image, but there were tiny differences in three pairs of characteristics out of 1900 studied that the authors felt were less easily explained - an example of one of these is shown in Figure 3.26. It could be argued that these differences may be a product of a difference in lighting in the two photographs, but whatever the cause, the authors concluded that the differences they did observe were so small, rare and insignificant that they did not affect their finding that characteristics were persistent.

Configurations of characteristics are much more variable than any first-level details. Support for this comes from the study of the way friction ridge skin develops, various statistical models that have been used to estimate the probability of different areas of skin sharing the same configuration of characteristics, and the observations of examiners and researchers.

The prevailing theory of friction ridge skin development asserts that ridges begin as small units that each form around a sweat gland and fuse into rows. As the ridges are forming, the growth of

(a)   (b)

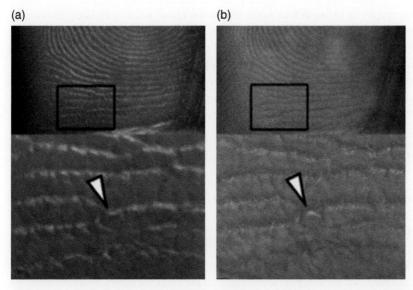

Figure 3.26 These photos were taken eight years apart, and the arrows point to what the authors of the study consider to be a dot in (b) that is not present in (a). *Source:* Reprinted with permission from Monson et al. (2019).

the foetus stretches the skin causing the ridges to increase in length and width. Swellings on the side of some ridges are stretched to the point where they pull away to form new ridges that remain connected to the original ridges – creating characteristics called bifurcations (Hale, 1952). The growth of the foetus also stretches the spaces between the ridges and results in new ridges forming in those spaces. Alternatively, some ridges will come to an end as a result of being sandwiched between two existing ridges – creating characteristics called ridge endings (Wertheim, 2011). Though hundreds of units may fuse together to create a single ridge, occasionally some units will not fuse with any others around them – creating characteristics called dots.

Whether a ridge unit forms a characteristic or a continuous ridge is determined at random by the tension or compression across the surface of the skin at that point at that time. These localised stresses are primarily the result of differences in growth rates across the skin and the interaction of separate developing fields of ridge, though they are also influenced by the various factors that determine pattern type and ridge count. Even slight variations in the stresses and strains across the surface of the skin or any of the other factors could result in different characteristics forming in different positions or orientations (Wertheim, 2011). This vast potential for variability for each tiny unit of ridge makes configurations of characteristics highly discriminating and means that the more that are clearly revealed in the mark, the less likely it is than anyone other than the donor of that mark could share a similar configuration. This extreme variability has been supported by various statistical models used to estimate the probability of finding the same specific configuration of characteristics on the skin of different people. Langenburg (2011) reviewed 17 of the models and found that, though the estimates vary, all demonstrated that the probability is staggeringly small. For example, when marks with configurations of 12 characteristics were put through one of the most sophisticated models, the average score was approximately 1 trillion – meaning that it would be 1 trillion times more likely to observe the same configuration of characteristics in a print if the donor of the print was also the donor of the mark, than if they were not (Neumann, Evett & Skerrett, 2012). The variability of characteristics is also supported by over a century of work by a global community of examiners and researchers who have only ever observed very small areas of ridge detail with configurations of characteristics that were similar in impressions made by different people. Further support comes from studies carried out to examine the similarities in the fingerprints of twins. Whilst twins have been found to be more likely to have the same pattern type and ridge count than non-twins (due to the genetic influence on the timing of ridge formation and volar pad merging), the studies found that the arrangements of characteristics were always different (Cummins & Midlo, 1961; Galton, 1892; Lin et al., 1982; Srihari et al., 2008).

Because of their variability, unlike first-level details, characteristics can be used 'on their own' to support a conclusion of identification (characteristics cannot exist without the first-level detail of ridge flow, so in reality, they are always used in conjunction with first-level detail).

3.5.1 Characteristics

Characteristics are also known as 'minutiae' or 'Galton details', and most sources agree that there are three types: ridge endings, bifurcations and dots (though the FSR (2017) definition only recognises bifurcations and endings as characteristics – dots are considered as one of several other 'ridge features').

3.5.1.1 Ridge Endings

A ridge ending is a characteristic in which a ridge ends, and the flanking ridges on either side of it converge to take its place (Figure 3.27).

There is considerable variation in the appearance of ridge endings. For example, they may occur at any orientation, and the ridges forming the characteristic may be curved or straight. If the ending is very clearly revealed, the examiner may also be able to observe that the ridge may come to an end abruptly, taper into a fine point, increase in size at its end or dissipate into a series of broken sections of ridge.[5] The examiner may also observe that the convergence of the flanking ridges occurs gradually over a distance or suddenly at the point of the ending, which itself may occur centrally between the flanking ridges or may end much closer to one ridge than the other. However, if the clarity is poor, the examiner may not be able to discern anything more than that the ridge comes to an end at a particular point. Sometimes they may not even be able to see whether a characteristic is a ridge ending or a bifurcation – just that an 'event' occurs in a location that could be due to the presence of either type of characteristic.

3.5.1.2 Bifurcations

A bifurcation is a characteristic in which a ridge splits into two, and the flanking ridges on either side of it diverge to make room for it (Figure 3.28).

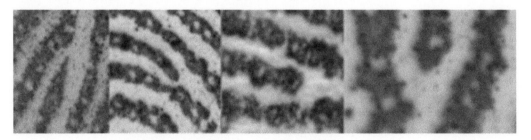

Figure 3.27 Four examples of ridge endings.

---

5 Technically the shapes and widths of the ridges are third-level details but where they are clearly revealed in both impressions the examiner may compare them simultaneously with second-level details – i.e. as a characteristic with a particular shape.

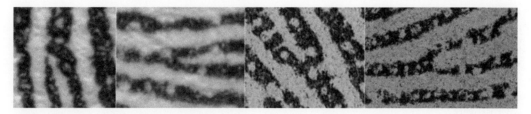

Figure 3.28   Four examples of bifurcations.

Like ridge endings, bifurcations can occur at any orientation, on straight ridges or curved ridges and can also vary considerably in appearance. For instance, a ridge may gradually widen until it splits into two, or the transition from one to two ridges may occur abruptly. The bifurcation may be formed by two ridges that both diverge from the path of the original ridge at the same angle, or by one ridge that diverges while the other maintains the path of the original ridge. As with ridge endings, the ridges may narrow or thicken at the point of the bifurcation.

3.5.1.3   Dots

A dot is a small, circular or oval characteristic that is approximately as wide as it is long (Figure 3.29). Dots vary in shape, size, and like ridge endings may appear closer to one flanking ridge than the other (Figure 3.29).

A characteristic may be surrounded by continuous ridges, or it may appear as part of a distinctive formation with other nearby characteristics. Some of the more commonly encountered formations can be referred to as:

- Short Independent Ridges
  A short independent ridge or 'island' is a formation featuring a ridge that ends in both directions as in (a) in Figure 3.30 (a short independent ridge is distinct from a dot in that it is longer than it is wide).

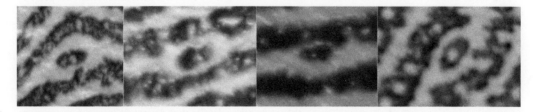

Figure 3.29   Four examples of dots.

(a)　　　　　　　　　(b)　　　　　　　　　(c)　　　　(d)

Figure 3.30   A short independent ridge (a), a lake (b), a spur (c) and a crossover (d).

- Lakes
  A lake or 'enclosure' is a formation in which a ridge bifurcates and then becomes a single ridge again when both branches of the bifurcation re-join each other as in (b) in Figure 3.30.
- Spurs
  A spur or 'hook' is a formation featuring a bifurcation in which one of the bifurcating ridges comes to an end as in (c) in Figure 3.30.
- Crossovers
  A crossover or 'bridge' is a formation featuring two bifurcations that connect adjacent ridges as in (d) in Figure 3.30. As well as appearing as it does in (d) the connecting ridge may be perpendicular to the other ridges.
- Opposed Bifurcations
  A formation in which two bifurcations appear in proximity on the same ridge opening in opposite directions as in (a) in Figure 3.31.
- Opposed Ridge Endings
  A formation in which two ridge endings occur facing each other as in (b) in Figure 3.31.
- Overlaps
  A formation in which two ridge endings overlap slightly as in (c) in Figure 3.31 (also known as 'over-unders').
- Changeovers
  A formation in which two ridges end in opposite directions and are separated by a single ridge that passes diagonally between them as in (d) in Figure 3.31.

Whilst these may be some of the more commonly encountered formations, there are many more – for instance, double bifurcations in which a ridge bifurcates and then one of the arms of the bifurcation itself bifurcates. Saviers (1989) looked at 40 texts and found they collectively mentioned 31 different formations (the author also observed that many of the same formations were known by different names). However, that list is by no means complete as, for example, the FBI (1972) illustrated 7 additional unnamed formations not considered by Saviers, and even these did not include the rare formation the Spanish Scientific Police refer to as a 'dock' (Gutiérrez-Redomero et al., 2011) or the unnamed one in Figure 3.32.

The value of naming a formation rather than describing it by the characteristics that comprise it (for example, a lake is made up of two bifurcations) is that the name allows the examiner to 'chunk' several characteristics together rather than having to remember their individual locations and orientations relative to each other when searching a print for a similar configuration.

As well as formations of multiple characteristics, the examiner will also encounter areas of ridge detail with no characteristics. These are known as 'open fields' of ridge (Figure 3.33).

Whilst there is general agreement that there are three types of characteristics, and that all formations are simply combinations of characteristics – there are other details that do not fit into either category (Hague in Saviers, 1987; Hawthorne, 2009; Rienti, 1988). For example, 'trifurcations' (in which one

(a) (b) (c) (d)

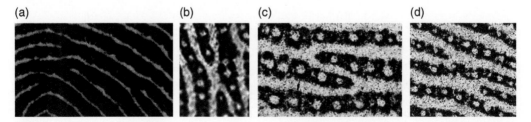

Figure 3.31 Opposed bifurcations (a), opposed ridge endings (b), an overlap (c), a changeover (d).

Figure 3.32 A rarely seen formation featuring a bifurcation in between two ridge endings.

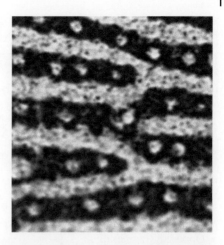

Figure 3.33 An 'open field' is an area of an impression that does not reveal any characteristics.

ridge splits into three at the same point) are occasionally seen and within the core of whorls, complete circles of ridge may be encountered. Rienti (1988) looked at 1000 whorls and found this detail in 50 of them. Hague (in Saviers, 1987) suggested 'angles' (small 'L' shaped formations of ridge) that can sometimes be found in deltas should be considered as characteristics. Templeton (1997) describes a detail in which two adjacent parallel ridges get closer together but without touching as a 'pinch'.

3.5.1.4 Target Group Selection
During the comparison stage, the examiner may need to compare the mark with many prints. By committing some of the details in the mark to memory, the examiner can efficiently discount prints that do not reveal similar details without comparing each one side by side with the mark. For example, if the mark reveals a whorl pattern then, allowing for the effects of distortion, it is not necessary to consider any prints that do not reveal a whorl pattern. As well as committing first-level details to memory, examiners may also do the same with some second-level details.

Estimates about the number of characteristics in a rolled impression vary from 80 (UK Home Office, 1969) to well in excess of 100 (Moenssens, 1975), so though it may be easy to hold a pattern type in their memory, in most instances the examiner will not be able to do the same with the type and orientation of all characteristics. Therefore, during the analysis of the mark, the examiner usually selects a small 'target group' of characteristics they can familiarise themselves with and look for in the print. So, if for example, a print that revealed the same pattern as the mark was found, the examiner would then look to see if it also revealed similar characteristics to those in the target

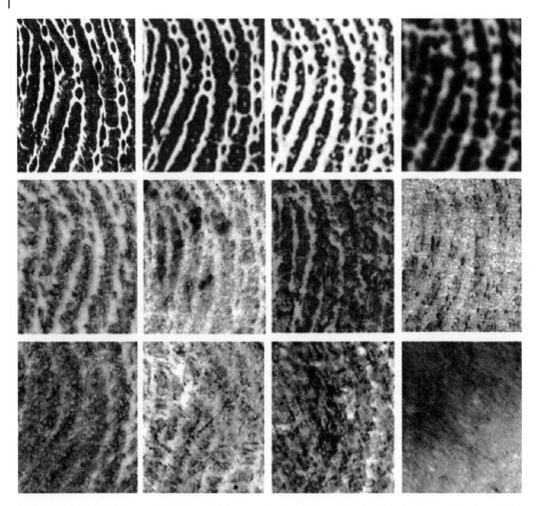

Figure 3.34 The same group of characteristics revealed at different levels of clarity. *Source:* Reprinted with permission from Hicklin et al. (2013).

group. Other details, such as creases, can also be used to form all or part of the group, but characteristics are the most frequently used detail.

With marks that reveal few details, the examiner may have little choice in the target group they select, but where there is more choice, there are several factors to consider – the most important of which is the clarity of the characteristics in the group. The examiner must be confident that the characteristics they select can be expected to appear similarly in a print made by the same area of skin. Figure 3.34 shows a series of impressions made by the same area of skin that illustrate how much more difficult it is to be confident of the details in an impression when the impression lacks clarity.

Another factor the examiner must consider is that the appearance of some characteristics can be more easily affected by the distortion factors discussed in Chapter 4 than others. For example, what appears to be a dot could be dirt or another contaminant in between the ridges on the surface the mark was on. Similarly, a contaminant on the surface or excess pressure during deposition of the mark could easily make a formation like an overlap appear as a slightly thicker continuous ridge, rather than two ridge endings. Ray & Dechant (2013) point out that such formations (and others, including lakes, spurs and short independent ridges) have another disadvantage in that unlike bifurcations and endings, their presence does not result in an increase or decrease in the number of ridges on one side of them. When using endings or bifurcations as part of a target group, this effect may allow the examiner to be confident of the presence of a characteristic even if the actual point a ridge ends or bifurcates is unclear. For example, a group that consists of a series of endings that all face in the same direction cannot easily be obscured by a contaminant or appear as continuous ridges because on one side of the group, there will be more ridges than there are on the other side. As well as favouring characteristics that affect the ridges around them, the examiner can reduce the likelihood of all the characteristics in a group being affected by distortion in a localised area by selecting a group that includes characteristics that are separated by several ridges.

As well as the type and orientation of the characteristics in the group, the examiner must also note their relative positions. A key component of this is the number of ridges between each characteristic. Skin is elastic and can stretch and contract, so the same characteristics can appear further away from each other, closer together or higher or lower in one impression than they do in another. However, considering the distortion factors discussed in Chapter 4, the examiner can expect that the number of ridges between characteristics in a mark and print made by the same area of skin will be the same.

Where possible, the examiner will usually select a group of characteristics in the mark that are near a 'fixed point' such as a core, delta or other distinctive detail. The comparison of the target group will begin with the examiner locating that fixed point in the print to ensure they are comparing characteristics from the same area in both impressions. If no suitable fixed point is revealed in the mark, and there are multiple groups of characteristics that could be used, the examiner should select one that is most likely to be revealed in a print. For example, if the mark was made by the tip of a finger, a group near the bottom of the mark may be more likely to appear in a print made by the same area of skin than a group at the top. This is because when prints are taken, the digits will be flat on the surface, which means that much of their tips may not contact the surface at all. As a target group at the bottom of a mark made by a tip would correspond to an area near the centre of a print, it therefore is more likely to have been recorded. Similarly, when choosing a target group in a mark likely made by the side of a finger or palm, the examiner may favour characteristics in an area that is closer to the centre of the finger or palm than the edge.

As well as their location, the rarity of the characteristics that make up a target group also needs to be considered. The rarer the group is, the less likely it is that a similar configuration will be found in any print other than one made by the same area of skin as the mark. Therefore, using a group comprising rare characteristics should make for efficient comparison. The rarity of characteristics can be influenced by several factors, including their type, location and orientation, as well as their combination with other characteristics in formations.

Whilst there have been several studies that examined how frequently different characteristics and formations of characteristics are encountered in impressions, the studies did not all consider the same characteristics or formations, and so as a result of that and other differences, only broad comparisons can be drawn between them. Gutiérrez et al. (2007) compared their own study of the Spanish population with 11 others from various population groups and found that collectively ridge endings made up between 40 and 60% of characteristics – a considerably higher proportion than any other formation (Gutiérrez-Redomero et al., 2011). Bifurcations were the next most commonly seen feature and made up between 18 and 40% with dots and all the formations of characteristics being encountered much less frequently (Gutiérrez-Redomero et al., 2011). Only two studies considered the frequency with which open fields are encountered, but both found that they were by far the most likely feature to be seen, reporting that between 76 and 83% of the time any area examined revealed an absence of characteristics or formations (NIST, 2012).

Frequencies for the characteristics and formations illustrated in this chapter reported by the most recent and largest study are shown in Table 3.1 (except for the spur, which was not considered). These frequencies were observed in 2000 prints from the Spanish population and included prints from all three pattern families (Gutiérrez-Redomero et al., 2011).

Table 3.1 Frequencies of characteristics and formations found by Based on Gutiérrez-Redomero et al. (2011).

Characteristic/formation	Frequency of occurrence
Ridge ending	60.3%
Bifurcation	29.0%
Short independent ridge	4.6%
Dot	1.5%
Lake	1.3%
Opposing ridge endings	1.0%
Overlap	0.7%
Crossover	0.3%
Opposed bifurcations	0.1

This study found that ridge endings were more than twice as common as bifurcations which were themselves more than six times more common than the next most frequently encountered detail (the short independent ridge). Only around 1% of characteristics were dots, lakes or opposing ridge endings. Overlaps, crossovers and opposed bifurcations each accounted for less than 1%. Opposing bifurcations were the least common formation, appearing at a frequency of only 1/1000 characteristics/formations making ridge endings more than 600 times more likely to be encountered (Gutiérrez-Redomero et al., 2011). Collectively, ridge endings and bifurcations were nine times as common as all the other characteristics or formations, which is similar to the 8 to 1 ratio reported by the UK Home Office (1969).

All the research clearly demonstrates that some characteristics and formations are considerably rarer than others. Some studies also reported that characteristics and formations are not distributed evenly across all areas of skin. For example, as can be seen in Figure 3.35, some recorded higher densities in the core and delta areas (where the ridges change direction most abruptly) than around the edges (Chen & Jain, 2009; Champod & Margot in Champod et al. 2016). One study found almost three times as many characteristics or formations in each square millimetre in the core and delta areas than outside those areas (Champod & Margot in Champod et al. 2016). This may explain why Gutiérrez et al. (2007) reported that the total number of characteristics in an

(a)   (b)   (c)   (d)   (e)

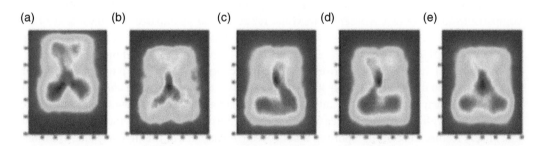

Figure 3.35 These images show the density of characteristics (the darker the area, the higher the density) in an arch pattern (a), a tented arch (b), a left loop (c), a right loop (d) and a whorl (e). *Source:* Reproduced with permission from Chen and Jain (2009).

impression varies according to its pattern family, with whorls (which have a core and at least two deltas) having the most, followed by loops (with one delta and a core) and arches the least (without a core or delta).

As well as there being more characteristics and formations in some areas than others, some studies have found that some characteristics or formations are more commonly encountered in some areas than others. The figures in Table 3.1 show the frequencies of characteristics and formations across all areas of impressions, but Champod & Margot (1996, in Champod et al. 2016) found that formations were more common in the core and delta than in the periphery of the impression. For example, a crossover is a rare formation anywhere in an impression, but Champod & Margot found that it was over 4 times more likely to be seen in the delta area than in the periphery. Gutiérrez-Redomero et al. (2011) compared the frequency of characteristics in the central area of prints with those on the periphery and found that ridge endings were encountered more frequently in the periphery (particularly at the bottom of the print). Conversely, Gutiérrez-Redomero et al. also observed that bifurcations were more frequently encountered in the central area (particularly around the top) than the periphery. The same study also found that the areas where the most ridge endings were seen were the bottom right and left of each print. These are the same areas that other studies have reported that, along with cores and deltas, have the highest densities of characteristics - as can be seen in Figure 3.35 (Chen & Jain, 2009 and Srihari, 2009). Gutiérrez-Redomero et al. (2011) found around 70% of the characteristics in these areas were ridge endings, and this is likely due to an effect known as 'pattern force'. The effect occurs because the ridge flow of most patterns typically results in ridges in these areas converging on each other and, as there is not enough space for all of them to continue, some are forced to end (Interpol European Expert Group on Fingerprint Identification (IEEGFI), 2004). The effect also means that the orientation of all the ridges that do end will be similar. For instance, the lower half of a loop pattern will usually reveal numerous ridge endings on the opposite side to the delta, and almost all of these will end pointing away from the delta (Figure 3.36). Where bifurcations are seen in this area, the bifurcation is more likely to open towards the delta (Gutiérrez-Redomero et al., 2011). Champod (1996, in Champod et al. 2016) found that endings or bifurcations with this orientation made up approximately 80% of the characteristics seen in this area of loops. The orientation of endings that point away from a delta or core and bifurcations opening towards the same detail are similar in that both feature ridge flow that opens or creates more ridges to the same side of the characteristic than the other.

Using such characteristics as a target group would likely result in the examiner encountering similar groups in that area of many prints. However, if a ridge ending was found in that area ending in the

Figure 3.36 The area within the funnel shape shows characteristics that have been influenced by 'pattern force'.

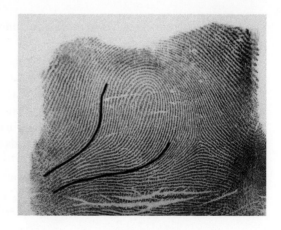

opposite direction, then it may be efficient to use it as part of a target group due to its rarity. Similarly, as a result of the formation itself, the examiner can expect to see multiple ridge endings or bifurcations that all open one way around the three arms of a delta due to the ridges spreading apart as they approach each arm. Neumann et al. (2015) observed that computer algorithms used to search marks against large databases of prints will return more similar responding prints when characteristics from a delta area are used than the other areas they tested. This means that though it is often useful to utilise a target group near a delta, using a target group that is part of an 'outflow' area from a delta is likely to mean that many prints may be encountered with similar configurations.

Another of Gutiérrez-Redomero et al. (2011) findings may also be attributable to pattern force in that they found that in the central area of a pattern, all characteristics opening left are more frequent on right hands and all characteristics opening right are more frequent on left hands. Champod (1996 in Champod et al., 2016) also made a similar finding, and it is likely because right sloping patterns are more frequently seen on right hands (and vice versa), and the delta (or one of the deltas) in those patterns can exert an influence on the orientation of the characteristics above the core. For example, if the delta is on the left of the pattern, most endings or bifurcations will usually open to the left – toward the delta (vice versa is true for patterns with the delta on the right) (Doak, 2004). Champod (1996, in Champod et al. 2016) found approximately 60% of characteristics in this area had this orientation.

Determining the rarity of characteristics and formations involves the consideration of a wide variety of factors, and the data available is limited by the relatively small number of studies and the differences between them. Ultimately it may be a determination that a statistical model is better able to quantify but for the purposes of choosing characteristics for a target group, considering the broad themes above will allow the examiner to reach a subjective determination of the rarity, and therefore suitability of a group of characteristics for their target group.

The observations the examiner documents about characteristics and formations may include their number, type, location, orientation, clarity and the number of ridges, or distance along a ridge between them. The examiner may also document both the confidence level that they have in the existence of a characteristic and the confidence they have in what type of characteristic that is. They may also trace the paths of specific ridges and record areas revealing open fields as well as noting the tolerance they have for characteristics or formations to appear differently in the mark than they do in a print made by the same area of skin as a result of distortion.

3.6 Third-Level Detail

Third-level details are the shapes and widths of individual ridges and the impressions made by the pores that line their summits.

The research conducted into how skin renews itself shows that the mechanism operates on a cellular level, i.e. each cell shed from the surface is replaced by another cell that replicates the former according to the basal layer template. As a result, even though third-level details are much smaller than other details, the research does support their persistence. However, whilst persistence of third-level details has been observed, there have been very few formal studies comparing impressions taken from the same area over periods of time and none that considered all aspects of third-level detail (Chatterjee in Moenssens, 1975; Faulds, 1913; Locard in Langenburg, 2011; Oklevski, 2011; Oklevski, 2012; Wilder & Wentworth, 1918).

Testing the idea that third-level details persist is problematic because of their size. Compared to details at the other levels, pores and edge shapes are much more susceptible to appearing differently in impressions made by the same area of skin or appearing in one and not the other as a result of a variety of different factors. For example, because they are so small, heavy downward pressure during the deposition of the mark can smooth out the shapes along the edges of the ridges. Similarly, tiny particles of the media used to develop the mark can 'fill in' pore or edge shape details. Third-level details may also be more susceptible to appearing differently as a result of even slight changes in the condition of the skin (Monson et al., 2019). As pores are active structures, their appearance may also be affected by whether the pore is open or closed (Monson et al., 2019). This known lack of 'reproducibility' of third-level details means it is difficult to tell whether the details are persistent because it is expected that they will appear differently (or not at all) in impressions that are made by the same area of skin.

Of the few studies that have attempted to assess the persistence of third-level detail, all those that examined it found that the size and shapes of pores were not persistent (Gupta & Sutton, 2010; Monson et al., 2019; Nagesh, Bathwal & Ashoka, 2011; Oklevski, 2011; Oklevski, 2012). However, when considering another aspect of pore detail, Oklevski (2011) found that the number of pores on each ridge was persistent, including in one example involving prints taken 48 years apart. Similarly, another study reported that there was no significant difference in the number of pores per centimetre of ridge in the impressions of groups of people of different ages – which would be expected if the number of pores changed over time (Nagesh, Bathwal & Ashoka, 2011). However, Monson et al. (2019) found that some pores that had been present in earlier impressions were not present in later impressions of the same area (Figure 3.37).

Regarding edge shapes, Monson et al. (2019) reported that in their study edge shapes were only persistent over a short period of time (one to four months), whereas Oklevski (2011) found persistent edge shapes in three-quarters of the impressions he examined, including in those taken at an interval of over 48 years (though the author did also report that the larger the interval was the fewer matching edge shapes there were). A separate study also observed persistence in some edge shapes over a period of 22 years (Oklevski, 2012). Lastly, the only study that specifically considered it found that ridge width was persistent (Monson et al., 2019).

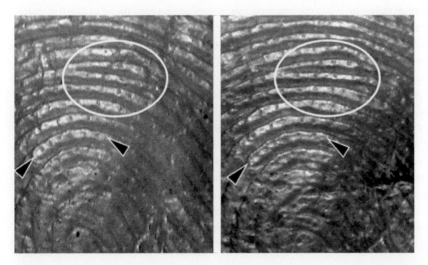

Figure 3.37 These are photographs of the same area of skin of a participant in the Monson et al. (2019) study taken two months apart. Whilst the authors consider that the pore presence and location are similar on the ridge indicated by the arrows, they consider the pores to be variable in the circled areas. *Source:* Reprinted with permission from Monson et al. (2019).

The persistence of third-level details has not been studied or demonstrated to the same extent as that of first and second-level details, and further studies are needed. A greater number of studies may make it easier to evaluate the contrary findings in those already carried out. The poor reproducibility of third-level details could be one explanation for the differences in the findings of these studies. For example, Oklevski (2011) did observe pores that were present in one impression but not in another made by the same area of skin but attributed the difference to distortion caused by the way the impression was recorded, and it could be argued that the differences in the photographs of the skin Monson et al. (2019) observed (as in Figure 3.37) could be similarly attributed to variations in contrast, lighting and focus across the rounded finger. However, Monson et al. felt that they were unable to rule out the possibility that the differences in the presence or absence of pores they encountered could be due to transient but reversible changes in the pores themselves. Another factor that may play a part is the extent to which third-level details are affected by the changes that occur as a result of the aging process, i.e. the detail itself may be the same, but the way it records in an impression may be different.

Like characteristics, third-level details are also extremely variable due to their formation as a result of the same localised tensions and compressions across the surface of the skin. These random forces act upon the three-dimensional ridge units resulting in variances in their width, height, length and shape. The number of the units and the vast potential for variation results in details that are extremely discriminating. There have been few studies that have examined the variability of third-level details, but support for their discriminating power was found using a statistical model which demonstrated that configurations of pores around one characteristic found in agreement in two impressions can provide strong evidence those impressions were made by the same person (Anthonioz & Champod, 2014). Further support comes from studies demonstrating that the use of third-level details in conjunction with details from other levels increases the accuracy of AFIS searches (2014 Jain, Chen & Demirkus, 2006; Parsons, Smith & Thönnes 2008; Roddy & Stosz, 1997; Vatsa et al., 2009).

Though they are considered variable enough to use 'on their own' to support a conclusion of identification, because of their lack of reproducibility, third-level details are usually only used in conjunction with other details. For the same reason, though they are present on all areas of friction ridge skin, third-level details are not recorded clearly enough to be usable by the examiner in every impression of it. In a study of prints taken in ink, Moenssens (1975) found that only 20% of them revealed useable pore detail. Similarly, Anthonioz et al. (2011) found less than 20% of the marks they examined revealed useable pore detail or edge shapes. Due to this and the fact that most impressions the examiner encounters that are comparable will reveal enough first and second-level details, though they may notice distinctive third-level details, the examiner will likely only make a record of them if they are very clear and there are very few other details revealed. Where that is the case, the observations the examiner documents about third-level details may include the number and locations of pores, their distances from each other, and spatial relationships to pores on adjacent ridges. The examiner may also trace distinctive edge shapes and record variances in the widths of the ridges. As with first and second-level details, the examiner may record the clarity of the details, the confidence levels they have that they would appear similarly in a print made by the same area of skin, and the tolerance they have for them to appear differently in a print as a result of distortion.

3.6.1 Pores

Along the top of the ridges are pore ducts that allow sweat from the eccrine glands to be excreted directly onto the friction ridge skin. The prevailing theory of friction ridge skin development

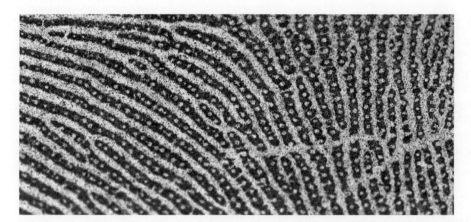

Figure 3.38 The pore ducts along the summits of the ridges often appear as voids in impressions.

indicates that the units that fuse into rows to form ridges each develop around what will become a sweat gland; as a result, there is one duct for each unit (Wertheim & Maceo, 2002). The ducts are located at the bottom of a conical structure that extends to the summit of the ridge, and it is the opening of this structure, rather than the duct itself, that produces the voids along the ridges shown in Figure 3.38 that are referred to as pores by examiners (Bush, 2002).

Pores vary in several ways, including:

- Size
  Two studies have reported that some pores can be over four times larger than others (Nagesh, Bathwal & Ashoka, 2011; Stosz & Alyea, 1994).
- Shape
  Studies have documented pores of various shapes, including circular, elliptical, rectangular, rhomboid, triangular, square and star (Bindra et al., 2000; Nagesh, Bathwal & Ashoka, 2011).
- Distance from each other
  Pores are generally spaced regularly along the ridges, and two studies have measured their frequency in finger impressions at between 5 and 18 per centimetre of ridge (Locard in Wentworth & Wilder, 1932; Nagesh, Bathwal & Ashoka, 2011). Bindra et al. (2000) noted significant variations in the spacing between pores in impressions made by different areas of skin, recording up to 25 pores per centimetre in the hypothenar area of the palm. Though their frequency is generally regular, each pores position within its ridge unit can vary relative to that of others, so some pores may be closer together and others further apart along the same ridge (Ashbaugh, 1999).
- Position
  Some pores may be centrally located in their ridge unit, whereas others may be closer to one side of it than the other (Figure 3.39). A pore that is close to the side of a ridge may create the appearance of a notch in the edge of the ridge in an impression.

Some of these variances have been found to be less reproducible than others. For example, where the same pore appears in two different impressions, its size and shape may appear very different, as shown in Figure 3.40 (Anthonioz et al., 2011; Ashbaugh, 1999; Faulds, 1913; Gupta, Buckley & Sutton, 2007; Gupta, Buckley & Sutton, 2008; Gupta & Sutton, 2010; Monson et al., 2019; Khan, 2011; Oklevski, 2011; Oklevski, 2012; Richmond, 2004; Roddy and Stotsz, 1997).

This means that pore shape and size have very limited value in a comparison. Similarly, the position of the same pore in relation to whether it appears to be in the middle or on the edge of a ridge

Figure 3.39 The size and shape of impressions made by the pore ducts can vary considerably, as can their position relative to the centre of the ridge. Similarly, though the spacing between pores is generally regular, some pores may be found much closer to each other than others.

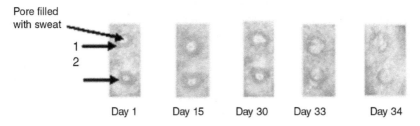

Figure 3.40 This image shows changes in size and shape of the same pores photographed on five different days. *Source:* Reprinted with permission from Gupta and Sutton (2010).

is also not particularly reproducible in different impressions (Anthonioz et al., 2011; Monson et al., 2019; Oklevski, 2011; Richmond, 2004). For example, Oklevski (2011) found that depending on factors such as the amount of downward pressure used to deposit the impression or the texture of the surface, a pore that appears as a notch because it is close to the edge of a ridge can appear as a fully enclosed pore in the centre of the ridge and vice versa.

In contrast, the distance between pores is considered to be less susceptible to distortion and therefore more likely to reproduce similarly in different impressions of the same area of skin (Anthonioz et al., 2011; Ashbaugh, 1999; Khan, 2011; Oklevski, 2011; Richmond, 2004). As a result, it is the distance between pores on the same ridge and their spatial relationships to those on adjacent ridges that are chiefly considered when pores are compared (Ashbaugh, 1999).

3.6.2 Shapes of the Ridges

The undulations that can be seen along the edges of the ridges are primarily a result of the shapes and alignment or misalignment of the ridge units (Figure 3.41) (Ashbaugh, 1999). Each ridge unit in an adult male is approximately half a millimetre long, and if each one had a visible edge shape on each side, then a five-millimetre length of the ridge would reveal twenty edge shapes (Ashbaugh, 1999).

Seven commonly seen shapes have been described by Chaterjee (in Ashbaugh, 1999) as; straight (flat), convex, concave, peak (a sharper, more angular convex shape), angle (a sharper, more

Figure 3.41 The edges of the ridges are not smooth and often reveal distinctive shapes.

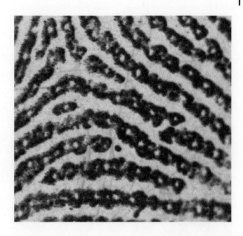

angular concave shape), pocket (the notch formed by a pore being close to the edge of a ridge) and table (the opposite of a pocket). If the examiner is documenting the shapes, they may find they broadly fit these types though they may also choose to describe a shape in their own terms. Sometimes a shape on one ridge will appear facing the opposite shape on the adjacent ridge, i.e. a peak in one ridge may be opposite an angle in the adjacent ridge (Ashbaugh, 1999; Richmond, 2004).

Like pores, the appearance of edge shapes in an impression can be particularly susceptible to distortion, so the edge shapes of the same ridge in two different impressions may vary considerably (Anthonioz et al., 2011; Richmond, 2004; Monson et al., 2019). For example, Richmond (2004) found increasing pressure usually flattens the shapes to produce a straight edge. Anthonioz et al. (2011) did observe some excellent correspondence of shapes in impressions of the same edges but also warned that it is difficult to differentiate between actual shapes made by the ridges and shapes that have been created as a result of the media used to develop the mark (for example particles of powder building up on the side of a ridge). Studies found that the largest, most pronounced shapes are the most likely to reproduce similarly in impressions made by the same area of skin (Anthonioz et al., 2011; Ashbaugh, 1999; Oklevski, 2011). Additionally, Anthonioz et al. (2011) found that shapes that were facing opposite shapes on adjacent ridges were among the most reproducible encountered in their study. Chatterjee (in Richmond, 2004) observed that edge shapes often appear more clearly near the bottom of impressions made by the distal phalanges. However, Richmond (2004) suggested that though the tip of the distal phalange leaves impressions that typically do not reveal shapes as pronounced as those at the bottom, the shapes that are visible are not as affected by distortion, so they may appear more similarly in impressions of the same area. Whilst edge shapes overall were not very reproducible, Anthonioz et al. (2011) did note that the shapes and widths of characteristics reproduced well in different impressions of the same area and maintained their appearance even when subjected to significant distortion.

As well as the shapes along the edges of the ridges, the examiner will also encounter small gaps in the ridges known as 'ridge breaks' (Figure 3.42).

The FBI (1972) described ridge breaks as 'interruptions' in a continuous ridge that are less than the width of the ridge. Occasionally it may be difficult to tell the difference between a ridge break and two opposed ridge endings, though in the latter case, the ridges on either side should react to the endings (Ashbaugh, 1999). Ridge breaks may be caused by two adjacent ridge units not fusing together but may also be caused by a slight dip in the height of a ridge which results in the lower section not

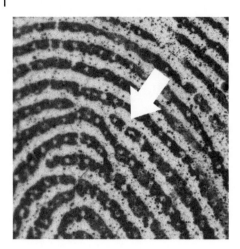

Figure 3.42   The arrow indicates a 'ridge break'.

always contacting the surface (Richmond, 2004). Ridge breaks could also be caused by damage to the skin or by a crease or alternatively by a pore that spans the width of the ridge. Like other third-level details, they suffer from a lack of reproducibility, and it is also possible for breaks to appear in the ridges of an impression that are not a feature of the skin of the donor. For example, a gap in the material on the surface of the skin may result in the appearance of a break in an impression. Similarly, even if there is not a gap in the material, the media used to develop the mark may only react with a constituent of the material it is made in that is not uniformly distributed along the ridges so may create the appearance of breaks in a ridge that are not present on the skin of the donor.

3.7   Creases, Subsidiary Ridges and Scars

Whilst most details are considered as part of only one level of detail, creases, subsidiary ridges and scars may reveal details that can be considered as first, second or third-level detail. For example, the presence and position of a scar can be considered as first-level detail, the specific path of the scar can be considered as second-level detail, and the actual ridge formations and shapes that form the scar can be considered as third-level detail.

3.7.1   Creases

Creases appear in impressions as linear breaks in the ridges and can be divided into two types: flexion creases and skin folds.

3.7.1.1   Flexion Creases

Flexion creases have a firm attachment to the underlying layers of skin and serve to accommodate the movement of the hands, digits and toes. Flexion creases can be categorised into three groups: major, minor and secondary (Loeffler in Schaumann & Alter, 1976).

Major flexion creases on the hands include the three most prominent creases on the palm: the Distal Transverse Crease, the Proximal Transverse Crease and the Thenar Crease (or Radial Longitudinal Crease) (Figure 3.43).

Generally, the paths followed by these creases are similar in most people though in a very small minority, the distal transverse crease and the proximal transverse crease are replaced with one

Figure 3.43 The three major flexion creases that run across the palm are the distal transverse crease (a), the proximal transverse crease (b) and the thenar crease (c).

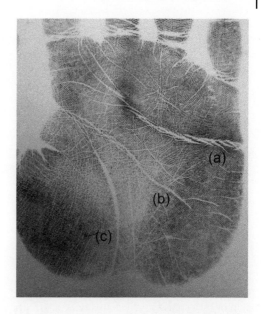

'Simian' crease that extends from one side of the palm to the other. Leadbetter (1976) examined twenty thousand palm prints and found only 39 (or 0.001%) that featured a simian crease. The 'Sydney' crease is another rarely seen variation in which the proximal transverse crease extends from one side of the palm to the other, rather than terminating before reaching the little finger edge of the hand (Schaumann & Alter, 1976). The only other major flexion creases found on the hands are the wrist crease and those that separate the phalanges of the fingers from each other and the palm – which are also known as 'flexures'.

The minor flexion creases are the Finger creases, the Accessory Distal Crease, the 'E' Lines and the Hypothenar Crease (Schaumann & Alter, 1976). These creases are found on the skin with much more variability than the major creases, i.e. some people's skin may not feature any of them, and if one or more of them is present, it may vary considerably in its prominence and length (Schaumann & Alter, 1976).

Finger creases may be found under one or all of the middle, ring and little fingers running vertically down the palm in line with the finger (Figure 3.44).

Where present, the accessory distal crease can be found under the fingers, running parallel to and above the distal transverse crease. 'E' lines may be seen on the edge of the palm between the distal transverse crease and the flexure that separates the little finger from the palm (Figure 3.45). If present, the hypothenar crease will be found running vertically down the little finger side of the palm curving towards the outer side of the hand.

All flexion creases that are not categorised as major or minor are known as secondary flexion creases. These are the smallest of the creases, and where they are present, their number, length and direction vary considerably in different people. Secondary creases include the major accessory creases (short creases which branch off the major creases), the short creases along the edge of the little finger and thumb sides of the palm and the crosshatched creases often seen in the thumb area (Ashbaugh, 1999).

Flexion creases form before birth, prior to, and at the same time as friction ridges. At this time, the basal layer is still forming, and it is believed that the presence of flexion creases disrupts it and prevents ridge development along the paths of the creases (Ashbaugh, 1999). As flexion creases are surrounded by ridges that have been shown to be persistent, it is considered that flexion creases

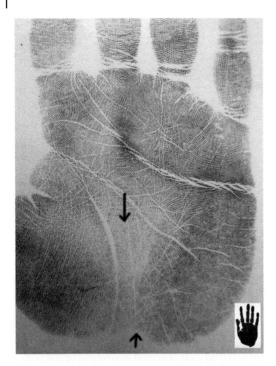

Figure 3.44 The arrows indicate the finger crease that relates to the right middle finger.

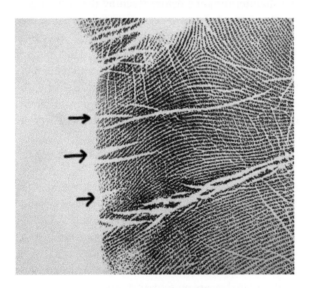

Figure 3.45 The arrows indicate three 'E' lines.

must also persist throughout life as changes to them would also change the ridges around them (Ashbaugh, 1999). Therefore, it is assumed that the same mechanism that maintains the persistent form of the ridges also ensures that the flexion creases persist. Persistence of flexion creases has been observed (as shown in Figure 3.46) though there have been very few studies carried out to systematically test it.

(a) (b)

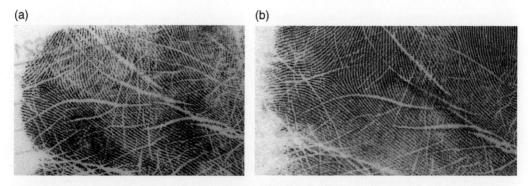

Figure 3.46 These palm prints were both made by the area of skin under the right index finger of the same person. The print in (a) was made 12 years before the one in (b), and both were recorded using the same technique. The most prominent creases are those of the end of the distal transverse crease, and their path is consistent in both impressions. Many, but not all, other creases are also present in both impressions.

One study was conducted by Oklevski (2012), who looked at the persistence of the distal transverse crease, the proximal transverse crease, the thenar crease and the major accessory creases that branch off them in 148 impressions. Using prints from the same palms taken over intervals of up to 24 years, Oklevski found that the path and location of all three major creases were persistent in all cases. However, the author also reported that some major accessory creases were absent in later prints, and some new ones were present. Oklevski observed that the longer the interval between the impressions was, the greater the chance of the differences occurring. The idea that secondary creases, such as the major accessory creases, could vary with age was also expressed by Schaumann & Alter (1976). However, Ashbaugh (1999) compared the major, minor and a selection of the secondary creases in fifty pairs of palm prints taken from the same people over a period ranging from one to sixty months and found that all palmar flexion creases were persistent. Ashbaugh did observe differences in the appearance of creases, but he attributed those to variances in the way one of the prints had been taken. For example, whether the thumb is close to the palm or extended when a palm print is made can affect the appearance of the crease. It is not clear whether the differences Oklevski (2012) observed could also have been due to this factor, a general lack of reproducibility of small creases in impressions, or could be due to the fact that there were much greater intervals between the impressions he examined than those Ashbaugh. The London Metropolitan Police Service (MPS) also carried out a study on sets of prints taken from the same people over time and, like Ashbaugh found that all major, minor and secondary palmar flexion creases persisted (Ashbaugh, 1999).

The flexion creases on the friction ridge skin of the feet (known as the plantar areas) do not have well-established names and are more variable than those of the palm. For example, some plantar impressions will reveal no creases (other than the flexures at the toe joints), whereas others may reveal many (Massey, 2004; Schaumann & Alter, 1976). The plantar flexion creases are formed in the same way at around the same time as the palmar flexion creases, and Massey (2004) compared 50 plantar prints taken from the same people over a period ranging from one to twelve years and found the creases to be persistent.

The power that flexion creases have to differentiate between different individuals varies according to their type because of differing influences on their formation. As well as the volar pads on the distal phalanges, there are also six volar pads on the palm at the time the flexion creases are forming. The general paths of most of the major creases are influenced by the size, shape and position of these pads (Ashbaugh, 1999). The pads appear in similar positions in most people, and as most of the major creases are channelled in-between the pads, there is only a small amount of variation possible in their paths. Studies conducted by Ashbaugh (1999) and the MPS (in Ashbaugh, 1999) indicated that the paths of the major flexion creases may occasionally appear similar in impressions made by different people. As a result, like first-level details, the discriminating power of the major flexion creases is limited. However, except for the finger creases, the paths of all the minor and secondary creases are not influenced in the same way by the volar pads and therefore are considered to form at random as a result of various influences, including the shape and flexing of the palm (Ashbaugh, 1999).[6] As a result, the minor and secondary creases are considered to be more variable and therefore have greater power to differentiate between different people. There is some support for this variability from a few studies involving the comparison of impressions of creases made by different people. Ashbaugh (1999) carried out a study in which 30 pairs of palm prints were paired at random and compared, and the MPS (in Ashbaugh, 1999) in a separate study compared one set of palm prints against 300 others. Both studies reported that other than the major flexion creases, all minor and secondary creases were different in different people. Ashbaugh (1999) also compared the creases of 90 pairs of twins and reported the same conclusion.

To document creases, the examiner may note their type and location and whether they appear as a single crease or are made up of several smaller creases. They may also record the path and length of the crease and any distinctive angles and shapes it forms when it crosses other creases. The examiner may also note the spatial relationships of the crease to other details and/or the number of ridges between them and the crease. The confidence level the examiner has that the crease will appear similarly in a print made by the same area of skin – bearing in mind the distortion factors discussed in Chapter 4 – may also be documented.

3.7.1.2 Skin Folds

Folds in the skin, also known as 'white lines', can form before or after birth as a result of the shape, movement and use of the hand (Figures 3.47 and 3.48) (Ashbaugh, 1999). The buckling of the surface

Figure 3.47 The creases the examiner encounters in impressions of the distal phalanges will be skin folds like the ones in this print. Licensed for use by the Mayor's Office for Policing and Crime under an Agreement between John Wiley & Sons Ltd and MOPAC dated 28th April 2016. *Source:* Courtesy of Mayor's Office for Policing and Crime, London.

---

6 Though the paths of the finger creases are influenced by the volar pads they pass between, these creases are often made up of multiple smaller sections of creases and these were found by Ashbaugh (1999) to vary in impressions made by different people even though the general path of the formation they comprised could appear similarly.

Figure 3.48   As well as flexion creases, this palm impression reveals numerous skin folds. *Source:* Courtesy of Mayor's Office for Policing and Crime, London.

that causes folds may also take place as a result of the aging process as skin becomes looser and folds in on itself, or due to damage caused by skin conditions or by recreational or occupational activities (for example those involving exposure to detergents or very hot or cold conditions) (Maceo, 2011).

Skin folds can be indistinguishable from some flexion creases, though as the latter only appear in specific places, the examiner may be able to differentiate between the two by considering the location of the crease. The main difference between the two is that skin folds develop after the basal layer has formed and do not have foundations deep in the structure of the skin. As a result, they are not supported by the same mechanism that maintains the persistent appearance of flexion creases. New folds can occur at any time (predominantly as a result of the ageing process), and whilst it is also possible for existing folds to change in appearance or cease to be present, once folds are present, they have generally been regarded as being 'more or less stable' (Ashbaugh, 1999; Champod et al., 2016; Schaumann & Alter, 1976). For example, Mierzecki and Miklaszewska (in Schaumann & Alter, 1976) found that if a skin fold is damaged, the skin can regenerate the crease as well as the ridges around it. However, the only study of the persistence of skin folds suggests that they may be less stable than was previously thought (Figure 3.49). Monson et al. (2019) examined 153 pairs of photographs of the same distal phalanges taken at intervals of over eight years and found that in 32% of the cases, new folds had formed in the later impression that were not in the former and in 27% of the cases folds that were in the earlier impression were not in the later impression.

Very few studies have been conducted that looked at skin folds, and none have involved extensive comparisons of the configurations of the skin of different people. However, because they develop at random as a result of the shape and movement of the digits, the aging process or activity, skin folds are considered variable enough to be used as part of the identification process (Ashbaugh, 1999).

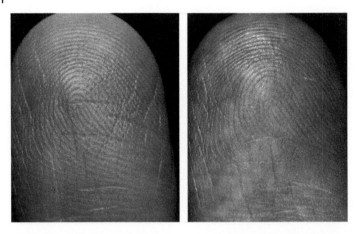

Figure 3.49 These photographs of the same digit taken eight years apart do reveal skin folds that are present in both but also reveal several that are present in one but not the other. *Source:* Reprinted with permission from Monson et al. (2019).

Where appropriate, the appearance of skin folds in the mark can be documented in the manner used for flexion creases.

3.7.2 Subsidiary Ridges

Subsidiary ridges (also known as incipient or interstitial ridges) are thin, often fragmented ridges that can appear in impressions in the furrows between the friction ridges (Figure 3.50).

Several studies have found subsidiary ridges present in one or more impressions of approximately 40% of people (Monson et al., 2019; Stücker et al., 2001). Stücker et al. (2001) did find impressions in which the subsidiary ridges were distributed throughout the impression but reported it was more common to see them either around the core or periphery of the impression only.

Ashbaugh (1992) found that subsidiary ridges may or may not have pore openings along their summits though Okajima (1979) and Penrose & Plomley (1969) both found that subsidiary ridges do not have sweat glands associated with them. The exact process that results in the development of subsidiary ridges is unknown, but one theory is that they develop as a result of a failure in the mechanism the body uses to signal when new ridges should stop forming, resulting in new ones forming which do not have the time to fully mature (Wertheim & Maceo, 2002).

Figure 3.50 Subsidiary ridges are thinner than friction ridges and where present, appear in the furrows between them. *Source:* Reprinted with permission from Cherrill (1954).

Subsidiary ridges are considered to be formed and maintained by the same mechanisms as characteristics, so the theory of skin development that supports the premise characteristics persist can also be applied to subsidiary ridges (Ashbaugh, 1992; Okajima, 1979; Penrose & Plomley, 1969; Wertheim & Maceo, 2002). Though subsidiary ridges may often appear similarly in a mark and print made by the same area of skin, the examiner must be aware that, like third-level details, they are not very reproducible, and their appearance can be highly susceptible to factors such as the amount of downward pressure used to deposit the impressions. The principal reason for this is that subsidiary ridges have a lower elevation than the friction ridges on either side of them. As a result, some or all of them may be absent in an impression made with light pressure but present in another made with heavier pressure. This makes assessing their persistence problematic. Also, several studies that looked at the frequency of subsidiary ridges in people of different ages have established that subsidiary ridges are more frequently found in impressions of older rather than younger people (Cohausz et al. in Okajima, 1979; Stücker et al., 2001). One explanation for this could be that new subsidiary ridges can develop over the course of a person's life. However, as ridges tend to flatten with age and reduce the differential between them and any subsidiary ridges between them, it may be that the subsidiary ridges are just more likely to be recorded in the impressions of older people. Irrespective of whether they were always present or developed over time, Stücker et al. (2001) concluded that once a 'new' subsidiary ridge appeared, it became persistent from that point.

The two studies of the persistence of subsidiary ridges reached different conclusions. Wentworth & Wilder (1932) reported that subsidiary ridges were persistent in impressions made every two years by one person between the ages of 4 and 14. Looking at the images, there are differences in the impressions; for example, every subsidiary ridge does not appear in every impression and is not the same length or width in those in which it does appear, so it is likely that the authors attributed these differences to a lack of reproducibility. However, in their study, which used both impressions and photographs of the skin, Monson et al. (2019) concluded that subsidiary ridges were not persistent. In impressions, Monson et al. (2019) found that their overall absence or presence in different impressions was similar, but they did notice the same sorts of differences that Wentworth & Wilder likely attributed to a lack of reproducibility. Monson et al. (2019) also observed impressions such as one taken 53 years after the first that had subsidiary ridges in it that were not present in the earlier impression. In photographs of the skin, whilst no new subsidiary ridges appeared, and none that were present in earlier photos disappeared, the authors considered that some of the photographs showed substantial differences in their appearance (as in Figure 3.51) though the differences in the photography make it difficult to conclude that the apparent differences do actually represent changes to the structure of the subsidiary ridges.

Because subsidiary ridges are immature ridges that are formed at the same time and by the same random and interdependent forces as characteristics, they are considered variable enough to be used in the identification process (Ashbaugh, 1999; Cowger, 1993; FBI, 1972; Wertheim & Maceo, 2002). However, the author is unaware of any studies that have compared the subsidiary ridges in impressions of different people to assess the extent of their variability.

The examiner may document the appearance of subsidiary ridges in the same way they document characteristics.

3.7.3 Scars

When a superficial injury to the skin (such as that shown in Figure 3.52) occurs, the mechanism that maintains a persistent configuration of ridges will ensure that new cells replace the damaged skin in a way that replicates its appearance before the injury.

Figure 3.51 These are photographs of the same area of skin of a participant in the Monson et al. (2019) study taken two months apart. The authors consider that the circled areas show subsidiary ridges which have changed in the interval between the photographs being taken though the differences in the lighting of the digits make it difficult to verify this. *Source:* Reprinted with permission from Monson et al. (2019).

(a) (b) (c)

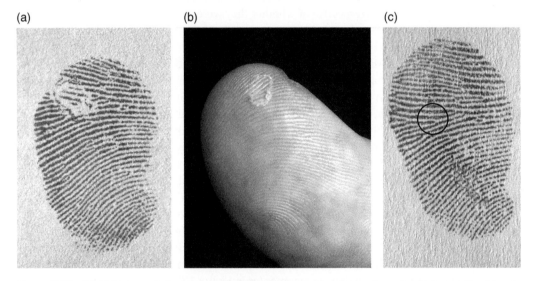

Figure 3.52 Print 'a' was made by the tip of the digit shown in 'b' and records a minor injury to the skin of that digit. Print 'c' was made by the same area of skin just over a month later and no longer shows any trace of the injury: the configuration of the ridges in this area is exactly as it was prior to the injury.

However, if the injury is deep enough to reach the basal layer (approximately 1.8 mm below the surface of the skin), the template that dictates the layout of the ridges on the surface will also be damaged (Cummins & Midlo, 1961). Once such an injury heals, the damage to the template will be reflected on the surface of the skin in a disruption to the layout of the ridges known as a scar (Figure 3.53). The scar itself will become a persistent detail on the skin as damage to the basal layer template will result in cells being produced by that layer that replicate the damage on the surface of the skin (Maceo, 2005). Therefore, once a scar has formed it is maintained by the same mechanism that maintains the persistence of the characteristics. As a result, considering the distortion factors discussed in Chapter 4 and providing both impressions were made after the injury had occurred and the scar had formed, the examiner can expect a scar to appear similarly in a mark and print made by the same area of skin (Figure 3.54).

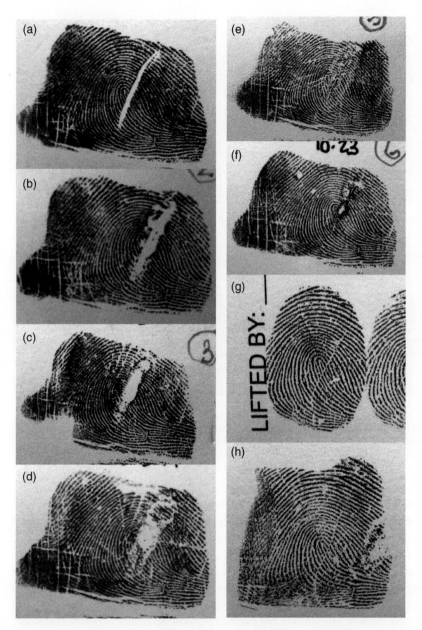

Figure 3.53 These prints illustrate an injury that resulted in a scar. Print 'a' was recorded immediately after the injury, and prints 'b', 'c', 'd', 'e' and 'f' were recorded with intervals of around a week to show the healing process. Prints 'g' and 'h' were recorded around 10 and 18 months respectively after the injury occurred and show the scar in its final form and demonstrate its persistence. Printed with permission from Pete Salicco.

The distinctive appearance of scars is a result of the healing process, which involves skin under the injured area contracting whilst basal layer cells at the edges of the damage spread across the injury to close it (Maceo, 2011). This contraction of the skin often results in a puckering of the ridges that is a feature of many scars (Maceo, 2011) (Figure 3.55).

Scars usually appear as linear breaks that delineate the path of an injury by interrupting the ridges at that point. Voids may also appear along with the break as a result of some of the newly

(a) (b)

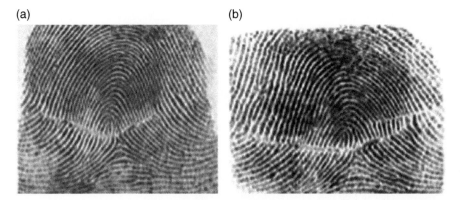

Figure 3.54 These prints were both made by the same area of skin. (b) was made 11 years after (a) and the same scar can be seen in both. *Source:* Reprinted with permission from, Dalrymple (2006).

(a) (b) (c) (d)

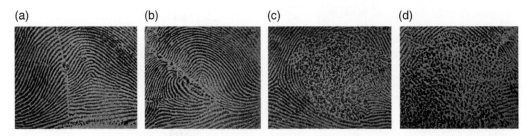

Figure 3.55 These prints all reveal scars, (a) was caused by a deep cut with a sharp knife, (b) was a laceration caused by broken glass, (c) was the result of a burn and (d) was caused by an injury that involved an appreciable loss of flesh. *Source:* Reprinted with permission from Cherrill (1954).

formed skin lying below the level of the ridges around it (Maceo, 2005). As well as puckering around the injury, the ridges often become narrower or wider as they approach the injured area (IEEGFI, 2004). Whilst some scars may be obvious, others may only result in slightly misaligned ridges or changes in the direction of the ridge flow at the point of the injury.

Scars are the result of random damage to the skin and can vary in location, length and appearance according to factors such as the type and severity of the injury. However, their principle discriminating power comes from the way the skin repairs itself (Schreel et al., 2018). Though they are guided, the cells that spread across the skin to repair it are not directed to a specific place but instead are subject to the random influence of the contracting skin. As a result, though there have been no studies to demonstrate it, it is considered that even if two areas of skin were injured in the same place and in the same way, the resulting scars would be different (Maceo, 2005).

3.8 Abnormalities

In addition to the patterns already described, the examiner may encounter extremely rare patterns that cannot be categorised into one of the three pattern families, such as that in Figure 3.56.

The examiner may also encounter impressions with common formations in unusual locations, such as a delta that is at the top of an impression made by a distal phalange.

Figure 3.56 This print reveals a 'cuspal' pattern. The pattern is extremely rare and does not fulfil the requirements to be considered as part of any of the pattern families (Moenssens, 1975). The ridges in the pattern flow vertically from the base of the distal phalange to the tip. *Source:* Reprinted with permission from Cherrill (1954).

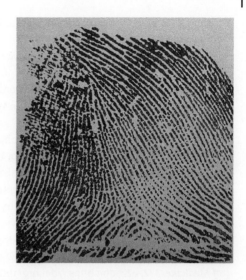

The examiner will also come across impressions revealing temporary damage to the skin caused by a wart, blister or condition such as eczema or psoriasis that may obscure some details or generally reduce the clarity of others (Drahansky et al. 2012). There are also several rare conditions that could affect the appearance of details, including Macrodactyly (where the digits may be unusually large), Brachydactyly (where the digits are unusually short), Syndactyly (where two or more digits are fused together), Polydactyly (where there are extra digits) and Ectrodactyly (where digits are completely or partially missing) (Parker, 1994). Two other conditions, Dissociated Ridges and Dysplasia, result in ridge units failing to fuse into rows and having a similar appearance in impressions. Another condition, Aplasia involves the complete lack of ridges on the skin (Wertheim, 2011). There are also several diseases, including Adermatoglyphia, that can cause the loss of ridge detail (Bordas & Bonsutto, 2020). Whilst these diseases are rare, the drug capecitabine, which is commonly used to treat some types of cancer, can cause a similar gradual loss of ridge detail (Al-Ahwal, 2012; Chavarrir-Guerra and Soto-Perex-de-Celis 2015; Bordas & Bonsutto, 2020; Wong et al. 2009). In one case, it took two years for the ridge detail to return after the patient stopped using the drug (Cohen, 2017). In those documented by Cohen, some ridge detail was still visible in most cases but was not as pronounced as it normally would be, and some patients reported that automatic fingerprint recognition software was unable to recognise their fingerprints.

References

Abraham, J, Champod, C, Lennard, C & Roux, C 2013, 'Modern Statistical Models for Forensic Fingerprint Examinations: A Critical Review', *Forensic Science International*, 232, pp. 131–150.

Al-Ahwal, M, S 2012, 'Chemotherapy and Fingerprint Loss: Beyond Cosmetic', *The Oncologist*, 17, pp. 291–293, doi: https://doi.org/10.1634/theoncologist.2011-0243.

Anthonioz, A. & Champod, C. 2014, 'Integration of Pore Features into the Evaluation of Fingerprint Evidence', *Journal of Forensic Sciences*, 59 (1), pp. 82–93.

Anthonioz, A, Egli, N, Champod, C, Neumann, C, Puch-Solis, R & Bromage-Griffiths, A 2011, 'Investigation of the Reproducibility of Third-Level Characteristics', *Journal of Forensic Identification*, 61 (2), pp. 171–192.

Ashbaugh D, R 1992, 'Incipient Ridges and the Clarity Spectrum', *Journal of Forensic Identification*, 42 (2), pp. 106–114.

Ashbaugh, D, R 1999, *Quantitative-Qualitative Friction Ridge Analysis*, CRC Press LLC, Boca Raton, Florida.

Bindra, B, Jasuja, O, P, & Singla A, K 2000, 'Poroscopy: A Method of Personal Identification Revisited', *Anil Aggrawal's Internet Journal of Forensic Medicine and Toxicology*, 1 (1), http://www.anilaggrawal. com/ij/vol_001_no_001/paper003.html.

Bordas, L. & Bonsutto, J. 2020, 'Adermatoglyphia: The Loss or Lack of Fingerprints and its Causes', *Journal of Forensic Identification*, 70 (2), pp. 154–162.

Bush, L 2002, Pores and Poreoscopy, unpublished.

Campbell, A 2011, *The Fingerprint Inquiry Report*, available at: http://www. thefingerprintinquiryscotland.org.uk/inquiry/3127-2.html.

Champod, C, Lennard, C, Margot, P & Stoilovic, M 2016, *Fingerprints and Other Ridge Skin Impressions*, Second Edition, CRC Press, Boca Raton.

Chavarrir-Guerra, Y. & Soto-Perex-de-Celis, E. 2015, 'Loss of Fingerprints', *The New England Journal of Medicine*, 372, e22, doi: https://doi.org/10.1056/NEJMicm1409635.

Chen, Y & Jain, A, K 2009, 'Beyond Minutiae: A Fingerprint Individuality Model with Pattern, Ridge and Pore Features', Advances in Biometrics, International Conference on Biometrics, June 2009, Alghero, Italy, pp. 523–533.

Cherrill, F, R 1954. *The Finger Print System at Scotland Yard*. Her Majesty's Stationery Office.

Cohen, P, R 2017, 'Capecitabine-Associated Loss of Fingerprints: Report of Capecitabine-Induced Adermatoglyphia in Two Women with Breast Cancer and Review of Acquired Dermatoglyphic Absence in Oncology Patients Treated with Capecitabine', *Cureus*, 9 (1), e969, doi: https://doi. org/10.7759/cureus.969.

Cowger, J, F 1993, *Friction Ridge Skin - Comparison and Identification of Fingerprints*, CRC Press, Boca Raton, Florida.

Cummins, H & Midlo, C 1961, *Finger Prints, Palms and Soles*, Dover Publications, INC, New York, USA.

Dalrymple, B, E 2006, Chapter 6 – Fingerprints, in Mozayani, A, and Noziglia, C (eds.), *The Forensic Laboratory Handbook*, Humana Press.

Drahansky, M., Dolezel, M., Urbanek, J., Brezinova, E. & Kim, T 2012, 'Influences of Skin Diseases on Fingerprint Recognition', *BioMed Research International, Article ID*, 626148, https://doi. org/10.1155/2012/626148.

Dror, I, E, Champod, C, Langenburg, G, Charlton, D, Hunt, H & Rosenthal, R 2011, 'Cognitive Issues in Fingerprint Analysis: Inter-and Intra-Expert Consistency and the Effect of a 'Target' Comparison', *Forensic Science International*, 208, pp. 10–17.

Dror, I, E, Péron, A, E, Hind, S & Charlton, D 2005, 'When Emotions Get the Better of Us: The Effect of Contextual Top-down Processing on Matching Fingerprints', *Applied Cognitive Psychology*, 19, pp. 799–809.

Dror, I, E & Charlton, D 2006, 'Why Experts Make Errors', *Journal of Forensic Identification*, 56 (4), pp. 600–616.

Doak, R 2004, 'Dominant Deltas – A Concept', *Fingerprint Whorld*, 30 (117), pp. 118–123.

Eldridge, H, DeDonno, M, Furrer, J & Champod, C 2020, 'Examining and Expanding the Friction Ridge Value Decision', *Forensic Science International*, https://doi.org/10.1016/j.forsciint.2020.110408.

Evett, I, W & Williams, R, L 1995, 'A Review of the Sixteen Points Fingerprint Standard in England and Wales', *Fingerprint Whorld* 21 (82), pp. 125–141.

Faulds, H 1912, *Dactylography, Or the Study of Finger-Prints*, Milner & Company, Halifax.

Faulds, H 1913, 'Poroscopy: The Scrutiny of Sweat-pores for Identification', *Nature*, 91 (2286), pp. 635–636.

FBI 1972, 'An Analysis of Standards in Fingerprint Identification', *FBI Law Enforcement Bulletin*, 41, pp. 1–6.

FBI 2006, *The Science of Fingerprints, Classification and Uses*, The Project Gutenberg EBook, #19022, United States Department of Justice, Federal Bureau of Investigation.

FSR 2013, Information, Fingerprint Examination – Terminology, Definitions and Acronyms, FSR-I-402

FSR 2017, Codes of Practice and Conduct, Fingerprint Examination – Terminology, Definitions and Acronyms, FSR-C-126, Issue 1.

FSR 2020, Codes of Practice and Conduct for forensic science providers and practitioners in the Criminal Justice System, FSR-C-100, Issue 5.

Galton, F 1892, *Finger Prints*, MacMillan & Co., London.

Gupta, A, Buckley, K & Sutton, R 2007, 'The Effect of Substrate on the Reproducibility of Inked Fingerprint Pore Dimensions Examined Using Photomicrography', *Fingerprint Whorld*, 33 (128), pp. 156–163.

Gupta, A, Buckley, K & Sutton, R 2008, 'Latent Fingermark Pore Area Reproducibility', *Forensic Science International*, 179, pp. 172–175.

Gupta, A & Sutton, R 2010, 'Pore Sub-Features Reproducibility in Direct Microscopic and Livescan Images – Their Reliability in Personal Identification', *Journal of Forensic Sciences*, 55 (4), pp. 970–974.

Gutiérrez, E, Galera, V, Martínez, J, M & Alonso, C 2007, 'Biological Variability of the Minutiae in the Fingerprints of a Sample of the Spanish Population', *Forensic Science International*, 172, pp. 98–105.

Gutiérrez-Redomero, E, Alonso-Rodríguez, C, Hernández-Hurtado, L, E & Rodríguez-Villalba, J, L 2011, 'Distribution of the *Minutiae* in the Fingerprints of a Sample of the Spanish Population', *Forensic Science International*, 208, pp. 79–90.

Hale, A, R 1952, 'Morphogenesis of Volar Skin in the Human Fetus', *The American Journal of Anatomy*, 91 (1), pp. 147–181.

Hawthorne, M. R 2009, *Fingerprints, Analysis and Understanding*, CRC Press, Boca Raton.

Herschel, W, J 1916, *The Origin of Fingerprinting*, Oxford University Press, London.

Hicklin, R, A, Buscaglia, J & Roberts, M, A 2013, 'Assessing the Clarity of Friction Ridge Impressions', *Forensic Science International*, 226, pp. 106–117.

Hicklin, R, A, Buscaglia, J, Roberts, M, A, Meagher, S, B, Fellner, W, Burge, M, J, Monaco, M, Vera, D, Pantzer, L, R, Yeung, C, C & Unnikumaran, T, N 2010, 'Latent Fingerprint Quality: A Survey of Examiners', *Journal of Forensic Identification*, 61 (4), pp. 385–418.

Home Office 1969, The Action Taken When Comparing Finger Impressions, Police Research and Development Branch, Report No. 1/69, January

IEEGFI 2004, Method for Fingerprint Identification II, http://www.latent-prints.com/images/ieegf2.pdf.

Jain, A, Chen, Y & Demirkus, M 2006, 'Pores and Ridges: Fingerprint Matching Using Level 3 Features', 18th International Conference on Pattern Recognition (ICPR'06), Hong Kong, pp. 477–480, doi: https://doi.org/10.1109/ICPR.2006.938.

John, J & Swofford, H 2020, 'Evaluating the Accuracy and Weight of Confidence in Examiner Minutiae Annotations', *Journal of Forensic Identification*, 70 (3), pp. 289–309.

Khan, H, N 2011, 'Identification from edgeoscopy and poroscopy in the examination of partial fingerprints and their significance in crime investigation', PhD thesis, Punjabi University, Patiala.

Laird, A & Lindgren, K 2011, 'Analysis of Fingerprints Using a Color-Coding Protocol', *Journal of Forensic Identification*, 61 (2), pp. 147–154.

Langenburg, G 2004, 'Pilot Study: A Statistical Analysis of the ACE-V Methodology – Analysis Stage', *Journal of Forensic Identification*, 54 (1), pp. 64–79.

Langenburg, G 2011, 'Scientific Research Supporting the Foundations of Friction Ridge Examinations', Chapter 14, National Institute of Justice, *The Fingerprint Sourcebook*, www.nij.gov.

Langenburg, G 2012, 'A critical analysis and study of the ACE-V process', PhD thesis, University of Lausanne.

Langenburg, G & Champod, C 2010, 'The GYRO System – A Recommended Approach to More Transparent Documentation', *Journal of Forensic Identification*, 61 (4), pp. 373–384.

Langenburg, G, Champod C & Genessay, T 2012, 'Informing the Judgments of Fingerprint Analysts Using Quality Metric and Statistical Assessment Tools', *Forensic Science International*, 219, pp. 183–198.

Langenburg, G, Champod, C & Wertheim, P 2009, 'Testing for Potential Contextual Bias Effects During the Verification Stage of the ACE-V Methodology when Conducting Fingerprint Comparisons', *Journal of Forensic Science, 54* (3), pp. 571–582, doi: https://doi.org/10.1111/j.1556-4029.2009.01025.x.

Leadbetter, M 1976, 'The 'Simian' Crease', *Fingerprint Whorld*, 2 (1),p. 12.

Lin, C, H, Liu, J, H, Osterburg, J, W & Nicol, J, D 1982, 'Fingerprint Comparison. 1: Similarity of Fingerprints', *Journal of Forensic Sciences*, 27 (2), pp. 290–304.

Maceo, A, V 2011, 'Anatomy and Physiology of Adult Friction Ridge Skin', in National Institute of Justice, *The Fingerprint Sourcebook*, www.nij.gov.

Maceo, A, V 2005, 'The Basis for the Uniqueness and Persistence of Scars in the Friction Ridge Skin', *Fingerprint Whorld*, 31 (121), pp. 147–161.

Massey, S, L 2004, 'Persistence of Creases of the Foot and Their Value for Forensic Identification Purposes', *Journal of Forensic Identification*, 54 (3), pp. 296–315.

Misumi, Y & Akiyoshi, T 1984, 'Scanning Electron Microscopic Structure of the Finger Print as Related to the Dermal Surface', *The Anatomical Record*, 208, pp. 49–55.

Mnookin, J, Kellman, P, J, Dror, I, Erlikhman, G, Garrigan, P, Ghose, T, Metler E & Charlton, D 2016, 'Error Rates for Latent Fingerprinting as a Function of Visual Complexity and Cognitive Difficulty', U.S. Department of Justice, 249890, May, 2009-DN-BX-K225.

Moenssens, A, A 1975, *Fingerprint Techniques*, Second Edition, Chilton Book Company, Radnor, Pennsylvania.

Monson, K, L, Roberts, M, A, Knorr, K, B, Ali, S, Meagher, S, B, Biggs, K, Blume, P, Brandelli, D, Marzioli, A, Reneau, R & Tarasi, F 2019, 'The Permanence of Friction Ridge Skin and Persistence of Friction Ridge Skin and Impressions: A Comprehensive Review and New Results', *Forensic Science International*, 297, pp. 111–131.

Nagesh, K, R, Bathwal, S & Ashoka, B 2011, 'A Preliminary Study of Pores on Epidermal Ridges: Are There Any Sex Differences and Age Related Changes?', *Journal of Forensic and Legal Medicine*, 18, pp. 302–305.

Neumann, C 2013, 'Statistics and Probabilities as a Means to Support Fingerprint Identification', in Ramotowski, R, S (ed.), *Lee and Gaensslen's Advances in Fingerprint Technology*, Third Edition, CRC Press, Boca Raton, Florida.

Neumann, C, Champod, C, Yoo, M, Genessay, T & Langenburg, L 2015, 'Quantifying the Weight of Fingerprint Evidence Through the Spatial Relationship, Directions and Types of Minutiae Observed on Fingermarks', *Forensic Science International*, 248, pp. 154–171.

Neumann, C, Champod, C, Yoo, M, Thibault, G & Langenburg G, 2013, *'Improving the Understanding and the Reliability of the Concept of "Sufficiency" in Friction Ridge Examination*, U.S Department of Justice, Document No. 244231.

Neumann, C, Evett, I, W & Skerrett, J 2012, 'Quantifying the Weight of Evidence from a Forensic Fingerprint Comparison: A New Paradigm', *Journal of the Royal Statistical Society, A*, 175 (Part 2), pp. 371–415.

New South Wales Police Force 2014, *Fingerprint Induction Course Learning Guide*, Fingerprint Training and Research Unit, Identification Services Branch, Forensic Services Group.

NIST 2012, *Latent Print Examination and Human Factors: Improving the Practice through a Systems Approach*, Report of the Expert Working Group on Human Factors in Latent Print Analysis, US Department of Commerce.

NIST 2013, *Markup Instructions for Extended Friction Ridge Features*, Special Publication 1151, http://dx.doi.org/10.6028/NIST.SP.1151.

NRC 2009, *Strengthening Forensic Science in the United States: A Path Forward*, National Academy Press, Washington, DC.

OIG 2006, *A Review of the FBI's handling of the Brandon Mayfield Case*, U.S. Department of Justice, https://oig.justice.gov/special/s0601/final.pdf.

Okajima, M 1979, 'Dermal and Epidermal Structures of the Volar Skin', *Birth Defects: Original Article Series*, XV (6), pp. 179–198.

Oklevski, S 2011, 'Poroscopy: Qualitative and Quantitative Analysis of the 2nd and 3rd Level Detail and Their Relation', *Fingerprint Whorld*, 37 (145), pp. 170–181.

Oklevski, S 2012, 'Utilization and Correlation Between Characteristics and Features from Different Levels in the Process of Identifying Latent Prints', *Fingerprint Whorld*, 38 (149), pp. 161–173.

OSAC 2020, Best Practice Recommendation for Analysis of Friction Ridge Impressions, Friction Ridge Subcommittee, Organization of Scientific Area Commitees for Forensic Science, Version 1.0, September.

Parker, C 1994, 'Digit Abnormalities', Texas Division, I.A.I newsletter, Oct-Dec., http://www.scafo.org/library/110404.html.

Parsons, N, R, Smith, J, Q & Thönnes, E 2008, 'Rotationally Invariant Statistics for Examining the Evidence from Pores in Fingerprints', *Law, Probability & Risk*, 7, pp. 1–14, doi: https://doi.org/10.1093/lpr/mgm018.

PCAST 2016, Forensic Science in Criminal Courts: Ensuring Scientific Validity of Feature-Comparison Methods, September.

Penrose, L, S & Plomley, N, J, B 1969, 'Structure of Interstitial Epidermal Ridges', *Zeitschrift für Morphologie und Anthropologie*, 61 (1), pp. 81–84.

Ray, E & Dechant, P, J 2013, 'Sufficiency and Standards for Exclusion Decisions', *Journal of Forensic Identification*, 63 (6), pp. 675–697.

Richmond, S 2004, 'Do fingerprint ridges and characteristics within ridges change with pressure?' Available at: http://www.latent-prints.com/images/changes%20with%20pressure.pdf.

Rienti, A, P 1988, 'The Ninth Ridge', *Fingerprint Whorld*, 13 (51), pp. 42–44.

Roddy, A, R & Stotsz, J, D 1997, 'Fingerprint Features – Statistical Analysis and System Performance Estimates', *Proceedings of the IEEE*, 85 (9), pp. 1390–1421.

Saviers, K, D 1987, 'Friction Skin Characteristics: A Study and Comparison of Proposed Standards', Proceedings of the International Forensic Symposium on Latent Prints', Laboratory and Identification Divisions, Federal Bureau of Investigation, July 7–10, Forensic Science and Training Center, FBI Academy, Quantico, Virginia.

Saviers, K, D 1989, 'Friction Skin Characteristics: A Study and Comparison of Proposed Standards', *Journal of Forensic Identification*, 39 (3), pp. 157–163.

Schaumann, B & Alter, M 1976, *Dermatoglyphics in Medical Disorders*, Springer-Verlag, New York.

Schreel, M, Stonehouse, A, Torres, A 2018, 'Assessment of Friction Ridge Skin and Scars with a Focus on Latent Print Examination', *Journal of Forensic Identification*, 68 (1), pp. 43–75.

Srihari, S, N 2009, 'Quantitative Assessment of the Individuality of Friction Ridge Patterns', report submitted to U.S. Department of Justice, No. 227842, Award No. 2005-DD-BX-K012.

Srihari, S, N, Srinivasan, H & Fang, G 2008, 'Discriminability of Fingerprints of Twins', *Journal of Forensic Identification*, 58 (1), pp. 109–127.

Stoney, D, A 2001, 'Measurement of Fingerprint Individuality', in Lee, H, C & Gaensslen, R, E (eds.), *Advances in Fingerprint Technology*, Second Edition, CRC Press, Boca Raton, Florida.

Stosz, J, D & Alyea, L, A 1994, 'Automated System for Fingerprint Authentication Using Pores and Ridge Structure', *SPIE-The International Society for optical Engineering*, 2277, pp. 210–223.

Stücker, M, Geil, M, Kyeck, S, Hoffman, K, Rochling, A, Memmel, U & Altmeyer, P 2001, 'Interpapillary Lines – The Variable Part of the Human Fingerprint', *Journal of Forensic Science*, 46 (4), pp. 857–861.

SWGFAST 2012, Standard for the Documentation of Analysis, Comparison, Evaluation, and Verification (ACE-V) (Latent), Document #8, Ver.2.0, 09/11/2012

SWGFAST 2013a, Standard for Examining Friction Ridge Impressions and Resulting Conclusions(Latent/Tenprint), Document #10, Ver. 2.0, http://clpex.com/swgfast/documents/examinations-conclusions/130427_Examinations-Conclusions_2.0.pdf.

SWGFAST 2013b, Standard Terminology of Friction Ridge Examination (Latent/Tenprint), Document #19, Ver. 4.1, http://clpex.com/swgfast/documents/terminology/121124_Standard-Terminology_4.0.pdf.

Swofford, H, Champod, C, Koetner, A, Eldridge H & Salyards, M 2021, 'A Method for Measuring the Quality of Friction Skin Impression Evidence: Method Development and Validation', *Forensic Science International*, 320, 110703.

Templeton, B 1997, 'The Schematic Fingerprint Chart', *Fingerprint Whorld*, 23 (87), pp. 17–18.

Triplett, M 2020, *Fingerprint Dictionary*, http://fprints.nwlean.net., accessed 8th July 2020.

Ulery, B, T, Hicklin, R, A, Roberts, M, A & Buscaglia, J 2014, 'Measuring What Latent Fingerprint Examiners Consider Sufficient Information for Individualization Determinations', *PLoS ONE*, 9 (11), e110179, doi: https://doi.org/10.1371/journal.pone.0110179.

Vatsa, M, Singh, R, Noore, A & Singh, S, K 2009, 'Combining Pores and Ridges with Minutiae for Improved Fingerprint Verification', *Signal Processing*, 89 (12), pp. 2676–2685.

Wentworth, B & Wilder, H, H 1932, *Personal Identification*, T. G. Cooke, Chicago.

Wertheim, K, V 2011, 'Embryology and morphology of friction ridge skin', Chapter 3, *The Fingerprint Sourcebook*, National Institute of Justice, www.nij.gov.

Wertheim, K & Maceo, A 2002, 'The Critical Stage of Friction Ridge and Pattern Formation', *Journal of Forensic Identification*, 52 (1), pp. 35–85.

Wilder, H, H & Wentworth, B 1918, *Personal Identification*, The Gorham Press, Boston.

Wong, M, Choo, S, P & Tan, E, H 2009, 'Travel Warning with Capecitabine', *Annals of Oncology*, Letters to the editor, Vol, 20 (7), p. 1281.

4

Are the Details in the Mark Likely to Appear Similarly in a Print Made by the Same Area of Skin?

All impressions will be subject to some form of distortion. Friction ridge skin is an elastic, three-dimensional organ that may contact a surface under differing amounts and directions of pressure that can cause it to stretch, compress, twist or slide across the surface. These and other factors mean that no two impressions of the same area of skin will ever be identical. In most cases, the resulting differences in appearance will be minor, and details in a mark and print made by the same area of skin will appear similarly; however, some forms of distortion can significantly alter the appearance of the details. Such distortion will usually result in indicators being left in the mark; however, sometimes the only indicator will be that the mark was found on a surface known to cause distortion or that it was developed using a technique known to do the same, so it is crucial that the examiner considers both the mark itself and its accompanying documentation when assessing whether distortion has occurred. Recognition of the indicators of distortion may allow the examiner to identify which details are affected, and an understanding of the effects of the different types of distortion may allow them to anticipate how those details may appear in a print made by the same area of skin. Failure to recognise significant distortion has occurred and make the appropriate allowances can affect the accuracy of the examiner's conclusion. For example, if the examiner does not consider a difference in two impressions could be due to distortion, they may conclude the impressions were not made by the same area of skin when they were. Conversely, if the examiner fails to consider a similarity is due to distortion, they may conclude two impressions were made by the same area of skin when they were not.

There is limited research available on the effects of distortion, and examiners predominately rely on knowledge gained from training and experience to recognise and interpret it. To date, research is also lacking in how effectively they can do this though one study found significant variability between the detection and interpretation of distortion in different examiners (Neumann et al., 2013).

With marks that do reveal indicators of significant distortion, the examiner may document the type, and extent of the distortion they think has occurred as a record of their views prior to seeing the print. In those cases, the examiner should record whether the distortion affects the entirety of the mark or define the area of the mark that it is limited to (for example, by outlining or shading it on a copy of the mark). Doing so establishes the area in which the examiner anticipates the details may appear significantly differently and separates that area from those where the details should appear similarly. The examiner may also record the tolerance that they have for how the affected details may appear differently in the print as a result of the distortion.

Those considered in this chapter do not encompass all possible causes of distortion but do include those the examiner is likely to encounter most frequently as well as those that can have the biggest effect on the appearance of the details.

The Forensic Analysis, Comparison and Evaluation of Friction Ridge Skin Impressions, First Edition. Dan Perkins.
© 2022 John Wiley & Sons Ltd. Published 2022 by John Wiley & Sons Ltd.

4.1 Downward Pressure

Friction ridge skin deforms on contact with a surface, and one of the factors that determines the extent of the deformation is the amount of downward pressure used to deposit the mark. Variances in downward pressure can create variances in the appearance of the mark and some of the details within it, including:

- The size of the mark
  As downward pressure increases the area of skin that contacts the surface also increases, resulting in a wider and longer mark (Maceo, 2009) (Figure 4.1).
- The width of the ridges and furrows
  Ridge widths vary in different people and in different areas of skin of the same person, but increasing pressure flattens ridges resulting in more of their sides contacting the surface (Ashbaugh, 1999; Maceo, 2009; Richmond, 2004). Therefore, in a mark made with light pressure, the ridges may appear widely spaced and narrower than the furrows between them, but in a mark made by the same area of skin with heavy pressure, the ridges and furrows may have the opposite appearance (Figure 4.1).
- Creases
  Some creases in a mark made with light pressure may not be visible in a mark made by the same area of skin with heavy pressure (Richmond, 2004) (Figure 4.2). Major creases, like those that separate the phalanges will likely still be visible but may appear much narrower.
- Characteristics
  Variances in downward pressure can cause a ridge ending in one impression to look like a bifurcation in another impression made by the same area of skin. Each ridge is not a separate structure from the ridges on either side of it. In fact, the ridges are all one continuous structure, so the downward sloping side of one ridge meets the upward sloping side of the ridge next to it in what is known as the furrow between those ridges (Ashbaugh, 1999). One reason that a characteristic may appear as an ending in

Figure 4.1 Both these impressions were made by the right thumb of the same donor. The mark on the left was made with light pressure and the one on the right with heavy pressure.

Figure 4.2 Both these impressions were made by the left middle finger of the same donor and illustrate the difference downward pressure (lighter in the impression on the left) can have on the appearance of creases; the smaller creases are much less apparent, and the distal flexure between the medial and distal phalanges is narrower.

one impression and a bifurcation in another is that increased downward pressure means more of the sides of the ridges contact the surface, and so the point where two ridges actually do meet gets closer to the surface and can be recorded in the impression. Where that occurs, what appears in an impression taken with light pressure as an ending may appear in an impression taken under heavy pressure as a bifurcation (Figure 4.3).

Varying downward pressure can also affect the appearance of the length of a ridge in an impression meaning that a characteristic can appear to be in a slightly different position in impressions made using different downward pressure. For example, the height of a ridge may vary slightly along its length, so light pressure could result in some parts with a lower elevation not contacting the surface (Ashbaugh, 1999). If a ridge happens to be lower at the point, it comes to an end, light pressure can make it appear as if the ridge comes to an end before it actually does.

- Subsidiary ridges

The entirety of a subsidiary ridge may have a lower elevation than the ridges on either side of it, so it can be absent in one impression and present in another made with heavier pressure (Maceo, 2009; Richmond, 2004) (Figure 4.4). Where they do appear in impressions, because they may vary in height along their length, subsidiary ridges in a mark made under light pressure may appear as short, disconnected sections of the ridge. Whereas in an impression made by the same area of skin under heavy pressure, the same details may appear as a single continuous subsidiary ridge. Richmond (2004) observed that, like the ridges on either side of them, subsidiary ridges also became thicker with increasing pressure.

- The shapes of the ridges

Edge shapes along the sides of the ridges that may be pronounced under light or medium pressure generally become smoother and less distinct under heavier pressure, as shown in Figure 4.5 (Maceo, 2009; Richmond, 2004).

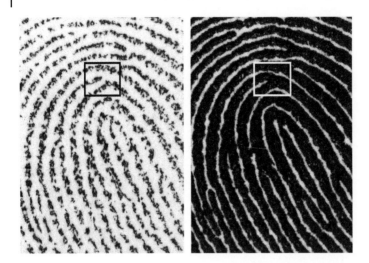

Figure 4.3 Both these impressions were made by the same area of skin, the one on the right with heavy downward pressure and the one on the left with light downward pressure. The highlighted areas show a characteristic that, due to the difference in pressure, appears to be a ridge ending on the left and a bifurcation on the right. The same effect has occurred with another ridge characteristic above and to the right of the highlighted one, though the other ridge endings visible appear as endings in both impressions.

Figure 4.4 Both these impressions were made by the same area of skin. Some of the subsidiary ridges revealed in the impression on the right, made with heavy downward pressure, are not revealed in the impression on the left that was made with light downward pressure. Also, some of the small separate sections of a subsidiary ridge that are revealed in the impression on the left appear connected in the impression on the right.

- Pores

  Richmond (2004) reported that the shape of impressions made by pore openings changes, and their size decreases as pressure increases (Figure 4.5). Pores that appear at the edge of a ridge in an impression taken under light pressure can appear further away from the edge in impressions taken under higher pressure because more of the side of the ridge has contacted the surface (Anthonioz et al., 2011; Oklevski, 2011; Richmond, 2004). Similarly, pores that appear joined or touching in impressions made with lighter pressure may appear to be separate in impressions made with heavier pressure (Richmond, 2004).

Figure 4.5 Both these impressions were made by the same area of skin. The heavier downward pressure used to deposit the impression on the right has resulted in a smoothing out of the undulations along the edges of the ridges and a reduction in the size of the details left by the pores on the summits of the ridges.

4.2 Movement

Movement, also known as lateral pressure distortion, refers to distortion caused by force applied to the skin across the surface the mark is being made on during the deposition of the impression that causes some or all of the ridges to slip. Lateral pressure can be linear or rotational and can result in the slight compression or stretching of some ridges or cause them to slide across a surface for some distance. As a result, humps or curves in ridge flow can be flattened or accentuated, straight areas ridges of flow can appear to be slightly curved, and ridges that slope in one direction can appear to slope in the other (Figures 4.6 and 4.7). Details can also appear to be closer to each other, further away, higher or lower or at different orientations in a mark than they do in a print made by the same area of skin (Figure 4.8). Movement can also obscure details and create the appearance of details that are not on the skin of the donor (Figure 4.9). In extreme cases, it can alter the appearance of the pattern in the mark and make it look like a different pattern (Figures 4.10 and 4.11).

Though much more research is needed, beginning with Maceo (2009), there have been a few studies that explored the extent to which movement can affect an impression (Fagert & Morris, 2015; Tate, Anderson & Eller, 2018; Wallis & Goulet, 2017). The extent of distortion caused by movement depends on a combination of interrelated factors, including the amount of downward pressure used, the ridge flow or pattern type on the skin and the direction or type of the movement (Fagert & Morris, 2015; Maceo, 2009). Studies show that when a distal phalange contacts a surface squarely, not all parts of the skin are under the same amount of downward pressure (Maceo, 2009; Richmond, 2004). The part under the greatest pressure is directly below the end of the distal phalange bone, which corresponds to an area of skin just above the core (Maceo, 2009; Richmond, 2004). The further away from that area, the less pressure an area is under. Maceo (2009) found that when pressure is applied to make a digit slide across a surface, the areas under least pressure will begin to move first whilst the area under greatest pressure remains in place – though that area also will eventually slip if the force is great enough. Because they move first, the details in areas furthest away from the core typically exhibit the greatest distortion to their orientation and location as a result of movement (Fagert & Morris, 2015; Maceo, 2009). Maceo noted that as downward pressure increases, so does the amount of pressure required to move the digit across the surface. As a result,

Figure 4.6 These impressions have been identified as having been made by the same area of skin, but movement during the deposition of the mark on the right has resulted in a compression of the details when compared to those of the print on the left. *Source:* Reprinted from Complete Latent Print Examination, http://clpex.com, Clough, G. FIG 006.

Figure 4.7 These impressions have been identified as having been made by the same area of skin, but the mark on the left appears to have a different slope to that of the print on the right as a result of movement distorting the ridge flow. *Source:* Reprinted from Complete Latent Print Examination, http://clpex.com, Cormier, J. FIG 81.

with heavy downward pressure, the digit is required to move further before the area under the most pressure finally slips on the surface, and that leads to greater distortion than would be seen with light downward pressure (Fagert & Morris, 2015; Maceo, 2009; Wallis & Goulet, 2017).

Whilst its often obvious that movement has occurred if the entire area of skin in contact with the surface slides across it, it can be more difficult to detect if some areas have remained in place while others have moved. Maceo (2009) & Fagert & Morris (2015) both experimented by applying force

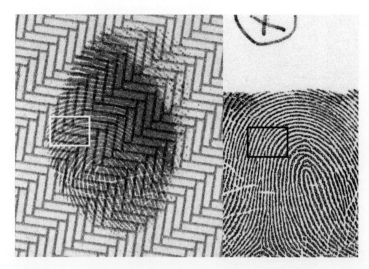

Figure 4.8 These impressions were both made by the right middle finger of the same donor but stretching and rotation of the skin in the mark on the left has resulted in some details appearing differently than they do in the print on the right. For example, the two ridge endings highlighted have a much more horizontal orientation in the mark than they do in the print.

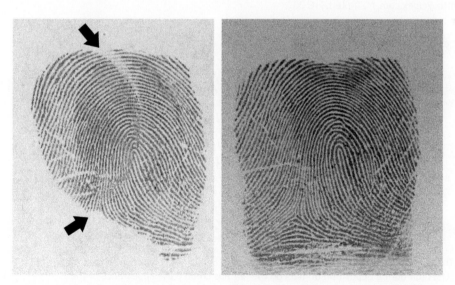

Figure 4.9 These impressions were both made by the right middle finger of the same donor, but there are several characteristics that appear to be in disagreement. For example, there is a ridge ending within the innermost recurve of the right impression, but there is not in the other impression. The impression on the left was deposited in a rolling motion from left to right, and halfway through this motion, the digit slipped downwards on the surface before continuing the motion to record the rest of the impression. The curved line delineated by the arrows is an indication of where the slippage took place. There is also smearing of the material on the ridges into the furrows along and to the left of this line. Though many of the ridges coincidentally align, near the top of the line, there are numerous misaligned ridges. Along and to the left of the line are the appearance of several characteristics which were created by the slippage and do not appear on the skin of the donor.

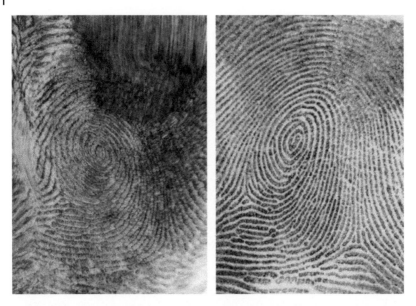

Figure 4.10 These impressions have been identified as having been made by the same area of skin but stretching, and movement of the skin has resulted in the slope of the pattern of the mark on the left appearing quite differently to that in the print on the right. *Source:* Reprinted from Complete Latent Print Examination, http://clpex.com, M. Triplett, FIG 002.

Figure 4.11 These impressions have been identified as having been made by the same area of skin, but the rotation of the digit during deposition of mark on the left has created the appearance of the impressions having different pattern types. *Source:* Reprinted from Complete Latent Print Examination, http://clpex.com, Manno, B. FIG 20.

to digits up until the point when their entire contact area slid across the surface and measuring the effects on the details in the resulting impressions before that occurred. Fagert & Morris (2015) reported that under heavy downward pressure, characteristic locations and their orientations were routinely altered by 1 mm and 10 degrees respectively before all of the digits slid across the surface, though extreme displacements of more than 3 mm and 30 degrees were also observed.

The greatest variability in the appearance of details as a result of movement occurs when the direction of movement is perpendicular to an area of parallel flowing ridges (Fagert & Morris, 2015;

Maceo, 2009). Maceo (2009) explains that furrows compress and expand depending on the force applied, so the more parallel ridges an area has in sequence, the more flexible that area is and the more its details can be distorted. Fagert & Morris suggest this is the reason why the details in plain whorls in their study exhibited a large degree of variability as a result of movement in any direction as they have ridges that flow parallel in every direction. Conversely, Fagert & Morris found that the less uniform nature of loops and arches meant that the variability of their details was more dependent on the direction or type of movement applied. Maceo and Fagert & Morris both reported that moving the finger distally (or 'up' in the direction of the nail) resulted in greater variability in the appearance of the details when compared to moving it left, right or down. Similarly, both reported significant distortion of details as a result of rotating the digit clockwise or anti-clockwise on the surface.

Indicators of possible movement include:

- Smearing
  Usually seen in the furrows and caused by the ridges sliding through the material transferred from the skin or a substance on the surface (Figure 4.9 and 4.13). Sometimes a darker line indicating a buildup of the material the mark was made in along the leading edge of the ridges may also be visible (Maceo, 2009).
- A line running through the impression
  If a digit contacts a surface and then some of the skin that was in contact breaks contact before the movement occurs, a line may be seen in the impression marking the point the movement began, as in Figures 4.9, 4.12 and 4.13.
- Misaligned ridges
  Usually, ridge flow will be straight or form a smooth curve, so a 'step' in the ridges or ridges that appear to end in furrows may indicate movement has occurred (Figures 4.9, 4.12 and 4.13).

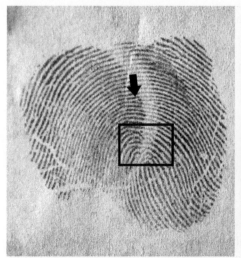 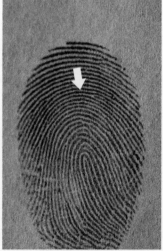

Figure 4.12 A similar motion to that described in Figure 4.9 was used to deposit the impression on the left, which was made by the same area of skin as the impression on the right. The arrow in the left image highlights the appearance of an apparent characteristic that is not on the skin of the donor. Similarly, within the rectangle are repeated areas of ridge flow and multiple misaligned ridges, some of which run at almost ninety degrees to each other. As in Figure 4.9, the point the slippage occurred is marked by a line through the impression.

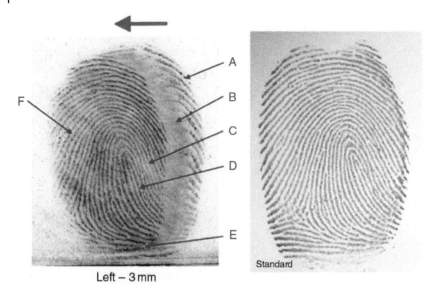

Left – 3 mm

Standard

Figure 4.13 These two impressions were made by the same area of skin. The mark on the left was deposited while the digit was being moved 3mm to the left under high downward pressure. 'A' indicates the initial contact of the digit with the surface. As the digit rolled to the left as it moved in that direction, the ridges initially deposited are clearly recorded along the right edge. 'B' indicates smearing caused by the ridges sliding through the material deposited on the surface during the initial contact. 'C' indicates the compression of the furrows into the core on the opposite side of the mark to the direction of movement ('F' indicates the widening of the furrows on the other side as a result of them being stretched away from the core). *Source:* Reprinted with permission from Maceo (2009).

- Kinks or waves in the ridge flow
  Undulations or sharp changes in the direction of the ridges may also indicate areas affected by movement (Figure 4.12).
- Differences in furrow width in areas of the mark
  Maceo (2009) observed the furrows on the side of the core that was in the direction of the pressure tended to be wider as a result of being stretched away from the core (the area under greatest pressure). Whereas the furrows on the opposite side tended to be squashed together as they were compressed into the core area (Figure 4.13).

4.3 Superimposition

Superimposition describes the deposition of one mark wholly or partly on top of another. Whilst superimposition may be obvious if the marks only slightly overlap, it can be more difficult to detect where the overlap is greater, particularly if the two marks have a similar ridge flow or their clarity is poor.

Superimposition can obscure details that are present on the skin of the donor and create the appearance of details that are not present on the skin of the donor. For instance, the superimposition of the marks in Figure 4.14 has resulted in the appearance of a single mark with a left sloping loop even though the upper mark was made by a right sloping loop.

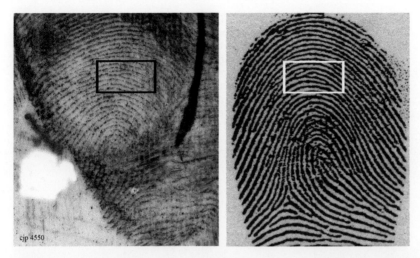

cjp 4550

Figure 4.14 At first glance, it may appear there is only one mark in the image on the left and that mark appears to be left sloping, in contrast with the print on the right. However, there are actually two marks in the image. The transition between the two occurs along a line running through the core where the light ridges from the upper mark meet the darker and thicker ridges of the lower one. The upper mark is clearly identifiable as having been made by the same area of skin that made the print on the right (the box highlights a group of characteristics in agreement in both impressions). The lower mark may also have been made by the delta area of the same digit, making it what is known as a 'double-tap' which is a term that describes two marks made by the same area of skin that are superimposed or touching. *Source:* Reprinted from Complete Latent Print Examination, http://clpex.com, Taylor, K. FIG 003.

The examiner may be able to recognise superimposition as the affected area may:

- Be unusually shaped
  If one of the marks is not entirely within the boundary of the other, their outlines may form a step or notch shape where they intersect, as in Figures 4.15 and 4.16.
- Include misaligned ridges
  Unnatural steps in ridges, ridges that cross each other or come to an end in furrows may be seen as in Figure 4.15.
- Be unusually large
  If the intersecting area lacks clear detail or if the ridges in the area coincidentally align, the result can appear to be a single large impression.
- Include ridges with differing tonal values
  Due to variances in the amount or constituents of the material the marks were made in and their reactions with a development media, the ridges of one mark may be lighter or darker than the other (Figure 4.16).
- Include variances in ridge and furrow width
  If the marks were made by different people or with the use of significantly different downward pressure, then the ridge and furrow widths in the two marks may be distinctly different, as in Figure 4.16.
- Include unusual areas of ridge flow
  Two superimposed marks may both be made by distal phalanges, but if their orientation is significantly different, then for example, the result could be what appears to be a single mark with a delta near the top.

Figure 4.15 This image is composed of two marks made by different donors that overlap. In places, the ridges coincidentally appear to flow almost continuously from one mark to the other. The image reveals several indicators of superimposition, including notches at the points the two marks intersect, ridges crossing over each other and differences in the ridge and furrow width. Pierce & Turnidge (2008) describe superimposed marks with this type of appearance as being 'butterfly' shaped.

Figure 4.16 Like Figure 4.15, this image is also composed of two marks made by different donors. Numerous indicators of superimposition are apparent, including areas of ridge with differing tonal values, misaligned ridges, a slight unnatural step in the outline of the marks where they intersect (hi-lighted by the black lines) and variances in ridge and furrow width. In this example, many of the ridges that intersect can be followed, and the characteristics from each mark can be differentiated from each other.

- Include repeated areas of ridge flow
  Superimposed areas are often made by the same digit contacting the same area of the same surface multiple times, for example, as the result of a drinking glass being picked up, put down and picked up again. In this scenario, the examiner may recognise the same distinctive group of details appearing multiple times in what could appear to be a single mark (Figure 4.17).

Once an area of superimposition is recognised the examiner may be able to follow both sets of ridges through it and be confident in the nature of the characteristics and other details revealed (Figure 4.16). However, if the superimposed area lacks clarity, the examiner must carefully consider the confidence they have that the characteristics within it would appear similarly in another impression made by the same area of skin.

Figure 4.17 The superimposition of the marks on the left has been created by the same area of the skin contacting the surface multiple times. *Source:* Reprinted from Complete Latent Print Examination, http://clpex.com, Texas. FIG 107.

4.4 Colour Reversal

In almost all prints of friction ridge skin, the ridges appear as dark lines. Whilst the same is true in most marks, in a reverse colour or 'tonally reversed' mark, the opposite is true; it is the furrows between the ridges that appear as the dark lines. As a result, such marks will reveal some details that are the opposite of those seen in a print made by the same area of skin, i.e. a bifurcation in a reverse colour mark will appear as a ridge ending in a print and vice versa (Figure 4.18).

Reverse colour marks can be produced in a variety of ways, including:

- A surface being contacted by skin with a contaminant (like blood) on it that has dried on the summits of the ridges but is still wet in the furrows. This can result in the contaminant only being transferred to the surface from the furrows (Langenburg, 2008). A similar scenario can occur if the contaminant is wiped off the ridges or removed from them by multiple contacts with a surface but remains in the furrows.
- The use of heavy downward pressure during deposition of the mark which squeezes a viscous substance (for example, sweat or grease) off the ridges into the furrows where it is left behind when the skin breaks contact with the surface as in Figure 4.18 (Langenburg, 2008).
- A contaminant on the surface being lifted off by the ridges leaving behind an impression of the furrows in the contaminant (Figure 4.19).

Reverse colour marks can also be produced if, rather than the skin itself, a cast of the skin is used to make an impression. In a cast of friction ridge skin, the ridges are recessed and the furrows protrude, so if the material is applied to the cast and used to make an impression, the material will transfer from the furrows to the surface and produce a reverse colour mark.

Pierce (1989) describes how the over-application of powder to 'true-colour' marks on some plastic surfaces (including plastic bags) can turn them into reverse colour marks. Pierce reported that the continued use of powder, past the point at which the mark became visible, resulted in powder

Figure 4.18 These two impressions were both made by the same area of skin, but the mark on the left is in reverse colour. As a result, the ridges are actually the lighter lines, and the furrows are the dark lines that contrast with the surface. This causes ridge endings to appear as bifurcations and vice versa, as can be seen in the circled area. This mark was made in a contaminant that was on the skin of the donor that has pooled along the right-hand edge of the mark.

adhesion to the surface and the removal of the powder that had initially adhered to the material transferred by the ridges.

Another study found that reverse colour marks could be produced by exposure to very high temperatures. Bradshaw et al. (2008) found that when marks on ceramic tiles that had been exposed to temperatures that may be encountered at arson scenes were powdered, the powder settled into the furrows rather than adhering to the ridges and produced reverse colour marks.

Moore et al. (2008) also reported reverse colour marks being produced as a result of exposure to high temperatures but by a different mechanism. The authors found that when marks in blood on ceramic tiles were exposed to temperatures of 700 and 900° C, though the blood was burned off the surface, the vacuum metal deposition process could still develop the mark but in reverse colour. It is thought that this is due to the process being able to detect chemical differences between the parts of the surface that were under (and therefore initially protected by) the mark and those that were not – which would include the furrows.

The use of vacuum metal deposition can result in reverse colour marks even if the marks have not been exposed to high temperatures. The process involves coating the surface under examination in a very thin layer of gold followed by a layer of zinc. Whilst the gold will deposit on and form a layer on the surface, it will diffuse into the material the mark was made in. As the zinc should only deposit on the gold, the ridges of the mark should be left distinct and undeveloped against a zinc-coloured surface. However, in practice, on some plastic surfaces, the zinc sometimes deposits onto the material the mark was made in, rather than the gold surface around it, resulting in zinc-coloured ridges rather than zinc-coloured furrows (Champod et al., 2016).

Figure 4.19 These impressions were both made by the same area of skin. The mark on the left was made on a metal surface that was covered in a contaminant, in this case, ink. When the digit came off the surface, the ridges removed some of the ink in contact with them, revealing the lighter coloured metal underneath but leaving the ink in the furrows undisturbed. These types of marks are known as 'takeaways' and have been encountered on surfaces contaminated with a variety of substances, including dust and dried liquids including blood and may also be seen on adhesive surfaces if the ridges have removed some of the adhesive (Ashbaugh, 1999; Geller, Volinits & Wax; 2017; Pierce, 1989). Whilst with some reverse colour marks, such as the one in Figure 4.18, the examiner may not realise the mark is in reverse colour until they compare it with a print made by the same area; it is often obvious with a takeaway mark that the details are in reverse colour. This is because in a takeaway mark, though the ridges may be light in the mark and dark in the print, they do contrast with the background in the same way they do in the print, i.e. the details that are different from the background are made by the ridges.

Several studies have reported reverse colour marks being produced as a result of blood being applied to true-colour latent marks on a surface. Under certain conditions, the blood can gather in the furrows and make the mark appear as a visible, reverse colour mark in blood on the surface (Huss et al., 2000; Lin et al., 2019; Praska & Langenburg, 2013).

Marks can be wholly or partly in reverse colour (Geller et al., 2018), and indicators that some or all of a mark may be in reverse colour include:

- Light lines in a mark with a dark background. Such marks can occur when powder adheres to a contaminant on the surface rather than the material the mark is made in (as in Figure 4.20) or when a contaminant on the surface is removed by the ridges (as in Figure 4.19). The dark background may be present around the mark as in 4.19 or may only be within the outline of the mark as in 4.20. The mark in Figure 4.18 also reveals a dark edge to one side of it that may have been due to the material the mark was made in (ink) being forced off the ridges by downward pressure.
- The dark lines in the mark being narrower than the spaces in between them (if the dark lines were ridges, they would normally be wider than the furrows between them) (Figures 4.19 and 4.20).
- Linear breaks in the mark consistent with the appearance of creases are dark rather than light lines (Figures 4.19 and 4.20).
- Details consistent with the appearance of pores can be seen in between the dark lines (Figures 4.19 and 4.20). The examiner must exercise particular care when using this indicator as particles of powder used to develop the mark or particles of a contaminant on the surface can appear like pores.

Figure 4.20 Both of these impressions were made by the same area of skin, but the impression on the left is in reverse colour and the one on the right is in true colour. The dark background and outline to the impression on the left are due to the powder used to develop the mark adhering to the background rather than the ridges. *Source:* Reprinted with permission from Castellon (2014).

Figure 4.21 Very closely spaced, parallel lines known as 'Split ridges' sometimes outline the furrows in marks that are in reverse colour. *Source:* Reprinted with permission from Castellon (2014).

Castellon (2014) reported that the presence of 'split ridges' can also be an indicator that a mark is in reverse colour. Split ridges (Figure 4.21) are thought to be caused by a thin line of the material on the skin accumulating on the shoulders of each ridge, resulting in the appearance of two thin parallel lines that represent either side of a furrow. This may be due to heavy downward pressure squeezing the material off the summits of the ridges to the shoulders (Richmond, 2004).

The examiner may encounter a further indicator during the comparison stage if they complete their analysis without realising the mark is in reverse colour and compare the details with a print made by the same area of skin. In this case, as the examiner would be comparing the furrows in the mark with the ridges in the print, though the details would have a similar location and orientation, the number of ridges between them would be different.

Reverse colour marks may not always reveal all the indicators discussed though usually there are multiple indicators present. In Castellon's (2014) experiments, at least three of the indicators were present in every reverse colour mark he created.

The examiner should be aware that many media used to develop marks, such as white powders or cyanoacrylate (superglue), are designed to render the ridges a light colour. Such marks would not be considered reverse colour as it is expected that marks developed by such techniques would have light ridges. If such marks have been developed by others and forwarded to the examiner, then the images of them will often be colour corrected, but if not, or if the examiner is viewing the actual mark *in situ* on the surface, then they should ensure they are aware of the colour 'signature' of the development media and make the appropriate allowances.

4.5 Direction Reversal

Occasionally a mark will appear as a mirror image of a print made by the same area of skin. For example, though they were both made by the same area, the mark in Figure 4.22 reveals a left sloping loop whilst the print reveals a right sloping loop.

Almost all marks the examiner encounters will have been deposited on the surface they are found on; however, most 'reverse direction' marks are deposited on one surface and transferred before being found on another surface (or a different part of the same surface). There has been very little research examining the conditions and circumstances under which this type of transfer can occur accidentally, but a study by Jabbal, Boseley & Lewis (2018) did identify several factors that contributed to whether a transfer would be successful with certain surface types, and how clear the mark would be if it was. In an experiment predominantly involving the transfer of marks from a non-porous surface to a porous surface, the authors found that increasing both the amount of pressure on the two surfaces and the amount of time they were in contact with each other improved the clarity of the transferred marks. Jabbal, Boseley & Lewis (2018) also reported that the contact between the surfaces needs to occur shortly after the deposition of the mark for the transfer to be

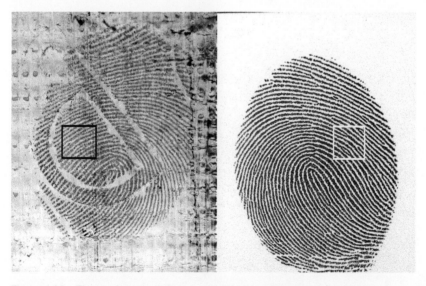

Figure 4.22 The mark on the left was made by the same area of skin that made the print on the right. The mark was deposited on a metal biscuit tin, whereupon a piece of tape was stuck to the tin over the mark. When the tape was removed, some of the mark transferred to the adhesive side of the tape and was developed on that surface. This resulted in a 'reverse direction' mark that appears to be a mirror image of the print - the same group of characteristics is highlighted in both impressions.

successful between non-porous and porous surfaces and that the quantity and composition of the material that made up the mark was also a factor with transfer between two porous surfaces. Where the transfer of the mark was successful, the authors reported that it was difficult to tell by looking at those marks whether they had been transferred or made on the secondary surface.

Though there has been very little published research into reverse direction marks, there have been several case studies.

Czarnecki (2005) describes a case where a mark was deposited on the adhesive side of a piece of tape that was then stuck to an envelope. When the tape was removed, enough of the mark had been transferred for it to be developed on the envelope. Martin (1994) describes a similar case involving the transfer of a mark from the adhesive side of the tape to the plastic frame of a cabinet. Whilst reverse direction marks can be identified as they were in these cases, unless the examiner anticipates a mark may be reversed, it is much less likely that it will be. This is because ordinarily there would be no reason for the examiner to compare the details within a mark revealing a left sloping loop to those in a print revealing a right sloping loop, or the details on the right of the core of a whorl in a mark with those on the left of the core in a print. In the cases described by Czarnecki and Martin, the examiners had access to the prints of the people who made the marks but initially did not identify them as having been made by the same areas of skin. The marks were only identified after the examiners noticed that neither mark extended beyond the area on the surface where the tape had been and considered the possibility that they could have been deposited on the tape and transferred to the surface. Zorich (1992) describes a similar case in which a mark was deposited on the adhesive side of a piece of tape and transferred to the non-adhesive side of another piece. In this case, the same mark was developed on both surfaces, and a comparison of the two established that they both revealed exactly the same area of skin but were mirror images of each other – suggesting that one was in the reverse direction.

As well as transfers occurring between adhesive and non-adhesive surfaces, case studies have also documented reverse direction marks being produced as a result of contact between two non-adhesive surfaces, as in Figure 4.23.

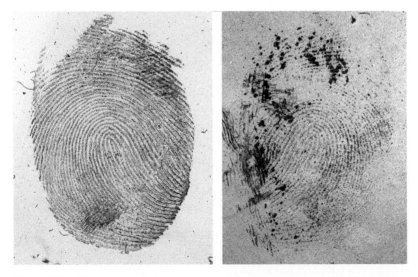

Figure 4.23 The mark on the left was made on a plastic bag, and after deposition, the area of the bag the mark was on was pressed onto another part of the bag. Some of the material the mark was made in was transferred to the other part of the bag allowing the same mark to be developed on both parts. When comparing the two, there are obvious similarities in their outlines. The fact that the powder used has developed much more detail in the one on the left is an indicator that this is likely to be the original mark.

Zampa et al. (2008) describe a mark being transferred from one area of a piece of paper to another, likely as a result of the paper being folded to fit into an envelope. Similarly, Czarnecki (2005) and Reneau (2003) both report instances of marks being deposited on plastic bags or wraps that were subsequently folded over on each other resulting in the transference of the marks to different areas. In both cases, the fold occurred right on the edge of the mark and created the appearance of a single mark with a whorl pattern that was in fact composed of the upper part of two separate marks made by digits with a loop pattern. Lane et al. (1988) describe cases involving the transfer of marks from one plastic bag to another (believed to be as a result of the bags being stacked on top of each other) and the suspected transfer of a mark from a metal surface to a plastic bag that was lying top of it, whereas Kershaw (2000) reports a case where a mark was transferred from a page of a textbook to the facing page. As in the case, Zorich (1992) describes, some of these marks were detected as being reversed as a result of both original and transferred marks being developed and the examiners noticing mirrored similarities in their shapes and the details they revealed. In such cases often the transferred, reverse direction mark will not have developed as strongly as the original mark as a result of less material being present in it than the original mark. However, in some cases, such as those described in Lane et al., only the transferred marks were found (the original marks were believed to have been destroyed by handling prior to examination). The examiners in one of these cases only considered that the marks could be in the reverse direction because of previous recent experience of reverse direction marks on a similar surface. In the other case Lane et al. describes, the mark had been mounted on transparent plastic backing, and an examiner discovered the mark was in the reverse direction after glancing at it from the reverse side and noticing its similarity to a print of a suspect. Similarly, Kershaw (2000) recognised details revealed in a mark as being a mirror image of those in a print of a person suspected of involvement in the offence as a result of his familiarity with that print. This led him to examine the facing page of the book the mark had been found on (which he had not been required to examine initially) and discover the original mark. Kershaw was able to provide further evidence for the transfer by comparing the relative locations and orientations of the marks on their respective pages.

As well as resulting from a transfer of a mark from one surface to another, reverse direction marks can also be produced on paper as a result of the mark being visible from both sides of the paper. After deposition, some of the material the mark is made in will penetrate a porous surface which may allow it to be developed on both sides. In such cases, if it is noticed that the mark is visible on both sides, the examiner will likely be provided with both reverse and correct direction images of the mark (Figure 4.24).

Brazelle et al. (2018) and Pickard (2012) both reported instances of the same marks being developed on both sides of some types of paper.[1] Brazelle et al. reported there was a difference in the appearance of the marks, with the reverse direction one having a less intense reaction with the development media and a hazy or diffused appearance. However, Picard reported that in their study, in most cases with marks made on tissue paper that could be seen from both sides, it could not be determined which side the mark had been made on without knowing what the correct direction of the mark should be.

The examiner may also encounter reverse direction marks if the mark has been made or mounted on a transparent surface and has been accidentally photographed or scanned through the surface from the wrong side. Similarly, whilst most tapes used to lift marks from surfaces are transparent

---

[1] Pickard (2012) also reported an instance in which a mark developed in a booklet of tissue paper had actually been deposited on one page and transferred through that page to be developed on the adjacent page – though the mark did not reveal clear enough detail for it to be identifiable. Pickard also describes another instance involving marks that were deposited on one side of tissue paper on a shop counter but were only developed on the other side.

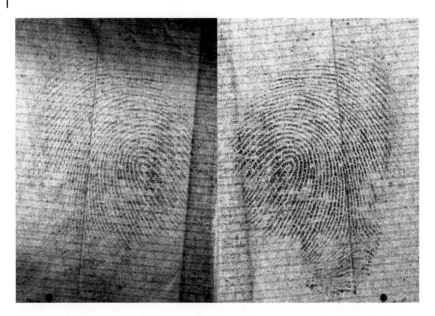

Figure 4.24 This mark was made on cigarette paper and as the paper is very thin the developed mark can be seen on both sides. It appears as a right loop on one side and a left loop on the other (the correct direction is shown in the image on the right).

and so allow the mark to be viewed through the tape in the correct direction, some lifting materials are opaque and so only allow the mark to be viewed from the underside in the reverse direction. Reverse direction marks can also be produced if, rather than the skin itself, a cast of the skin is used to make an impression.

Most marks the examiner encounters will be in the correct direction, and it is not usually practical or necessary to consider them as being anything other than the direction they appear in. However, if the mark was found on a surface that reverse direction marks are known to have been found on and attempts to identify it in the direction it appears to be in have failed, the examiner should consider reversing its direction.

Reverse direction marks have been found on surfaces that:

- were in contact with another surface (for example, adhesive tape);
- are flexible and could have folded over on themselves (for example, a plastic bag);
- are transparent or very thin.

As well as being aware of those surfaces, the examiner may also detect a reverse direction mark if both original and transferred marks have been developed by observing that the marks are mirror images of each other in terms of their shape, the details they reveal and their position and orientation on their relative surfaces. In this case, if one mark is 'weaker' than the other, that mark is more likely to be in reverse direction. If only the transferred version of the mark has been developed, then there may be little or no indicators that it is in the reverse direction. Saviano (2003) observed that some transferred marks (such as the one in Figure 4.23) have a 'blotchy' overall appearance that is in part due to the material the mark was made in being trapped between and re-distributed by the two surfaces resulting in differences in concentrations in different areas. However, marks that have not been transferred can also have a blotchy appearance, and so if both original and transferred marks have not been developed, then the only way an examiner may detect a reverse direction mark may be as a result of their familiarity with a particular print that reveals similar but mirrored details to the mark. In a study of marks transferred from one piece of paper to another, Flanders, Moyer & Fisher (2021) found that some reverse direction marks displayed a 'corona' or 'halo' effect caused by the edges of the mark reacting more strongly to the development technique than the centre. The authors also reported that the majority of the reverse direction marks they created were faint and/or 'fuzzy' in that they appeared out of focus.

The identification of the donor of a reverse direction mark is a relatively rare occurrence, but as marks may reveal few or no indicators they are in the reverse direction it is likely some such marks are never considered as being in the reverse direction and therefore never identified. If a reverse direction mark is identified, it is crucial that information is documented and reported due to the potential implications on its evidential value (i.e. that the donor may not have touched the surface the mark was found on).

4.6   The Surface the Mark Was On

Flexible surfaces like plastic bags and adhesive tape can wrap around a digit and create unusually shaped marks that are significantly larger than those left by the same digit on a less flexible surface (Figure 4.25). The size of the mark can be an indicator of the digit that made it, so distortion of this type can be an important consideration with a mark on a flexible surface.

If a mark is deposited over a small fold or crease on a flexible surface, it may be necessary to unfold the surface to photograph the mark. This may split the mark into two separate areas of ridge detail and may also result in one or both of the marks having an unnatural straight edge marking the location of the fold or crease (Figure 4.26).

Such marks may end up some distance from each other on the surface and therefore may be provided to the examiner in separate images. As a result, the examiner may be unaware that the

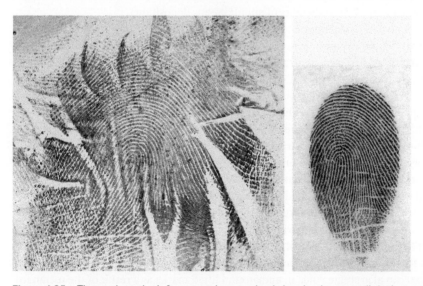

Figure 4.25 The mark on the left was made on a plastic bag by the same digit that made the print on the right (a left little finger). The bag wrapped around the digit as the mark was being deposited, resulting in a mark with an unusual shape and one that appears far larger than would be expected to be produced by that digit on a less flexible surface.

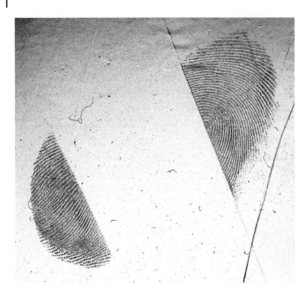

Figure 4.26 These two marks were deposited by the same digit at the same time across a fold in a plastic bag. The bag was unfolded to allow the marks to be photographed, and as a result, two marks were produced. The presence of a mark with an unnatural straight edge on a flexible surface should alert the examiner that it could have been deposited simultaneously with another mark.

marks were made together, so the observation of the unnatural edge may lead them to consider that possibility. For example, Zampa (2008) describes a case involving two such marks on a napkin that was handled by suspects in a firearms offence. Neither mark on its own revealed enough detail for the examiners to identify the donor, but as the marks revealed indicators of being deposited simultaneously over a crease in the napkin, the examiners were able to consider them as one mark and identify them (see Chapter 7 for indicators of simultaneous deposition).

Separated areas of ridge detail can also be created if a mark is deposited across the point where two different surfaces meet, for example, where one piece of paper overlaps another piece underneath it. Similarly, an uneven surface can create gaps in the ridge flow as a result of the skin only contacting the elevated parts of the surface (Figure 4.27).

Even if the skin does contact recessed areas of a surface, if the mark is lifted from the surface using adhesive tape, gaps may still result as the tape may not be able to contact all the same areas. As well as gaps, if the powder is used to develop the mark, uneven surfaces can also create lines through the mark that correspond with the uneven edges of the surface as a result of the powder adhering to those edges (or contaminants along those edges).

Curved surfaces can cause a mark to appear differently to a print of the same area of skin made on a flat surface (Figure 4.28). The angles and spatial relationships between details can be altered as a result of the skin wrapping around the surface. For example, if marks are made on a pole by the fingers wrapping around it, when the marks are lifted, those made by the distal phalanges may appear in the lift as being very close to a mark made by the triradiate area of the palm. When compared to a print on a flat surface made by the same hand, these areas would be significantly further apart. Curved surfaces can also create voids in and around lifted marks as a result of the rectangular lifting tape being unable to fully conform to the surface.

Dirty or patterned surfaces (for example, banknotes) can create the appearance of details that are not present on the skin of the donor as well as obscuring those that are (Figures 4.29 and 4.30).

Figure 4.27 This mark was made on the metal lid of a box of chocolates. The mark crosses over one of the letters of the brand name that protrudes from the surface. The uneven surface has divided the mark into three sections and created gaps in between each section as a result of the lifting tape used to retrieve the mark from the surface not contacting the sides of the raised letter.

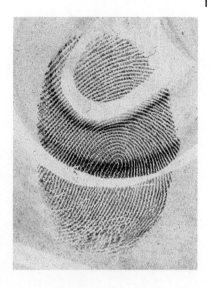

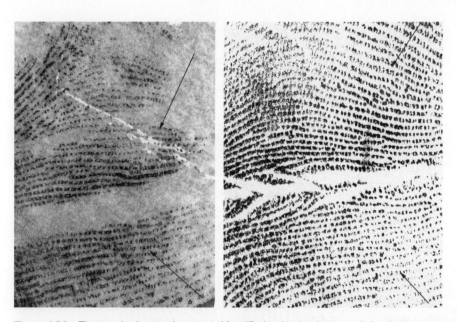

Figure 4.28 These palm impressions were identified as having been made by the same area of skin. The distance between the two arrowed groups of characteristics appears greater in the print on the right as a result of the mark being made by the hand wrapping around a curved surface (a large jug). *Source:* Reprinted from Complete Latent Print Examination, http://clpex.com, Austin Police Department, FIG 26.

If the surface does obscure a detail, the examiner may sometimes be able to infer its presence in a location even though they cannot see it. For example, if the examiner observes three ridges flowing into a small area of surface interference yet only observes two flowing out, then providing the flanking ridges can be followed either side of the interference, the presence of a characteristic can be inferred as in Figure 4.31.

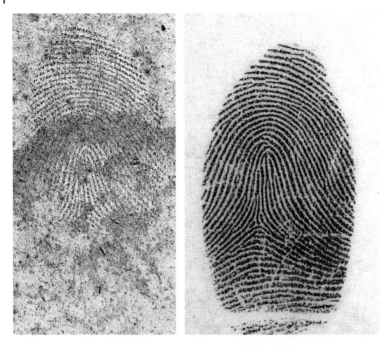

Figure 4.29 These impressions were both made by the right forefinger of the same donor. The mark on the left was made on the bonnet of a car that was contaminated with dirt and other substances. The powder used to develop the mark has also adhered to the contaminants on the surface and makes it difficult to see some of the details in the mark.

Figure 4.30 These impressions have been identified as having been made by the same area of skin. The mark on the left was made on a door frame, and the grain of the wood interrupts the flow of the ridges throughout the mark. *Source:* Reprinted from Complete Latent Print Examination, http://clpex.com, S. Siegel, Tx, FIG 95.

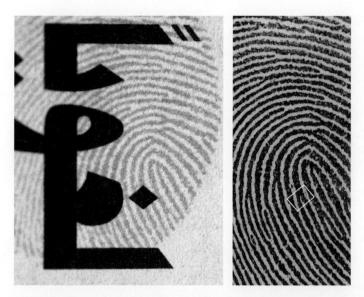

Figure 4.31 The mark on the left was made on a newspaper by the same digit that made the print on the right. Some of the text has obscured a bifurcation in the mark, but as three ridges can be seen going into the text and only two coming out, the presence of a characteristic in the mark (though not its type) can be inferred.

4.7 The Material the Mark Was Made In

A large amount of material on the skin at the time of deposition may thicken the appearance of the ridges in the mark, filling in and obscuring finer details in much the same way that heavy downward pressure can. Conversely, small amounts of material may result in marks with very thin or discontinuous ridges as a result of the material not covering the entire width or length of the ridges.

In addition to the natural secretions of the body, there are many contaminants that may be found on the friction ridge skin of a donor, including dirt, traces of foodstuffs, cleaning products, cosmetics, lotions, grease, blood etc. and the appearance of the mark may vary depending on the constituents of the material it is made in. For example, the details revealed in marks made in the blood are often less clear and do not accurately reveal the finer details (Figure 4.32).

Even if the details in the mark do appear similarly to those in a print at the time the mark is deposited, the material the mark is made in may deteriorate significantly between then and the time the mark is found. The extent of the deterioration that occurs will depend on the amount of time the mark is on the surface, the nature of the surface, the environmental conditions and the constituents of the material the mark is made in (see Appendix 5 on Activity level propositions for more information). For example, the deterioration may involve evaporation of a significant part of the material that can result in the developed mark appearing faint with very thin ridges. Alcaraz-Fossoul (2016) described distortion ('ridge-drift') considered to be associated with the amount of time a mark had been on a surface that resulted in a ridge that bifurcated with the one on its left appearing to bifurcate with the one on its right after the mark had remained undeveloped on the surface for several weeks.

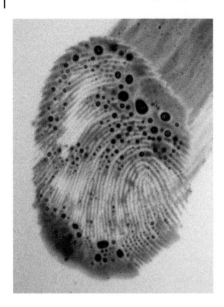

Figure 4.32 This mark was made on a ceramic tile in the blood that was on the digit.

4.8   The Media Used to Develop the Mark

Chemical treatments used to develop marks react with a particular constituent of the material the mark is made in, and so if that constituent is not uniformly distributed along the ridges, some areas of the ridge will have a stronger or weaker reaction than others. This can result in the ridges having the appearance of being made by a series of dots rather than a continuous line, particularly where the constituent is concentrated in the pores along the summits of the ridges (Figure 4.33). It can also result in a ridge appearing to end before it actually does and a bifurcation appearing as a ridge

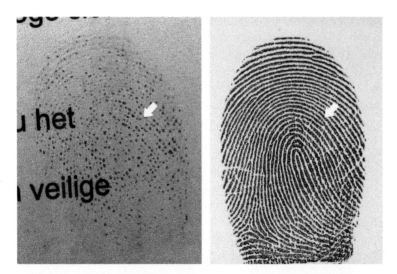

Figure 4.33   These two impressions were both made by the right middle finger of the same donor. The mark on the left was made on paper and chemically developed using nynhydrin. The chemical reacts with a particular constituent of sweat that, in this case, was not evenly distributed over the ridges and has resulted in a series of dots marking the paths of the ridges. The arrows highlight the same bifurcation in both impressions.

ending if there is little or no reaction between the substance and the chemical treatment at the point the ridge ends or bifurcates.

Some development media like powders can fill in fine details or thicken and alter the appearance of others depending on how heavily or lightly they are applied.

4.9 Changes to the Skin

The interval between the mark and the print being made maybe hours, days, weeks, months, years or decades and changes to the skin during that period can cause the details in the two impressions to appear significantly differently.

For example, if one impression was made when the person was very young, it may be significantly smaller than an impression made by the same person in later life. Sinclair and Fox (2007) describe an instance where a plantar print taken from a baby (in order to prevent children being mixed up in the hospital) was used to identify the adult donor and establish the place of their birth.

Over the course of a person's life, their ridges will flatten, become coarser, and their skin will lose its elasticity resulting in more creases (Maceo, 2011).

As well as the effects of ageing the activity of the donor may affect the appearance of the details. For example, a regular physical activity involving the hands such as manual labouring, or the use of chemicals or other liquids may wear down the heights of the ridges on their skin, making it difficult to discern the details in impressions (Schaumann & Alter, 1976) (Figure 4.34).

Gibbs (2012) describes a case in which a person's ridge detail is thought to have been affected by an unknown chemical they used in the course of their work that gave all the ridges on their distal phalanges a wavy and broken appearance. As well as occupational damage, activities such as playing the guitar or rowing can also affect the appearance of ridge detail by causing small callouses to develop in particular areas that can obscure ridge detail.

Some diseases or skin conditions can also cause damage that can result in impressions made by the same area appearing differently. For example, eczema, psoriasis, dermatitis or warts can obscure details and result in scaling, wrinkling and creasing of the skin (Drahansky et al., 2012; Samischenko, 2001). Some procedures used to treat skin conditions can also have an effect, such as

Figure 4.34 These prints were both made by the same right thumb, but the one on the left was taken the day after the donor had been using sandpaper.

dermabrasion which results in smooth skin that is devoid of ridges (Moenssens, 1975). A similar effect can also be produced as a side effect of the drug Capecitabine, prescribed to cancer sufferers (Al-Ahwal, 2012; Chavarrir-Guerra and Soto-Perex-de-Celis, 2015; Bordas & Bonsutto, 2020; Harmon, 2009; Wong et al., 2009).

Damage to the skin in the form of cuts, abrasions and burns that occur between the deposition of the impressions can also cause their details to appear differently and may result in the formation of a scar. Injuries can obscure details or create the appearance of details not present on the skin of the donor and can also alter the appearance of the spatial relationships between details (Figure 4.35). With a minor injury, generally, the ridges the injury interrupts will only be affected at the point of interruption; however, a more severe injury can affect the skin away from the injury site (Maceo, 2005).

In extreme cases, the scar formed by a significant injury can alter the appearance of the pattern to the extent where it may appear to be a different pattern type or family, as in Figures 4.36 and 4.37 (Samischenko, 2001).

Serious injury to friction ridge skin can occasionally result in skin being taken from another area and grafted over the injury (Samischenko, 2001). Mazurek (1994) describes a case in which skin was taken from a forearm and grafted onto a thumb resulting in a ridge-less and heavily creased area bordered by scars where the core of the pattern had been.

There have also been instances where friction ridge skin from one digit has been grafted over an injury to the friction ridge skin area of another digit (Putter, 1983; Stammers, 1977). In one case, this resulted in the digit the skin was grafted onto being able to leave impressions that could be identified

Figure 4.35 Both these prints have been identified as having been made by the same area of skin and the one on the right was made a year after the one on the left. A small injury (the path of which is indicated by the arrows) in that intervening period has created the appearance of several characteristics that were not previously on the skin of the donor. *Source:* Reprinted from Complete Latent Print Examination, http://clpex.com, Triplett, M. FIG 32.

Figure 4.36 These impressions have been identified as having been made by the same area of skin, but damage to the skin between the times the impressions were made has altered the appearance of the pattern type. *Source:* Reprinted from Complete Latent Print Examination, http://clpex.com, FIG 09 & FIG 23.

Figure 4.37 These impressions have been identified as having been made by the same area of skin, but damage to the skin between the times the impressions were made has altered the appearance of the pattern type. *Source:* Reprinted from Complete Latent Print Examination, http://clpex.com, FIG 09 & FIG 23.

as having been made by both digits (Stammers, 1977). There have also been single and double entire hand transplants (at least one of which involved using the donor hand of a person whose fingerprints were held on file by the police) (Szajerka et al., 2010). As well as procedures carried out for medical reasons, a print made by the same area of skin as the mark could also appear differently to it as a result of intentional mutilation to the skin in an attempt to avoid identification, as discussed in Chapter 8.

References

Al-Ahwal, M, S. 2012, 'Chemotherapy and Fingerprint Loss: Beyond Cosmetic', *The Oncologist*, 17, pp. 291–293, doi: https://doi.org/10.1634/theoncologist.2011-0243.

Alcaraz-Fossoul, J, De Roberts, K, A, Feixat, C, B, Hogrebe, G, G & Badia, M, G 2016, 'Fingermark ridge drift', *Forensic Science International*, 258, pp. 26–31.

Anthonioz, A, Egli, N, Champod, C, Neumann, C, Puch-Solis, R & Bromage-Griffiths, A 2011, 'Investigation of the Reproducibility of Third-Level Characteristics', *Journal of Forensic Identification*, 61 (2), pp. 171–192.

Ashbaugh, D, R 1999, *Quantitative-Qualitative Friction Ridge Analysis*, CRC Press LLC, Boca Raton, Florida.

Bordas, L. & Bonsutto, J. 2020, 'Adermatoglyphia: The Loss or Lack of Fingerprints and its Causes', *Journal of Forensic Identification*, 70 (2), pp. 154–162.

Bradshaw, G, Bleay, S, Deans, J & NicDaeid, N 2008, 'Recovery of Fingerprints from Arson Scene: Part 1 - Latent Fingerprints', *Journal of Forensic Identification*, 58 (1), pp. 54–82.

Brazelle, S, Inlow, V & Leitner, M, L 2018, 'Laterally Reversed Latent Prints Developed Using 1, 2 – Indandione', *Journal of Forensic Identification*, 68 (2), pp. 161–170.

Castellon, S 2014, 'Clues in Friction Ridge Comparisons: Tonal Reversals', *Journal of Forensic Identification*, 64 (3), pp. 223–237.

Champod, C, Lennard, C, Margot, P & Stoilovic, M 2016, *Fingerprints and Other Ridge Skin Impressions*, Second Edition, CRC Press, Taylor & Francis Group, Boca Raton.

Chavarrir-Guerra, Y. & Soto-Perex-de-Celis, E. 2015, 'Loss of Fingerprints', *The New England Journal of Medicine*, 372, e22, doi: https://doi.org/10.1056/NEJMicm1409635.

Czarnecki, E, R 2005, 'Laterally Inverted Fingerprints', *Journal of Forensic Identification*, 55 (6), pp. 702–706.

Drahansky, M., Dolezel, M., Urbanek, J., Brezinova, E. & Kim, T. 2012, 'Influences of Skin Diseases on Fingerprint Recognition', *BioMed Research International*, 626148, https://doi.org/10.1155/2012/626148.

Fagert, M & Morris, K 2015, 'Quantifying the Limits of Fingerprint Variability', *Forensic Science International*, 254, pp. 87–99.

Flanders, J, Moyer, A & Fisher, C, P 2021, 'Potential Characteristics to Aid Latent Print Examiners in Analyzing Possible Laterally Reversed Images on Porous Surfaces', *Journal of Forensic Identification*, 71 (1), pp. 49–59.

Geller, B, Volinits Y, Wax, H 2017, 'Can Dry Bloodstains Provide a Source for a Blood-Contaminated Fingermark?', *Journal of Forensic Identification*, 67 (3), pp. 355–360.

Geller, B, Leifer, A, Attias, D & Mark, Y 2018, 'Fingermarks in Blood: Mechanical Models and the Color of Ridges', *Forensic Science International*, 286, pp. 141–147.

Gibbs, P 2012, 'Metamorphosis of Friction Ridge Skin', *Journal of Forensic Identification*, 62 (3), pp. 191–1913.

Harmon, K 2009, 'Can You Lose Your Fingerprints?', *Scientific American*, May 29, https://www.scientificamerican.com/article/lose-your-fingerprints/.

Huss, K, Clark, J & Chisum, W, J 2000, 'Which Was First – Fingerprint or Blood?', *Journal of Forensic Identification*, 50 (4), pp. 344–350.

Jabbal, R, S, Boseley, R, E & Lewis, S, W 2018, 'Preliminary Studies into the Secondary Transfer of Undeveloped Latent Fingermarks Between Surfaces', *Journal of Forensic Identifiction*, 68 (3), pp. 421–437.

Kershaw, M, H 2000, 'Laterally Reversed', *Journal of Forensic Identification*, 50 (2), pp. 138–140.

Langenburg, G 2008, 'Deposition of Bloody Friction Ridge Impressions', *Journal of Forensic Identification*, 58 (3), pp. 355–389.

Lane, P, A, Hilborn, M, Guidry, S & Richard, C, E 1988, 'Serendipity and Super Glue: Development of Laterally Reversed, Transferred Latent Prints', *Journal of Forensic Identification*, 38 (6), pp. 292–294.

Lin, S, Luo, Y, Xie, L, Yu, Y & Mi, Z 2019, 'Faux Blood Fingermark on Pistol: Latent Fingerprint Developed by Whole Blood', *Journal of Forensic Sciences*, 64 (6), pp. 1913–1915.

Maceo, A, V 2005, 'The Basis for the Uniqueness and Persistence of Scars in the Friction Ridge Skin', *Fingerprint Whorld*, 31 (121), pp. 147–161.

Maceo, A, V 2009, 'Qualitative Assessment of Skin Deformation: A Pilot Study', *Journal of Forensic Identification*, 59 (4), pp. 390–440.

Maceo, A, V 2011, 'Anatomy and Physiology of Adult Friction Ridge Skin', *The Fingerprint Sourcebook*, National Institute of Justice. www.nij.gov.

Martin, K, F 1994, 'Laterally-reversed Transfers of Latent Fingerprints Upon Non-porous Surfaces', *Journal of Forensic Identification*, 44 (5), pp. 530–541.

Mazurek, D 1994, 'Disfigurement of Fingerprint Patterns', *Fingerprint Whorld*, 20 (75), pp. 9–11.

Moenssens, A, A 1975, *Fingerprint Techniques*, Second Edition, Chilton Book Company, Radnor, Pennsylvania.

Moore, J, Bleay, S, Deans, J & NicDaeid, N 2008, 'Recovery of Fingerprints from Arson Scene: Part 2 – Fingerprints in Blood', *Journal of Forensic Identification*, 58 (1), pp. 83–108.

Neumann C, Champod, C, Yoo, M, Genessay T & Langenburg, G 2013, 'Improving the Understanding and the Reliability of the Concept of "Sufficiency" in Friction Ridge Examination', National Institute of Justice – Office of Justice Program, Award 2010-DN-BX-K267.

Oklevski, S 2011, 'Poroscopy: Qualitative and Quantitative analysis of the 2nd and 3rd level detail and their relation', *Fingerprint Whorld*, 37 (145), pp. 170–181.

Pickard, R 2012, '*A study into the likelihood & possible mechanisms of ridge detail transference onto adjacent pages of a booklet*', MSc. King's College London.

Pierce, D, S 1989, 'Tonally Reversed Friction Ridge prints on Plastics', *Journal of Forensic Identification*, 39 (1), pp. 11–22.

Pierce, D, S & Turnidge, S, S 2008, 'The Significance of Butterflies' *Journal of Forensic Identification*, 58 (6), pp. 696–711.

Praska, N & Langenburg, G 2013, 'Reactions of Latent Prints Exposed to Blood', *Forensic Science International*, 224, pp. 51–58.

Putter, P, J 1983, 'Finger/Thumb Transplantation', *Fingerprint Whorld*, 9 (33), p. 26.

Reneau, R, D 2003, 'Unusual Latent Print Examinations', *Journal of Forensic Identification*, 53 (5), pp. 531–537.

Richmond, S 2004, 'Do fingerprint ridges and characteristics within ridges change with pressure?' Available at: http://www.latent-prints.com/images/changes%20with%20pressure.pdf.

Samischenko, S, S 2001, *Atlas of the Unusual Papilla Patterns*, IUrisprudentsiia, Moskva.

Saviano, J 2003, 'The Significance of Using Level 1 Detail in Latent Print Examinations', *Journal of Forensic Identification*, 53 (2), pp. 209–218.

Schaumann, B & Alter, M 1976, *Dermatoglyphics in Medical Disorders*, Springer-Verlag, New York.

Sinclair, R & Fox, C 2007, 'Infant-to-Adult Footprint Identification', *Journal of Forensic Identification*, 57 (4), pp. 485–492.

Stammers, J 1977, 'Fingerprint Transplant', *Fingerprint Whorld*, 2 (7), p. 47.

Szajerka, T, Jurek, B & Jablecki J 2010, 'Transplanted Fingerprints: A Preliminary Case Report 40 Months Posttransplant', *ScienceDirect*, 42 (9), pp. 3753–3755.

Tate, D, Anderson E & Eller, J 2018, 'Assessing the Appearance of Latent Print Distortion on Absorbent and Nonabsorbent Substrates', *Journal of Forensic Identification*, 68 (3), pp. 403–420.

Wallis, J & Goulet, J 2017, 'Quantitative Analysis of the Distortion of Friction Ridge Impressions According to Three Deposition Pressure Levels and Horizontal Movement', *Journal of Forensic Identification*, 67 (2), pp. 259–277.

Wong, M, Choo, S, P & Tan, E, H 2009, 'Travel Warning with Capecitabine', *Annals of Oncology*, Letters to the editor, 20 (7), p. 1281.

Zampa, F, Cappiello, P, Vaccaro, G, Carullo, V, Cervelli, F, Mattei, A & Garofano, L 2008, 'Latent Print Examination on Foldable and Porous Surfaces: Analysis of Three Cases', *Journal of Forensic Identification*, 58 (1), pp. 46–53.

Zorich, S, R 1992, 'Laterally Reversed Cyanoacrylate Developed Prints on Tape', *Journal of Forensic Identification*, 42 (5), pp. 396–400.

5

Is the Mark Suitable?

Once the examiner has considered the details in the mark, they can decide whether the mark is suitable or 'of value' for the remaining parts of the analysis to be completed and for the mark to be progressed to the comparison stage. Not all marks will be suitable, and for those that are not suitable, work will cease at this point.

There are broadly two different types of suitability: suitability for identification and suitability for comparison. Marks not suitable for either would be deemed 'no value/insufficient', but a mark that is suitable for comparison but not for identification would be one that could only be used to exclude someone as having been the donor. An example of such a mark would be one that only revealed a pattern type but no other details. Though it would not usually be possible to identify that mark's donor, as different people can have the same pattern type,[1] if the prints of a person it was compared with all revealed different pattern types, that person could be excluded as being its donor.

Often exclusions are only useful to the investigation of an offence if the mark has significant evidential value. For instance, if someone has been shot and a mark is found on a surface near where the shooting took place, excluding that mark as having been made by a person suspected of being involved may not be particularly useful. However, if a mark is found on the trigger of the firearm used in the offence, then excluding a suspect does at least indicate that there is another person whose finger has been on that trigger. Because most marks do not have that kind of evidential value and comparing them requires time and resources, some organisations favour the suitability for identification approach – under which a mark not suitable for identification would be deemed 'insufficient/no value' and not progressed. As a compromise, an organisation may vary the approach used based on the crime type, for example only using the suitable for comparison approach with marks from serious offences. Alternatively, an organisation may retain marks that are only suitable for exclusion but only compare them if requested.

There is no objective standard for suitability, so the examiner makes a subjective decision based on the details in the mark that can also be influenced by other factors including their training and experience. The decision reflects the examiners opinion of the likelihood of being able to reach a conclusion of identification or exclusion with the mark (Langenburg, 2012).

Often the assessment of a mark's suitability can be done rapidly as many will reveal either an abundance of or lack of clear detail. Generally, marks in which enough clear ridge flow is revealed to form a recognisable configuration like a pattern, core or delta will be suitable for at least exclusion. To be suitable for identification, a mark would usually need to reveal multiple characteristics (if only

---

1 It may be possible to identify the donor of such a mark in a largely hypothetical situation where it could only have been made by one of a small group of known individuals, only one of whom has an area of skin revealing that pattern (Langenburg, 2012).

The Forensic Analysis, Comparison and Evaluation of Friction Ridge Skin Impressions, First Edition. Dan Perkins.
© 2022 John Wiley & Sons Ltd. Published 2022 by John Wiley & Sons Ltd.

a few are revealed, the mark may still be identifiable if their type or configuration is rare). In most cases, a decision about whether the mark is suitable can be reached by considering these details; if not, the examiner should assess the contribution that other details such as creases, subsidiary ridges, scars or third level details could have.

Studies have demonstrated variability in suitability decisions when different examiners consider the same marks (Eldridge et al., 2020; Langenburg, 2012; Neumann et al., 2013; Peterson & Markham, 1995; Smith, 2004 in Haber & Haber, 2009; Ulery et al., 2011; Ulery et al., 2013). In a study using marks selected to be broadly representative of those found in casework, Ulery et al. (2011) found that examiners only reached a unanimous conclusion that a mark was or was not of value for identification on 43% of marks. The authors observed that it was the very poor quality or very good quality marks at either ends of the scale that resulted in unanimous conclusions. Whilst Eldridge et al. (2020) also reported the suitability decisions involving marks at the very good quality end were fairly consistent, they found that there was much more variability at the other end of the scale. Eldridge et al. reported no consensus no value decisions and found that with many marks, decisions were nearly equal across all options (Figure 5.1).

Neumann et al. (2013) conducted a study only using 'challenging' marks selected to maximise variability between examiners and found that 12 of the 15 marks resulted in split decisions from the examiners, including some such as the one in Figure 5.2 where the number of examiners who thought the mark was suitable for identification were similar to the number who thought it was not.

One study also found that the same examiner can reach a different decision when re-presented with the same mark (Ulery et al., 2012). In the study examiners maintained their no value decisions 85% of the time, their suitability for identification decisions 93% of the time and their suitability for exclusion decisions 55% of the time. Complete reversals between no value and suitability for identification were also seen and occurred at the rate of 1%. Ulery et al. (2012) concluded that much of the variability in their study appeared to be due to examiners having to make categorical decisions with borderline cases, and Champod et al. (2016) argue that in those situations, the decision of the examiner becomes almost a random assignment.

Several studies have shown that the number of characteristics observed by the examiner is the single biggest factor in whether they decide a mark is suitable or not (Eldridge et al., 2020;

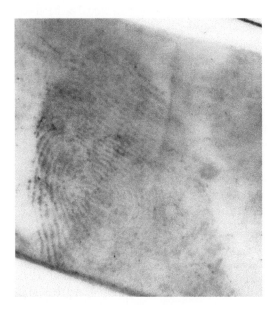

Figure 5.1 This mark was used in a study in which the examiners were given five options to describe its suitability. The options and the number of examiners who selected each option were: no value (9), some investigative value but insufficient for identification or exclusion (5), value for exclusion only (9), value for identification only (2) and value for both identification and exclusion (10). *Source:* Reprinted with permission from Eldridge et al. (2020).

Figure 5.2   Thirty-four examiners decided this mark was value for identification, 44 decided it was value for exclusion only and 26 decided it was no value. *Source:* Reprinted from Neumann et al. (2013).

Langenburg, 2012; Neumann et al., 2013; Ulery et al., 2013; Ulery et al., 2014). Some organisations may require examiners to find a specific number for a mark to be deemed suitable, but where that is not the case, several studies have shown that the threshold for suitability appears to be around 7–8 characteristics (Eldridge et al., 2020; Langenburg, 2012; Ulery et al., 2013; Ulery et al., 2014).[2] The use of the number of characteristics as the principal means of determining suitability may explain some of the variability in the suitability decisions of examiners as several studies have demonstrated that given the same mark, different examiners will report the presence of different numbers of characteristics (Dror et al., 2011; Langenburg, 2004; Neumann et al., 2013; Swofford et al., 2013). For example, Langenburg (2012) tasked 144 examiners with counting the number of characteristics in the same 12 impressions, which varied in the amount and clarity of detail revealed. The smallest difference between the minimum and maximum numbers of characteristics found for any of the marks was 11 (min. 8, max. 19), the largest was 42 (min. 3, max. 45) and the average difference was 22 characteristics (Langenburg, 2012). Dror et al. (2011) conducted a similar study of 20 examiners with 10 marks and found an average difference of 11 characteristics. As well as not seeing the same number of characteristics, Eldridge et al. (2020) also reported that examiners vary in how they see them. For instance, the authors found that on many occasions similar numbers of examiners would designate the same characteristic a high confidence ridge ending as would designate it a high confidence bifurcation and that such differences were seen even in clear areas of marks.

However, though characteristics may be the principal factor, there are other factors that influence judgements of suitability as several studies have found examiners deciding marks with few characteristics are of value and those with many are not (Langenburg, 2012; Ulery et al., 2013; Ulery et al., 2014). For instance, Ulery et al. (2013) reported examiners deciding marks with up to

---

2 Eldridge et al. (2020) reported a similar threshold when considering characteristics the examiners indicated they were confident in, but found a slightly higher one if all the characteristics examiners had used to make their decisions, including those marked as uncertain, were considered.

27 characteristics were suitable for exclusion only, those with up to 12 being no value and those with 0 being suitable for identification. Eldridge et al. (2020) reported factors such as how sure the examiner is of the pattern type revealed in the mark, how confident they are in the characteristics they can see, the clarity and amount of distortion present and the specificity of the characteristics revealed also impacted suitability decisions. Eldridge et al. (2020) suggest that though the number of characteristics had the greatest impact, factors such as these may help to tip the balance of a decision, particularly with marks revealing ambiguous details.

In addition to the details in the mark, there are also other factors that can influence suitability decisions. One study found that laboratory practitioners were more likely to conclude the same mark was suitable when it was presented to them as relating to an offence of murder rather than an offence of theft (Earwaker et al., 2015). Fingerprint examiners may also be affected by the seriousness of the offence or by other contextual information, such as whether there are many marks or just one in a case. For instance, they may be more likely to conclude a borderline suitable/no value mark is suitable if it represents the only fingermark evidence than they would be if there were many other marks with the same evidential value. Fraser-Mackenzie et al. (2013) found that examiners' suitability decisions could be influenced by knowledge of another examiner's decision about the same mark. The study reported that if examiners thought another examiner had already decided a mark was 'unsuitable', they were more likely to decide it was unsuitable than if they made the decision without that knowledge. However, the study also found that examiners were not more likely to decide a mark was suitable when they thought another examiner had already made that decision. Another part of the same study found that suitability decisions were influenced if the examiner made the decision whilst a print was alongside the mark (Fraser-Mackenzie et al., 2013). The authors found that the effect varied depending on whether the print was or was not made by the same area of skin as the mark. Compared to the decision they made in the absence of a print, where the impressions were made by different areas of skin, the examiner was more likely to decide the mark was suitable, and where they were made by the same area, the examiner was less likely to decide the mark was suitable and more likely to decide it was 'questionable'. The latter finding may fit in with another study that reported that the presence of a print made by the same area of skin as the mark during the analysis of some marks can reduce the number of characteristics examiners observe, which therefore potentially could influence a suitability decision (Dror et al., 2011).

In general, the examiners' suitability decision is less likely to be influenced by these contextual factors where the mark reveals an abundance of, or significant lack of, clear detail. However, where the detail revealed lies somewhere in-between, the examiners' decision may be more vulnerable. With these marks the use of tools like the Defense Fingerprint Image Quality Index (DFIQI) or Universal Latent Workstation (ULW) (see Chapter 3) that can quantitatively assess the value of a mark can be used to standardise and provide support for the examiners' decision and to make it more transparent.

The report of the Expert Working Group on Human Factors in Latent Print Analysis (NIST, 2012) recommended that organisations provide their examiners with guidelines that are as explicit as possible regarding how to determine whether a mark is suitable or not to encourage consistent decision making and to enhance the transparency of their analysis. The working group also stressed the importance of documentation of suitability decisions, particularly any that are changed after the examiner has seen the print the mark is to be compared with. Langenburg (2012) suggests that some suitability decisions could be made using characteristics whose presence has been agreed on by a consensus of examiners. This approach would compensate for variability by having multiple examiners independently analyse the same mark and only use the characteristics they all agreed on to decide suitability.

In addition to suitability for identification or exclusion determinations, the examiner may also be required to determine whether the mark is suitable for searching through a computer database of prints using an Automated Fingerprint Identification System (AFIS). If a mark is suitable for identification but not suitable for searching, then it usually could only be identified if a specific person were nominated to have their prints compared with it. The AFIS may have minimum acceptance criteria, such as a specific number of characteristics that must be designated in the mark before a search can be set. Alternatively, the system may only be capable of searching marks made by certain areas of skin, for example those made by the distal phalanges rather than those made by the medial or proximal phalanges.

A study of casework in one laboratory found that examiner decisions about AFIS suitability were associated with their ability to establish the most likely area of skin to have made the mark. Gardner, Kelley & Neuman (2021) found that examiners were more likely to decide a mark was suitable for searching if they could confidently determine which area of skin made it. The same study also reported examiners were slightly more likely to decide a mark was suitable for searching if the offence it is related to involved an offence against a person, which may suggest examiners unintentionally set a lower threshold for marks from offences they consider more important or serious. The Gardner, Kelley & Neuman (2021) study also found considerable variation in the percentage of marks different examiners considered suitable for search. The study examined casework over the course of one year and reported some examiners considered only one in three marks were suitable (35.8%), whereas others considered more than half were (56.5%).

Eldridge et al. (2020) recommend the threshold used to deem a mark suitable for search should be considered separately from that required to deem a mark suitable for identification. This is because whilst a mark that is suitable for identification could be identified after comparison with the prints of one suspect in the offence, if that mark is to be searched against a large database, then the examiner is much more likely to encounter a print not made by the same donor that happens to reveal a similar configuration of details. As a result, the examiner may consider whether a greater number of details are necessary to deem a mark suitable for search than identification, and/or as Eldridge et al. (2020) suggest the organisation may employ additional quality control measures for identifications made using an AFIS such as more verifications or blind verification. To assist in determining suitability for search, examiners may also use the ULW's capability to generate a latent quality score to indicate the probability that a search of the mark through an AFIS system would return a print made by the same donor if that donor's prints were held in the database.

Determining the suitability of the mark is a critical decision as if a mark that is suitable is erroneously determined to be of no value, then the opportunity to identify the person who made it, or exclude a person who did not, will be missed. Conversely, if a mark that is of no value is determined to be of value, then time and resources can be wasted in the attempt to compare it and such efforts may introduce greater risks of erroneous conclusions.

As well as the mark itself, whether a mark judged suitable for identification or exclusion can be identified or excluded also depends on the details revealed in the print it is compared with. The examiner will likely not have seen the print when they are determining the suitability of the mark, so they will make that determination on the assumption of what would be possible if the best print of the corresponding area was available. However, if the print is poor quality or does not reveal the required area, the examiner may not be able to identify a mark that is suitable for identification. Similarly, it may also not be possible to exclude someone even though the mark is suitable for exclusion – if, for example, the mark only reveals a pattern type and no characteristics, and that same pattern type is revealed in the print. Both scenarios may result in the examiner reaching an inconclusive conclusion rather than an identification or exclusion. However, irrespective of the print, the examiner may encounter a situation where they realise their suitability determination was incorrect. In this scenario, it is crucial the examiner is able to modify their decision and document the reasons for the change. For example, Grieve (1988) describes a scenario in which an examiner who concludes a mark is suitable for identification may feel their credibility would be damaged if they changed that view after seeing the print. Grieve suggests that in this scenario the examiner may be influenced to identify the mark as having been made by the same donor as the print, even if there is insufficient detail to support that conclusion. The opposite scenario would involve the examiner determining the mark to be of no value, then reversing that decision after seeing a print that they consider was made by the same donor. The right to change the decision could prevent the wrong person being identified or allow the right person to be identified, but it also brings with it a risk of the examiner being inappropriately influenced by the print. Therefore, where the examiner has documented they have changed their view, the organisation may implement additional quality control measures (for example, the involvement of additional examiners or additional verifications) to reduce the risk of an erroneous conclusion.

Prior to determining a mark to be of no value, the examiner should assess whether there are any other actions they could take or factors they could consider that may affect their decision. For example, as well as revisiting the options for enhancing the image of the mark addressed in Chapter 2, Langenburg (2004) reported that examiners observed significantly more characteristics in an image of a mark that was enlarged (15 times) than they did in a life-size image of the same mark. Additionally, if other marks are consistent with having been made simultaneously with the mark in question, the examiner should consider whether the details in those marks could be used collectively to determine the suitability of all the marks.

When deciding a mark is no value, the examiner should also be aware that statistical models have been able to provide strong evidence of an association between no value marks and their donor (Neumann et al., 2011; Stoney et al., 2020). Fingerprint evidence exists along a continuum; at one end is a finger-shaped smudge revealing no detail and at the other is a mark revealing an abundance of clear detail. In between the two are marks revealing varying levels of support for who the donor was, and so the practice of only considering marks to have any value if they reach an arbitrary point designated as the beginning of a category of suitable for comparison or suitable for identification will result in potentially useful evidence being lost (Champod et al., 2016; Eldridge et al., 2020; Langenburg, 2012; Ulery et al., 2011). For example, Stoney et al. (2020) were able to use a model to provide strong evidence of association with the donor of marks that revealed at least three characteristics but had been judged of no value for identification (Figure 5.3).

Using a different model Neumann et al. (2011) were able to demonstrate an association with the donor of 24% of marks that had been judged no value for identification by examiners and reported that some of those associations carried similar evidential weight to that seen in identifications. For example, some marks revealing three characteristics were found to have a similar weight to those revealing 12 characteristics (Neumann, Evett & Skerrett, 2012). Currently, very few examiners use such models, and none of the ones that are in use have been generally accepted but, provided an organisation is willing to report conclusions that do not meet their criteria for identifications, their use can mean that some marks judged to be of no value under the current categorical framework may yield considerable value.

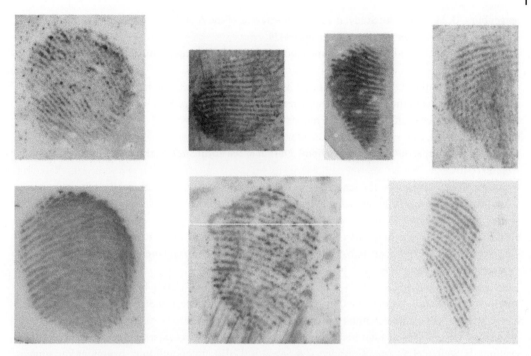

Figure 5.3 Examples of marks judged no value by examiners that may be able to provide strong associative value. *Source:* Reprinted with permission from Stoney et al. (2020).

References

Champod, C, Lennard, C, Margot, P & Stoilovic, M 2016, *Fingerprints and Other Ridge Skin Impressions*, Second Edition, CRC Press, Boca Raton.

Dror, I, E, Champod, C, Langenburg, G, Charlton, D, Hunt, H & Rosenthal, R 2011, 'Cognitive Issues in Fingerprint Analysis: Inter- and Intra-Expert Consistency and the Effect of a 'Target' Comparison'. *Forensic Science International*, 208 (1–3), pp. 10–17.

Earwaker, H, Morgan, R, M, Harris, A, J, L & Hall, L, J 2015, 'Fingermark Submission Decision-Making Within a UK Fingerprint Laboratory: Do Experts Get the Marks They Need?', *Science & Justice*, 55 (4), pp. 239–247.

Eldridge, H, De Donno, M, Furrer, J & Champod, C 2020, 'Examining and Expanding the Friction Ridge Value Decision', *Forensic Science International*, https://doi.org/10.1016/j.forsciint.2020.110408.

Fraser-Mackenzie, P, Dror, I & Wertheim, K 2013, Cognitive and Contextual Influences in Determination of Latent Fingerprint Suitability for Identification Judgments (ojp.gov). NCJRS, 241289.

Gardner, B, O, Kelley, S & Neuman, M 2021, 'Latent Print Comparison and Examiner Conclusions: A Field Analysis of Case Processing in One Crime Laboratory', *Forensic Science International*, 319, 110642.

Grieve, D, L 1988, 'The Identification Process: Attitude and Approach', *Journal of Forensic Identification*, 38 (5), pp. 211–224

Haber, L & Haber, R, N 2009, *Challenges to Fingerprints*, Lawyers & Judges Publishing Company, Inc., Tucson.

Langenburg, G, M 2004, 'Pilot Study: A Statistical Analysis of the ACE-V Methodology – Analysis Stage', *Journal of Forensic Identification*, 54 (1), pp. 64–79

Langenburg, G 2012, 'A critical analysis and study of the ACE-V process', PhD thesis, University of Lausanne.

Neumann C, Champod, C, Yoo, M, Genessay T & Langenburg, G 2013, 'Improving the Understanding and the Reliability of the Concept of "Sufficiency" in Friction Ridge Examination', National Institute of Justice – Office of Justice Program, Award 2010-DN-BX-K267.

Neumann, C, Evett, I, W & Skerrett, J 2012, 'Quantifying the Weight of Evidence from a Forensic Fingerprint Comparison: A New Paradigm', *Journal of the Royal Statistical Society, A*, 175 (Part 2), pp. 371–415.

Neumann, C, Mateos-Garcia, I, Langenburg, G & Kostrowski, J 2011, 'Operational Benefits and Challenges of the Use of Fingerprint Statistical Models: A Field Study', *Forensic Science International*, 212, pp. 32–46.

NIST 2012, *Latent Print Examination and Human Factors: Improving the Practice through a Systems Approach*, Report of the Expert Working Group on Human Factors in Latent Print Analysis, US Department of Commerce.

Peterson, J, L & Markham, P, N 1995, 'Crime Laboratory Proficiency Testing Results, 1978–1991, II: Resolving Questions of Common Origin', *Journal of Forensic Sciences, JFSCA*, 40 (6), 1009–1029.

Stoney, D, A, De Donno, M, Champod, C, Wertheim, P, A & Stoney, P, L 2020, 'Occurrence and Associative Value of Non-Identifiable Fingermarks', *Forensic Science International*, 309, 110219.

Swofford, H, Steffan, S, Warner, G, Bridge, C & Salyards, J 2013, 'Inter- and Intra-Examiner Variation in the Detection of Friction Ridge Skin Minutiae', *Journal of Forensic Identification*, 63 (5), pp. 553–570.

Ulery, B, T, Hicklin, R, A, Buscaglia, J & Roberts, M, A 2011, Accuracy and Reliability of Forensic Latent Fingerprint Decisions, *Proceedings of the National Academy of Sciences of the United States of America*, 108 (19), 7733–7738, https://doi.org/10.1073/pnas.1018707108.

Ulery, B, T, Hicklin, R, A, Buscaglia, J & Roberts, M, A 2012, 'Repeatability and Reproducibility of Decisions by Latent Fingerprint Examiners', *PLoS ONE*, 7 (3), e32800. doi:https://doi.org/10.1371/journal.pone.0032800.

Ulery, B, T, Hicklin, R, A, Kiebuzinski, G, I, Roberts, M, A & Buscaglia, J 2013, 'Understanding the Sufficiency of Information for Latent Fingerprint Value Determinations', *Forensic Science International*, 230, pp. 99–106.

Ulery, B, T, Hicklin, R, A, Roberts, M, A & Buscaglia, J 2014, 'Measuring What Latent Fingerprint Examiners Consider Sufficient Information for Individualization Determinations', *PLoS ONE* 9 (11), e110179. doi:https://doi.org/10.1371/journal.pone.0110179.

6

What Is the Most Likely Orientation of the Mark?

To facilitate comparison the mark should be oriented so it will correspond with the way a print made by the same area of skin will appear on the screen and on a fingerprint form. Failing to orient the mark in this way could lead the examiner to compare it with prints made by the wrong area of skin or the right area the wrong way up. Also, if the mark is to be searched through a computer database, then the accuracy of the search may be compromised if the orientation of the mark is significantly different from that of a corresponding print on the database.

The majority of marks the examiner encounters will have been made by the skin of the distal phalanges (UK Home Office, 1969), and this chapter describes the principal indicators the examiner should consider to determine the likely orientation of such marks. If the mark's size, shape, ridge flow or the details it reveals are consistent with it having been made by a different area of skin, then the information in Chapter 7 should be considered to assist with orientation.

6.1 Shape and Ridge Flow

Marks made by the skin of the distal phalanges often have a horseshoe-shaped outline in which the curved end corresponds with the top of the digit.

Whilst the ridges at the top of the mark will follow the curvature of the digit, those at the bottom will usually run straight across the width of the mark perpendicular to the sides. Just as the curved shape of the top of the mark echoes the curvature of the tip of the digit, the sides of the mark will usually reflect the straight, parallel sides of the finger and indicate one of two possible orientations (Figure 6.1).

Even if only the tip of a finger contacted the surface to leave a mark, the shape and ridge flow of that mark may reveal indicators of its orientation. Such marks are often triangular in shape, and where that is the case, the most acute angle will usually be found at the bottom of the mark. As a result, that area will reveal the shortest sections of the ridge in the mark (as in the leftmost mark in Figure 6.2). Marks made by tips may also be circular or oval but irrespective of the shape, whilst the ridges at the top of such a mark will follow the gradual curve of the tip of the digit, those at the bottom will often curve much more sharply as they flow around the top of the core of a pattern (Figure 6.2). If several marks are made by the tips of digits of the same hand at the same time, they may form an arc shape in which the bottom of the arc will correspond to the bottom of the marks, as in Figure 6.2.

The Forensic Analysis, Comparison and Evaluation of Friction Ridge Skin Impressions, First Edition. Dan Perkins.
© 2022 John Wiley & Sons Ltd. Published 2022 by John Wiley & Sons Ltd.

6.2 Pattern Family

Each pattern family has distinguishing features that can assist with orientation. For example, the rise in the ridge flow of an arch pattern will be towards the top of the mark (Figure 6.3).

In a loop pattern, the ridges that form the recurve will often be almost horizontal at the bottom of the mark before sloping upwards towards the core and turning back on themselves.

Whilst it may be circular around the core, the ridge flow of some whorl patterns may become more egg shaped towards the edges of the mark (Figure 6.4). Where this shape is seen, the widest area will almost always be at the bottom of the mark.

If they do not reveal an egg shape, whorls can be difficult to orient because of their rotational symmetry. However, where deltas are revealed, they can assist as they are almost always found in the bottom half of marks and the ridge flow below them is usually close to horizontal. Even if a delta cannot be seen, multiple diverging ridges may indicate the presence of one below them, as in Figure 6.5.

Figure 6.1 This impression reveals numerous indicators that can be used to verify the orientation it is presented in is correct, including a sharply curved top edge, a flat bottom edge (with ridges from the medial phalange below it) and relatively straight sides that, in this example, can also be seen to continue along the sides of the part of the mark made by the medial phalange.

6.3 Distal Flexion Crease

The bones of the digits are connected by interphalangeal joints that are marked by flexion creases on the surface of the skin. The distal flexion creases are the ones that correspond to the joints between the distal phalanges and the medial phalanges, so when revealed in a mark made by a distal phalange, they are always found at the bottom. The crease usually appears as a linear break in the ridges across the width of the mark, as seen in Figure 6.1. The distal flexion crease may take the form of a single, solid crease, several overlapping creases or two or more parallel creases a short distance apart.

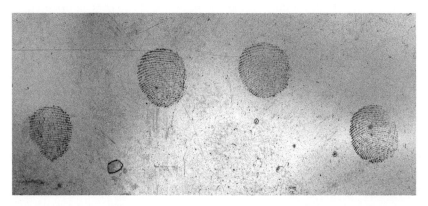

Figure 6.2 These marks were made by the tips of the fingers of a right hand and reveal several indicators that can be used to verify that the orientation they are presented in is correct, including the shapes of some of the marks, their ridge flow and the arc shape they form.

6.4   Other Marks

If there are other marks near the mark in question, the examiner can consider whether they are consistent with having been made by the same hand at the same time (indicators of simultaneous deposition are described in Chapter 7). If other marks are consistent with being deposited simultaneously, then they can be examined for the orientation indicators they reveal; however, the examiner must exercise caution as if those marks were not deposited with the mark, they may have a very different orientation.

6.5   The Surface the Mark Was Found on

Information provided by those who found the mark may indicate the surface it was found on and where it was on that surface. This type of information can be helpful when considering the possible orientation of a mark, as some surfaces are likely to have been contacted by friction ridge skin that was oriented in a particular way. For example, when the rear-view mirror in a car is adjusted, the adjustment is often made by someone sitting in the driver's seat, reaching up and touching the reflective surface.

The examiner may also be provided with information about the offence the mark relates to, and this may also assist with its orientation. For instance, if a burglary has been committed and entry to the premises has been gained by climbing in through a window and the mark has been found on or around that window, considering how someone may have climbed in through it may suggest a possible orientation for the mark (providing that mark was actually left by someone climbing in through the window). Whilst many surfaces can be contacted by areas of skin at any orientation, considering how the surface could have been touched may suggest a likely or possible orientation for the mark that can be particularly useful if it reveals few indicators of its orientation.

6.6   Summary

Though determining the orientation of most marks is relatively easy, those revealing small areas of ridge detail with few indicators can be more challenging. The best way for the examiner to determine the most likely orientation of a mark is to identify all the indicators it reveals and consider them in aggregate. In considering every indicator, the examiner may observe ones that suggest conflicting orientations

Figure 6.3   The rise or hump in the ridges of an arch pattern made by a distal phalange will usually be in the direction of the top of the mark. labsas/Getty Images.

Figure 6.4   In whorls that reveal an egg shape, the widest part of the shape will usually correspond with the bottom of the mark. Arkadiusz Fajer/ Dreamstime LLC.

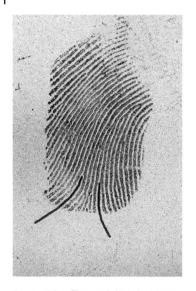

Figure 6.5 The straight left side to this mark indicates one of the two likely orientations. No delta is visible, but the presence of one can be inferred at one end of the mark by a divergence of ridges that is typically found above a delta. As deltas are almost always found in the bottom half of marks, this divergence suggests the orientation the mark is presented in is the most likely one.

and will therefore need to evaluate which is more likely. For example, the way the surface is likely to have been contacted may suggest an orientation that, if correct, would mean that a delta is at the top of a mark, which is consistent with being made by the skin of one of the distal phalanges. In this instance, whilst a delta appearing in this position is not impossible, unless the surface can only be touched in one way, it is more likely that the surface has been contacted in a different way and that the delta is at the bottom of the mark.

The examiner will occasionally encounter marks that reveal few, if any, indicators of orientation. In these instances, it may be necessary to compare the mark at different orientations with the same print. For example, a mark that only reveals a small area with a delta in it may need to be compared at all three possible orientations.

When considering the orientation of a mark, the examiner must be aware that patterns and deltas also appear in areas of friction ridge skin other than the distal phalanges. So even though the mark may be a size and shape that is consistent with it being made by a distal phalange, it may have been made by an area in which these details cannot be used as orientation indicators in the same way. For example, the skin of the medial and proximal phalanges can feature upside-down arch-type patterns, and deltas can appear at the top and bottom of the palm (see Chapter 7 for more information on how to orient marks made by other areas of skin).

As the likely orientation of most marks is obvious, the examiner may not routinely document the justification for their chosen orientation. However, if the orientation is not obvious, particularly where the mark is complex, the examiner may record what they consider to be the orientation of the mark and the indicators that support it.

Reference

UK Home Office 1969, Police Research and Development Branch, Report No. 1/69, January, J. W. Godsell.

7

Which Area of Skin Is Most Likely to Have Made the Mark?

In most cases, the only way to be certain which area of skin made the mark would be to have witnessed it being deposited. However, consideration of the factors discussed in this chapter may allow the examiner to form an opinion as to the area that is most likely to have made it. Determining the most likely area and comparing the mark with prints made by that area first is usually the most efficient way to identify its donor. The UK Home Office (1969) conducted a survey involving 967 'lone' marks and found that examiners were confident they could determine the particular digit that made a mark over 65% of the time, and when they did nominate a digit, they were correct 91% of the time.

The first part of this chapter deals with marks made by the skin of the distal phalanges though some of the factors discussed, such as how the surface may have been touched and whether there are other marks around the mark in question, should also be considered for marks made by any area of skin.

The UK Home Office (1969) carried out a study of 3903 cases and reported over 80% of marks found were made by the skin of the distal phalanges, so in the absence of indicators to the contrary, it is usually most efficient, to begin with the hypothesis that the mark was made by that area. If that hypothesis does not lead to the identification of the mark's donor, then the examiner should consider the likelihood that the mark was made by a different area.

Where appropriate, for example, with a complex mark, the examiner may document the area they think is most likely to have made the mark and note the indicators they have observed to support this determination.

7.1   The Surface the Mark Was Found On

Considering how the surface may have been touched can indicate whether the area of skin that made the mark is likely to be on the left or right hand. For example, the tap on the right side of a sink, or the handle on a door that is hinged on the right are more likely to have marks from the right hand on them. With other surfaces, a particular digit or digits may be more likely to have touched them than others; for example, the reflective surface of a rear-view mirror in a car is often contacted by a thumb when it is being adjusted.

If the surface does not suggest a particular hand or digit, considering the marks orientation on the surface may assist. For example, a cup can be contacted by digits from either hand, but marks on them are usually oriented towards 9 o'clock if made by the left hand or 3 o'clock if made by the right hand – the opposite would mean either the hand or the glass had to have been inverted (Figure 7.1).

The Forensic Analysis, Comparison and Evaluation of Friction Ridge Skin Impressions, First Edition. Dan Perkins.
© 2022 John Wiley & Sons Ltd. Published 2022 by John Wiley & Sons Ltd.

Figure 7.1 When made by a person holding it in a way that is consistent with using it to drink from, marks made by fingers of the left hand on a cup will usually be oriented towards 9 o'clock, whereas ones made by fingers of the right hand will usually be oriented towards 3 o'clock. The other side of the cup will often reveal a single, larger mark made by the thumb with the opposite orientation.

7.2 Other Marks

The area around the mark in question may reveal another mark or marks that are consistent with having been made by the same hand (or foot) at the same time. The study of 3903 cases by the UK Home Office (1969) reported that just over 48% revealed such marks. These 'simultaneous' impressions (also known as 'sequences' of marks) may be located above, below or to one or both sides of the mark. They may also be on the opposing side of a surface that has been gripped, like the key to a door or a piece of glass removed from a broken window (Figure 7.2). Marks like these may have been lifted or photographed separately from the mark in question, so the examiner may only discover their relative location by studying the documentation produced by those who found the marks.

Simultaneous impressions can also be produced by a single contact of one digit. For example, a finger that contacts a surface across a recess may leave two separate areas of ridge detail – one on either side of the recess. Alternatively, contact may occur across a fold in a surface or across the point at which one surface overlaps another – both instances will result in two separate areas of ridge detail if the former surface is unfolded, or the later surfaces are separated. Similar separate marks can also be created if a mark is split into two as a result of the middle part of it being wiped from the surface or if it is deposited across multiple surface types and the development medium used does not develop the mark on all of them.

If other marks have been deposited simultaneously, they may reveal indicators that can assist in establishing the most likely area of skin to have made the mark in question. Consideration of the following factors may allow the examiner to arrive at an opinion as to whether the marks are or are not consistent with simultaneous deposition:

- The spatial relationships of the marks
  Generally, if the marks were deposited while the fingers were relatively straight, their spatial relationships should correspond to the different lengths of the fingers (Figure 7.3). The easiest

(a)

(b)

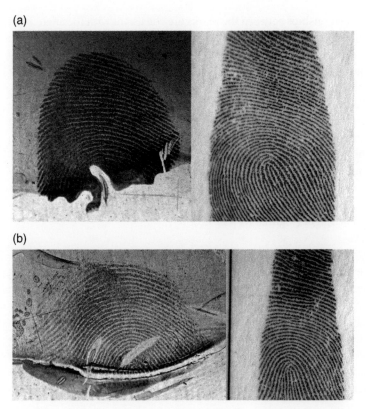

Figure 7.2 The marks on the left were made by the right thumb (a) and right index finger (b) in conjunction with each other, meaning that they were made by the same hand at the same time on opposite sides of a surface – in this case, the fuel cap of a car. The prints on the right were made by the same areas of skin.

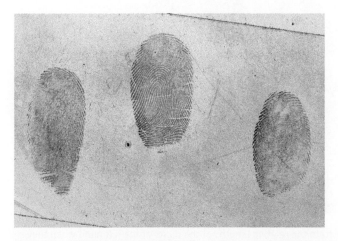

Figure 7.3 These marks were deposited by the same hand at the same time on a flat surface and reveal numerous indicators of simultaneous deposition. Most notably, even though two of the marks reveal little detail, their spatial relationship to the other mark indicates that the mark is most likely to have been made by a middle finger. As discussed later in this chapter, the slope of the pattern of that mark indicates it was most likely made by a left middle finger.

way to establish whether the relationships between the marks are consistent with the anatomical possibilities is for the examiner to place their fingers over a life-size image of the marks to observe which of their distal phalanges correspond with the locations of the marks. In doing this, the examiner must allow that the donor of the marks may have significantly larger or smaller hands than them and may be able to spread their digits further apart. Additionally, though most peoples' middle and little fingers are their longest and shortest, respectively – there is more variation in the lengths of the other digits. Comparing the lengths of the index and ring fingers Peters et al. (2002) reported that in men, the ring finger tends to be longer, though in women, the results are more variable.

If the marks were deposited while the fingers were wrapped around something, the relationships of the marks may not correspond to the lengths of the fingers. This is because as the fingers close around an object, the relative difference in their lengths decreases and the ring finger often extends further than the middle finger (Figures 7.4 and 7.5).

- The orientations of the marks
Generally, marks made by the same hand at the same time will have similar orientations. However, those made on different sides of an item that has been gripped will usually have opposing orientations, and thumb marks made by a hand placed flat on a surface can have an orientation almost perpendicular to that of the fingers.
- The parts of the distal phalanges that made the marks
Simultaneous impressions made by the distal phalanges will usually have been made by a similar area of each phalange (Figure 7.6). For example, they each may reveal more of the area to the left of the core than to the right. If one or more of the marks reveals significantly more of the opposite area, then it is unlikely they were made simultaneously. However, when simultaneous

(a)   (b)

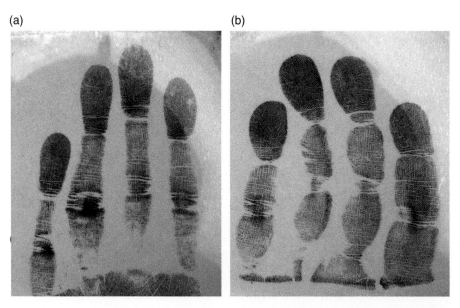

Figure 7.4 The impressions in (a) illustrate the relative lengths of the digits of the left hand when they are placed flat on a surface. The impressions in (b) illustrate the difference in the relative lengths when impressions are made by the same digits on a curved surface (in this instance, a cup). The relative differences are reduced, and the ring finger now appears to be the longest digit. The impressions on the right also appear curved and are wider, and when seen in marks, these are both indicators that the marks may have been left on a curved surface.

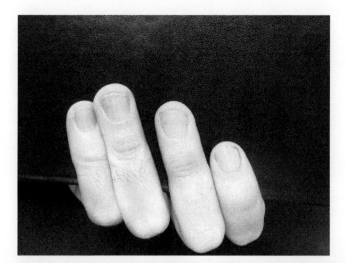

Figure 7.5 Though this surface is flat, as a result of the hand being wrapped around it, the relative positions of marks made simultaneously by these digits would appear very different from those the examiner may expect to see on a flat surface.

Figure 7.6 These marks were made by a right hand on a cup. When marks are made by fingers wrapping around a curved surface, they usually reveal more detail on the side of each digit closest to the thumb than the other.

impressions of the fingers and thumb are made on a flat surface, the fingers may reveal a similar amount of detail on both sides of the core while the thumb only reveals detail on the left side of the right thumb or right side of a left thumb. Also, if the marks are made by a hand wrapping around an object, then the thumb mark will usually reveal detail on the opposite side of the core to that revealed in the finger marks.

- Indications of movement in the marks
  If there are signs of the skin moving during the deposition of simultaneous impressions, those signs will often indicate that the type, direction and extent of the movement were similar for each mark (Figure 7.7). However, movement in different directions is possible as, for example, fingers can spread apart to an extent when contacting a surface. It is also possible for a mark to

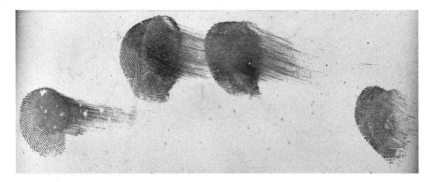

Figure 7.7 These marks reveal numerous indicators that they were made by the same hand at the same time, including movement that has resulted in smearing of all four marks to a similar extent and in a similar direction.

reveal no indicators of movement even though it has been deposited simultaneously with others that do – for instance, if the digit that made that mark breaks contact with the surface before the movement began.

- The appearance of the details within the marks

  The amount of downward pressure used to deposit each mark will generally be similar if the marks have been deposited simultaneously on a flat surface, so the ridge and furrow widths should be relatively consistent (though simultaneous impressions can be left with differing amounts of pressure). In impressions made by a single contact with one area of skin that have resulted in separate areas of ridge detail, the ridge flow and spatial relationships of the details in the separate areas should, allowing for the gap, be consistent.

  If one area of ridge detail is very heavily creased and others are not, this may be an indication that different donors deposited the marks. However, when considering this factor, the examiner will have to take into account that some areas of skin are often much more heavily creased than others (for example, the thenar area of the palm).

  If a development media has been used to make the marks visible, the reaction between that media and constituents of the material the marks were made in will usually result in marks that have been deposited simultaneously having a similar appearance. However, as a result of variations in those constituents across the surface of the hand, some marks that were deposited simultaneously may have reacted with greater or lesser intensity to the development media than others.

In many cases, marks will either be clearly consistent or inconsistent with being made by the same hand at the same time; however, there will be instances where the examiner is unable to form an opinion either way.

If the examiner reaches the conclusion that marks are consistent with being made simultaneously, then, as well as being able to use those marks to establish the likely area of skin that made the mark in question, some examiners will also use the details in all the marks collectively in the comparison and evaluation stages. For example, the examiner may consider that the four marks in Figure 7.8 each reveal areas of detail that are insufficient for identification on their own, but if their appearance is seen as consistent with being made simultaneously, the examiner may consider using all the details collectively to identify the donor (Figures 7.9, 7.10, 7.11 and 7.12).

Whilst this practice is not new and was endorsed by the Scientific Working Group on Friction Ridge Analysis, Study and Technology (SWGFAST, 2008) and numerous authoritative publications (Ashbaugh, 1999; Cowger, 1993; FBI, 1972; Vanderkolk, 2011), some consider that the available

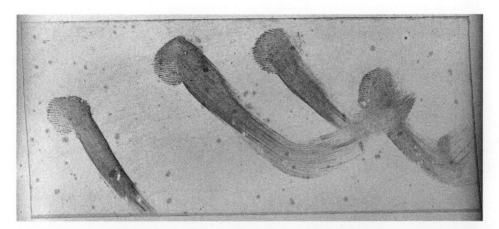

Figure 7.8 These marks were deposited simultaneously by a right hand on glass and developed with aluminium powder. The orientations of the marks and their spatial relationships are consistent with having been made by the same hand at the same time. Also, the marks all appear to have been made by a similar area of each digit (the right side of the tip) and their appearance (for example, ridge width) and reaction with the development media is similar. Each mark also reveals indicators of a similar type of movement during deposition.

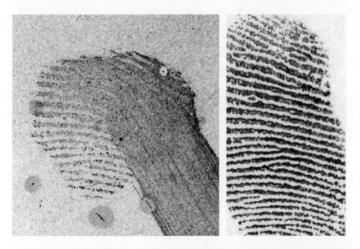

Figure 7.9 Figures 7.9, 7.10, 7.11 & 7.12 show each of the marks in Figure 7.8 with the corresponding part of a print made by the same area of skin. Though there may not be enough detail in any of the marks on their own for the examiner to identify them, if the examiner considers the prints are consistent with being deposited simultaneously, there may be enough detail in all the marks collectively for them to reach that conclusion.

research does not provide a scientific basis for it (OIG, 2011). For similar reasons, a US court ruled an identification made using the practice inadmissible in 2005. In Commonwealth vs. Terry L. Patterson (2005), the Supreme Judicial Court of Massachusetts decided that the practice was not generally accepted or recommended, there were no studies or articles that validated or demonstrated an error rate for it, and there was no formalised procedure or standards to govern its use.

In response to this ruling the first, and so far, the only study was undertaken to assess whether examiners could reliably determine whether marks were made by the same hand at the same time. The study found that, after a thorough analysis incorporating all available indicators,

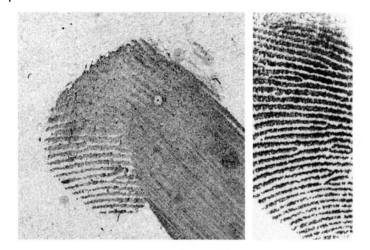

Figure 7.10

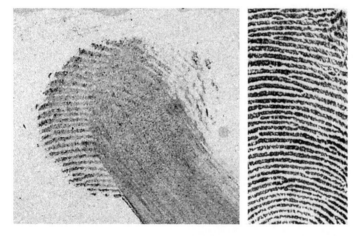

Figure 7.11

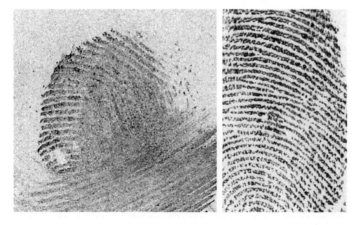

Figure 7.12

examiners were able to do this with clear marks nearly 88% of the time (Black, 2006). Also, since the ruling, SWGFAST has addressed the lack of formalised procedures by publishing requirements for determining whether marks are consistent with being simultaneously deposited and requirements for determining conclusions from the comparison of those marks (SWGFAST, 2008).

If the examiner uses this practice, they must ensure the conclusion that marks are or are not consistent with having been made by the same hand at the same time is reached before the print is analysed. This is necessary to minimise the risk of the examiner being influenced by the details in the print. For example, the observation that two marks both have a small amount of detail in agreement with two prints of the same person, but only enough to identify that person as the donor if they are used in aggregate, could influence the examiner to reach the conclusion they must have been made simultaneously (IEEGFI, 2004). The determination that marks are consistent with being made simultaneously must be based on their appearance and other information gathered during the analysis of the mark, independent of the details in a print.

To reach a conclusion about whether marks are or are not consistent with being deposited simultaneously the examiner must carefully observe and evaluate all indicators. The extent to which they document this conclusion depends on the procedures of the organisation they are employed by and the marks themselves. For example, if none of the marks reveals enough clear details to support a conclusion on their own and the examiner considers them to be consistent with being deposited simultaneously, their justification for this conclusion should be thoroughly documented. However, if all the marks reveal abundant clear detail and the simultaneous aspect is being considered only to arrive at the most likely digit to have made the mark in question, less extensive documentation of the justification for their conclusion may be appropriate.

7.3 Lone Marks

The UK Home Office (1969) study of 3903 cases found that just over 65% of them revealed marks that were found on their own, without other marks around them that were consistent with being made by the same hand at the same time. Such marks are more likely to have been made by thumbs or index fingers than any other digits. The same Home Office study examined 931 lone marks that had been identified and found that thumbs accounted for over 53% and index fingers over 23%, with middle fingers at 13% being the next most encountered digit (UK Home Office, 1969). However, it is important to note that, though they were encountered less frequently, lone marks made by all the other digits were also seen.

Though the majority of people are right-handed, the study found that the proportion of right digits identified was almost the same as the proportion of left digits. Another study of over 3000 marks also reported that the percentage of those made by right or left hands was virtually even (Brooks, 1975).

Marks made by thumbs often have a distinctive appearance that can be used to differentiate them from those made by fingers. As the distal phalanges of thumbs are significantly wider and longer than those of the fingers, the examiner may be able to recognise that a mark is likely to have been made by a thumb by its size. However, the examiner must also consider that as the size of people's digits can vary considerably, a mark made by one person's thumb could appear smaller than one made by another person's finger (Figure 7.13). The size of a mark can also be affected by a number of other factors, including the amount of the digit that contacted the surface (i.e. a thumb contacting a surface at an angle may leave a smaller mark than a finger that squarely

(a)   (b)   (c)

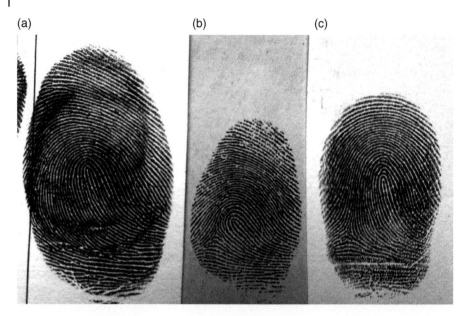

Figure 7.13 Though a large mark is often an indicator that the digit that made it was a thumb, there is considerable variation in the size of digits. These impressions were all made by adults and are all the same scale but were made by a right thumb (a), a right thumb (b) and a right middle finger (c). (b) *Source:* Courtesy of Mayor's Office for Policing and Crime, London.

contacts the same surface). Also, as heavier downward pressure usually produces a larger mark, a thumb that only lightly contacts a surface can leave a smaller mark than a finger.

Due to their size, some thumb marks will also be distinctive because of the large number of ridges between the core and the top of the mark or because of their high ridge count. In a study of in excess of 22 million sets of fingerprints, over 53% of loops with a ridge count of 21 or greater were found on thumbs (Washington, 1993)

Another distinctive aspect of some thumb marks is their shape. In comparison with finger marks, their sides will often be more bulbous (Figure 7.14), and some thumb marks may have a pear shape. Thumb marks may also be more likely to reveal subsidiary ridges than marks made by the other digits (Penrose and Plomley, 1969; Stücker et al., 2001).

Marks made by thumbs that reveal the tip of the digit will almost always include a distinctive arrangement of ridges known as 'fault ridges'. Whilst the ridge flow at the extreme tips of the fingers typically runs parallel to the fingernail and then slopes downwards on both sides, the fault ridges at the extreme tips of thumbs typically slope steeply downwards to the left or right (Figure 7.15).

When seen in impressions, these fault ridges flow down to the right on right thumbs and down to the left on left thumbs irrespective of the pattern type (Figure 7.16). Two studies both found 100% of the fault ridges they examined exhibited this trend (Mandrah & Kanwal, 2015; Singh et al., 2005).

Fault ridges are not found on any other digit and as the direction of their slope correlates to the hand that made them, their presence is a strong indicator that the mark was made by a thumb, and which thumb it was. However, the examiner must be aware that the ridges at the top of marks made by fingers can sometimes appear like fault ridges (Figure 7.17). This will not usually be the case if a finger squarely contacts a surface, but where the mark is made by the side of the digit, the only differences may be that the sloping ridges made by the finger are shorter in length and less numerous than fault ridges.

Figure 7.14   In contrast to the straight sides seen in many finger marks, the sides of thumb marks like this one are often more bulbous.

Figure 7.15   The print in (a) was made by a left thumb and displays fault ridges at the top that slope sharply down to the left. The print in (b) was made by a left index finger and displays ridges at the top that are virtually parallel to the fingernail and have a much less pronounced directional slope.

(a)   (b)

(a)   (b)

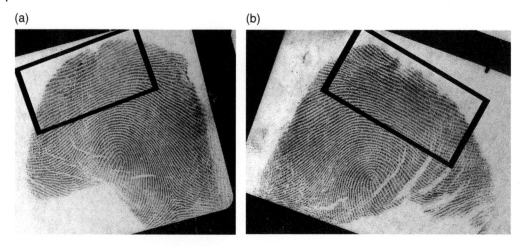

Figure 7.16 The print in (a) was made by a left thumb and the print in (b) by a right thumb. The boxes frame the fault ridges that slope down and to the left in the left thumb print and down and to the right in the right thumb print.

(a)   (b)

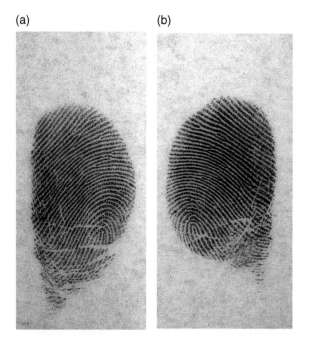

Figure 7.17 These two marks were made by the sides of left (a) and right (b) middle fingers and reveal ridges with a similar appearance to fault ridges.

7.4   The Part of the Digit That Made the Mark

When a hand contacts a surface squarely, subtle differences in the parts of the distal phalanges that touch the surface may be seen in the marks. Whilst fingers often have a slight inclination or rotation towards the little finger side of the hand; thumbs have a much more pronounced and opposite inclination (Bansal and Craigen, 2007; Tan et al., 2008). As a result, often only the extreme side of a thumb touches a surface when the hand contacts it squarely (Figure 7.18). Consequently, marks on flat surfaces that reveal detail made by the extreme side of a digit are often found to be made by the sides of thumbs that are furthest away from the fingers.

When more of one side of a digit contacts a flat surface than the other, one side of the resulting mark will usually be straighter and the other more bulbous (Figures 7.17 and 7.18). The straighter side will usually be the side that the digit is inclining towards, therefore where they have been made by the hand contacting a surface squarely, marks made by right thumbs often have straighter left sides and more curved right sides due to their natural inclination (and vice versa).

If a hand does not contact a surface squarely or that surface is not flat, then the areas revealed in any resulting marks may be different. For example, when a hand is wrapped around something like a cup, usually more of the side of each finger closest to the thumb, and the side of the thumb closest to the fingers, contacts the surface (Figure 7.19). So, marks made by the left-hand fingers around a cup may often reveal more detail to the right of the core, and the mark made by the thumb on the other side may often reveal more detail to the left of the core.

When the thumb and index finger are used in conjunction with each other to grip a small surface, whilst the thumb is often relatively flat against the surface, the index finger will usually be inclined away from it, resulting in much more of its thumb side being revealed in the mark.

Figure 7.18 This mark was made by the right thumb when the donor's whole hand contacted a flat surface. Due to the natural inclination of the thumb, relative to the fingers, much more of the side of the thumb furthest from the fingers has contacted the surface, which has resulted in more detail to the left of the core being revealed (the opposite would be expected in a mark made by the left thumb). The left side of the mark is noticeably straighter than the right side, which is also typical of marks made by digits inclined in this direction.

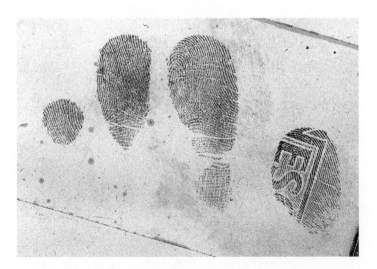

Figure 7.19 These marks were made on a cup by the fingers of a left hand and subsequently reveal more detail to the right of the core than to the left.

7.5   The Pattern

The slope of the pattern revealed in the mark can indicate whether a right or left hand is most likely to have made it. Additionally, some pattern types are found on particular digits more frequently than others.

Unless otherwise stated, the pattern frequency distribution figures used in this section are based on data published by Andrés Washington available on http://www.dermatoglyphics.com. This data is the result of the examination of in excess of 22 million FBI fingerprint forms (over 17 million male and over 4 million female) which were collated by the National Institute of Standards and Technology (NIST) in 1993. Whilst differences in frequency have been observed in different populations (Babler, 1977; Rife, 1953; Swofford, 2005), the same general pattern frequency trends in the data published by Washington have also been found in data collected from 5000 people in Australia (Murray, 2008) and 4 million people in the United Kingdom (Metropolitan Police Service (MPS), 1989 – unpublished).

Though some patterns are much more likely to be found on some digits than others, it is crucial to acknowledge that any pattern can be found on any digit. In the data published by Washington, there were no instances where a particular pattern was never found on a particular digit.

Of the three pattern families, the data indicates that loops comprised 65% of patterns encountered, whorls 28% and arches 6%. Table 7.1 shows the frequency of occurrence of the seven pattern types that data was collected for.

7.5.1   Loops

Loops made up approximately 65% of all patterns, were not significantly more likely to be found on one hand than the other and were commonly encountered on all digits (Table 7.2). Consequently, the presence of a loop pattern in a mark does not provide a significant indicator of the hand or digit that made it. However, whether the loop slopes to the left or the right is an excellent indicator of the hand that made the mark as 93% of right loops were on right hands and 94% of left loops were on left hands (Figure 7.20).

Table 7.3 shows that whilst they are very rare on all other digits, loops that slope right in impressions made by the left hand or left in impressions made by the right hand are relatively common on the index fingers (almost one third have that slope). So, it is not unusual to see a mark revealing

Table 7.1   Frequency of occurrence of the patterns.

Loops	65%
Plain whorls	20%
Central pockets	4%
Plain arches	4%
Twinned loops	3%
Tented arches	3%
Accidentals	0.001%

Table 7.2   Distribution of loop patterns.

Thumbs	Index fingers	Middle fingers	Ring fingers	Little fingers
17%	17%	22%	18%	26%

(a)   (b)

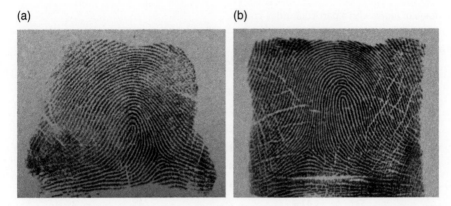

Figure 7.20 Left (a) and right (b) sloping loop patterns. (a) *Source:* Courtesy of Mayor's Office for Policing and Crime, London.

Table 7.3 Percentage of loops on each digit that slope left in impressions made by the left hand or right in impressions made by the right hand.

	Thumbs	Index fingers	Middle fingers	Ring fingers	Little fingers
Loops sloping left in left-hand impressions or right in right-hand impressions	99%	69%	98%	98%	99.7%

a right sloping loop that was made by a left index finger and vice versa. The examiner will often encounter simultaneous impressions revealing loops that, apart from one, all slope in the same direction. Because loops that oppose the general tendency are rare on the other digits, the loop that has the opposite slope is likely to have been made by an index finger.

The ridge count can also provide an indicator of the most likely digit to have made a mark revealing a loop pattern. Table 7.4 shows that ridge counts of fewer than five were not often seen on thumbs but were relatively common on index fingers. Also, over half of impressions with ridge counts of over 21 were made by thumbs and just under a quarter were made by ring fingers.

7.5.2 Plain Whorls

Plain whorls made up approximately 20% of all patterns, were not significantly more likely to be found on either hand and were commonly seen on all digits (Table 7.5).

If a whorl reveals a slope, then that can be a good indicator of the hand that made the mark as right sloping whorls are more likely to have been made by right hands and left sloping whorls by left hands (Figure 7.21). One study of American Caucasian males found that 90% of right sloping whorls were made by right hands and over 90% of left sloping whorls were made by left hands though a study of prints from a Central Indian (Marathi) male and female population reported slightly lower figures at 79 and 73% respectively (Brazelle & Brazelle, 2018; Kapoor & Badiye, 2015). Brazelle & Brazelle (2018) also observed that whorls with a slope that opposed this tendency were most commonly seen on index fingers.

If both deltas can be seen in a mark revealing a plain whorl, then its tracing can also be used to indicate which hand is most likely to have made it. The majority of outer tracing whorls were

Table 7.4 Percentage of loops with a particular ridge count.

Ridge count (% of loops with that count)	Thumbs	Index fingers	Middle fingers	Ring fingers	Little fingers
1–5 (14%)	7%	32%	25%	15%	21%
6–10 (23%)	11%	21%	26%	14%	27%
11–15 (36%)	14%	14%	26%	17%	28%
16–20 (24%)	24%	16%	13%	21%	25%
21–25 (4%)	52%	6%	4%	25%	13%
26–30 (0.004%)	69%	5%	1%	22%	3%
31+ (0.0002%)	51%	11%	2%	34%	2%

Table 7.5 Distribution of plain whorl patterns.

Thumbs	Index fingers	Middle fingers	Ring fingers	Little fingers
26%	23%	13%	29%	9%

(a) (b)

Figure 7.21 Left (a) and right (b) sloping whorl patterns. (a) ar-chi/Getty Images, (b) Juri Bizgajmer/Dreamstime LLC.

found on right hands (75%), whereas the majority of inner tracing whorls were found on left hands (64%) (Figure 7.22). 66% of meeting tracing whorls were found on digits of the right hand.

Table 7.6 shows that though outer tracing plain whorls were more commonly seen on right hands (in particular thumbs and ring fingers), they were also frequently found on the left index finger. In fact, plain whorls on the left index finger, unlike all the other digits of the left hand, were more likely to be outer tracing than inner tracing.

Table 7.7 shows that though inner tracing plain whorls were more commonly seen on left hands, they were also frequently found on the right index finger (only the left ring finger had a higher

(a) (b)

Figure 7.22 Whorls with an inner tracing like the print in (a) are more commonly seen in marks made by the left hand, whereas outer tracing whorls like the print in (b) are more commonly seen in marks made by the right hand. (a) 4x6/Getty Images, (b) kristo74/Getty Images.

Table 7.6 Percentage of outer tracing plain whorls on each digit.

	Thumbs	Index fingers	Middle fingers	Ring fingers	Little fingers
Right hand	26%	5%	6%	26%	11%
Left hand	6%	14%	3%	1%	0.001%

Table 7.7 Percentage of inner tracing plain whorls on each digit.

	Thumbs	Index fingers	Middle fingers	Ring fingers	Little fingers
Right hand	7%	17%	5%	6%	0.006%
Left hand	16%	7%	9%	24%	8%

occurrence of plain whorls with this tracing). In the same way that outer tracing whorls were more common on the left index finger than inner tracing whorls, whorls on the right index finger were more likely to be inner tracing than outer tracing.

Inner and outer tracing plain whorls were more common than meeting tracing ones and Table 7.8 shows that meeting tracings were most often seen on right hands.

In addition to considering the slope of the major axis and the tracing, with some whorls, there is another way to establish the likely hand that made the mark. If the ridges that form a plain whorl, elongated whorl or central pocket pattern unwind from the core in a spiral formation, the direction the spiral unwinds will usually indicate whether a left or right-hand digit is most likely to have made the mark (Figure 7.23). Marks revealing spiral whorls that unwind clockwise are more likely to have been made by left hands, whereas those that unwind anti-clockwise are more likely to have been made by right hands.

Table 7.8 Percentage of meeting tracing plain whorls on each digit.

	Thumbs	Index fingers	Middle fingers	Ring fingers	Little fingers
Right hand	16%	14%	10%	22%	4%
Left hand	6%	11%	7%	8%	1%

(a)   (b)

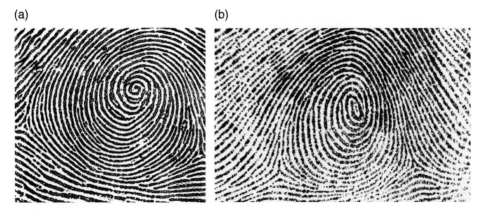

Figure 7.23 Both these prints reveal a spiral formation of ridges. If the spiral is followed from the ridge ending at the centre of the core, it unwinds in a clockwise direction in the print on the left and an anti-clockwise direction in the print on the right. (a) SCIENCE SOURCE/SCIENCE PHOTO LIBRARY, (b) babyblueut/Getty Images.

Singh (2005) looked at in excess of one thousand spiral whorls and found that over 97% of those on the right hand unwound anti-clockwise, and over 98% of those on the left hand unwound clockwise. Singh (2005) also reported that the digits most like to oppose this trend were the index fingers, in which approximately 2–3% were found to have patterns that unwound the opposite way.

7.5.3 Central Pockets

Central pocket patterns made up approximately 4% of all patterns and were not significantly more likely to be found on either hand. As Table 7.9 illustrates, they were much more likely to be found on ring fingers than any other digit.

Central pocket patterns may have a slope that indicates the most likely hand that made them in the same way it does for plain whorls, but another indicator is the position of the closest delta to the core. If that delta is on the right, then the mark was most likely to have been made by a right hand and vice versa (Brazelle, 2015). Tracing can also indicate the hand that made the mark, and the same tracing tendencies for plain whorls also apply to central pockets patterns, i.e. outer tracings were more commonly seen in impressions made by the right-hand digits, and inner tracings were more commonly seen in impressions made by the left-hand digits.

7.5.4 Twinned Loops

Twinned loops made up approximately 3% of all patterns and were not significantly more likely to be found on either a right or left hand. As Table 7.10 illustrates, they were much more likely to be found on thumbs than any other digit but were also fairly commonly encountered on index fingers.

Table 7.9   Distribution of central pocket patterns.

Thumbs	Index fingers	Middle fingers	Ring fingers	Little fingers
5%	21%	12%	44%	17%

Table 7.10   Percentage of twinned loop patterns on the digits.

Thumbs	Index fingers	Middle fingers	Ring fingers	Little fingers
64%	18%	7%	7%	3%

(a)   (b)

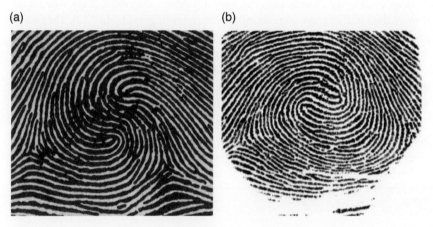

Figure 7.24   The ascending loop in impressions revealing a twinned loop pattern made by the left hand is more likely to be left sloping as in (a), whereas in impressions made by a right hand, it is more likely to be right sloping as in (b). (a) SCIENCE SOURCE/SCIENCE PHOTO LIBRARY, (b) VCNW/Getty Images.

As with other patterns in the whorl family, slope or tracing can be used to indicate the most likely hand to have made a mark revealing a twinned loop pattern. If the ascending loop is left sloping, the mark was most likely to have been made by a left hand and vice versa (Figure 7.24). An unpublished study by the MPS of around 4 million people in the United Kingdom in 1989 found this was the case approximately 81% of the time. The same study reported that the majority of twinned loops that did not conform to this tendency were found on the index fingers and that those digits revealed patterns with the opposite trend 47% of the time. The study also found that the right index finger was more likely to reveal a pattern with the opposite tendency (62%) than the left index finger (33%).

The same tracing tendencies for plain whorls also apply to twinned loops, i.e. outer tracings were more commonly found on right hands (84% were outer tracing) and inner tracings on left hands (83% were inner tracing). Meeting tracings were slightly more common on right hands (57%). The vast majority of inner or outer tracing twinned loops that do not conform to these tendencies occur on the index fingers. Twinned loops were actually more likely to oppose the tendency when seen on the index fingers – 67% of twinned loops on the index fingers opposed the tendency, but there was a significant difference between the left and right digits in that 82% of twinned loops on the right index opposed the tendency whereas only just over 50% did on the left index.

7.5.5 Accidentals

Accidentals made up approximately 0.001% of all patterns, were not significantly more likely to be found on either hand but were much more frequently encountered on index fingers than any other digit (76% were found on index fingers).

7.5.6 Plain Arches

Plain arches made up approximately 4% of all patterns, were not significantly more likely to be found on either hand and were most commonly seen on the index fingers, middle fingers and thumbs (see Table 7.11).

7.5.7 Tented Arches

Tented arches made up approximately 3% of all patterns, were not significantly more likely to be found on either hand, but more than half of them were found on the index fingers (Table 7.12). They were also frequently seen on middle fingers and were much more rarely seen on thumbs than plain arches.

If an arch pattern does reveal a slope, left sloping ones are more likely to have been made by left hands and right sloping ones by right hands (Figure 7.25). Murray (2008) examined 5000 Australian sets of fingerprints and found that over 78% of left sloping arches were found on left hands, and over 73% of right sloping arch patterns were found on right hands.

Table 7.11   Distribution of plain arch patterns.

Thumbs	Index fingers	Middle fingers	Ring fingers	Little fingers
23%	34%	29%	8%	5%

Table 7.12   Distribution of tented arch patterns.

Thumbs	Index fingers	Middle fingers	Ring fingers	Little fingers
3%	54%	27%	9%	6%

(a)   (b)

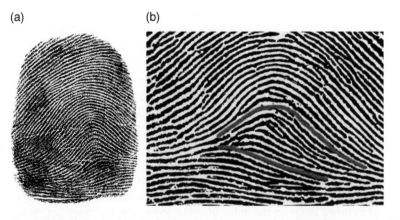

Figure 7.25   Left (a) and right (b) sloping arch patterns. (a) labsas/Getty Images, (b) CSA-Printstock/ Getty Images.

7.5.8 Scarred or Mutilated Patterns

Completely scarred or mutilated patterns were most commonly seen on index fingers (37%) followed by middle fingers (29%). Thumbs were the digits in which scars or mutilated patterns were seen with the lowest frequency (7%).

7.5.9 Indeterminate Patterns

The examiner will encounter marks that do not reveal enough of a pattern to classify its type, and in those situations, they must be careful to make all possible allowances for the pattern that could be on the skin of the donor. For example, a twinned loop can appear to be a plain loop if only the ascending loop is revealed in the mark.

If only the upper part of a mark is revealed, then the examiner may not be able to discern the pattern type, but it is sometimes still possible to ascertain the slope of the pattern and therefore at least suggest the hand, if not the digit, that is most likely to have made the mark. For instance, the ridges above the core will often turn downwards more steeply on one side than the other, and this often occurs on the left in a right sloping mark and the right in a left sloping mark (Figure 7.26).

Another indicator that can be useful if only the upper part of a mark is revealed is the presence of groups of ridge endings and/or bifurcations above and within approximately 12 ridges of the core (Figure 7.27). If the majority of ridges that end flow from the left and end towards the right and/or the bifurcations point to the right, the pattern on the digit that made the mark is likely to be right-leaning (Doak, 2004; Stoney and Thornton, 1987). If the majority of endings are ending to the left and/or the bifurcations are pointing to the left, then the pattern on the digit that made the mark is likely to be left-leaning.

If only the lower part of the mark is revealed, and a delta can be seen, then certain patterns can be excluded as having made the mark. For example, arches do not have deltas so the pattern could not be an arch, and if the delta is on the left side of the mark, then it also could not be a left loop (and vice versa) (Figure 7.28).

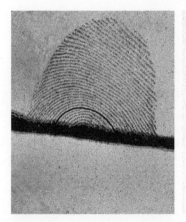

Figure 7.26 The mark on the left was made on the edge of a cardboard box. The mark does not reveal a pattern type, but the ridges turn down more steeply above and to the left of where the core would be than they do to the right. This is an indicator that the mark may have been made by a digit with a right sloping pattern (if the gradient was steeper on the right, then the mark is more likely to have been made by a digit with a left sloping pattern). The print on the right was made by the same area of skin and shows the almost vertical ridges on the left side of the core and the less steeply angled ridges on the right.

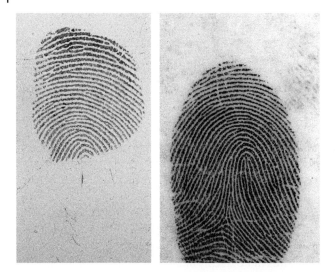

Figure 7.27 The mark on the left does not reveal a pattern type, but within 12 ridges directly above where the core would be, there are five ridge endings or bifurcations. All three bifurcations point to the right and one of the ridge endings ends towards the right. The presence of the majority of characteristics in this area with that orientation indicates the mark is likely to have been made by a digit with a right sloping pattern (a left sloping pattern would be likely if the majority of endings were ending to the left and/or the bifurcations were pointing to the left). The print on the right was made by the same area of skin and reveals the slope of the pattern.

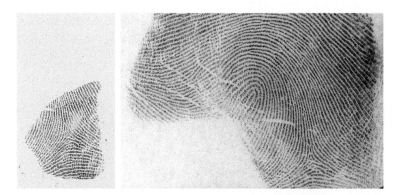

Figure 7.28 The mark on the left does not reveal a pattern but does reveal a delta which can be used to indicate the type of pattern on the digit that made it. Once the most likely orientation for the mark is established, it can be seen that the delta is on the right side of the mark, indicating that the most likely pattern types are left sloping loops or those of the whorl family. This mark was made by the left thumb that made the print on the right.

When using pattern as an indicator to establish the likely hand or digit that made a mark, the examiner must bear in mind that many patterns found on the digits can also be found on the palm, phalange or plantar surfaces, and in rare instances, a small mark left by one of these areas may appear very similar to that made by the distal phalange of a digit.

7.6   Summary of Digit Determination

To accurately determine the most likely digit to have made a mark, the examiner must consider and evaluate all indicators. Relying too heavily on one indicator can lead to the wrong area of skin being compared first. For example, though over 40% of central pocket patterns are found on ring fingers, a mark that reveals that pattern is not necessarily most likely to have been made by a ring finger. Similarly, the size of a mark may suggest it was made by a thumb, but it could also have been made by a person with large fingers or have been made on a surface that has distorted the size of the mark. However, if that large mark also reveals a twinned loop pattern and fault ridges, then a thumb would be the most likely area of skin to have made it.

Some marks will reveal indicators that are consistent with being made by multiple digits. For example, marks made simultaneously by either the left little and left ring fingers or the right index and right middle fingers can have similar spatial relationships. If two marks like this are found, the examiner may consider the pattern types of the marks to establish which combination is more likely. The mark on the left (made by the left little or right index finger) may reveal a left sloping loop which is consistent with patterns found on either digit, but if the mark on the right also reveals a left sloping loop, then it is much more likely that the digits that made the impressions were the left little and left ring fingers as left sloping loops are rarely seen on right middle fingers. Some examiners (Battley, 1930; Cherrill, 1954) suggest that another subtle indicator that could be considered in such a case is that the ridge detail in index fingers is usually more clearly revealed in marks than the ridge detail in little fingers.

In considering all indicators examiner will also encounter conflicting ones and so must be able to justify which ones to attach more emphasis to. For example, the mark in Figure 7.29 was found on a can of drink and was oriented towards 9 o'clock. The mark is large and was accompanied by marks

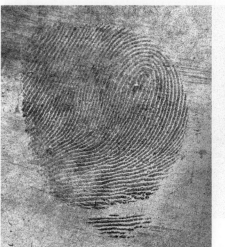

Figure 7.29   The mark on the left was found on a can of drink. The size of the mark is consistent with it being made by a thumb and its orientation on the can (pointing to 9 o'clock) suggests a right thumb is the most likely digit. However, the pattern is left sloping and more of the area of skin to the left of the core has contacted the can, which would be the opposite of what would result from the can being contacted by a right thumb if the can was being held in a conventional way. Therefore, the most likely digit to have made the mark is a left thumb that has contacted the can while either it or the donor's hand was inverted. The print on the right was made by the left thumb of the same donor who made the mark.

with the opposite orientation that were consistent with being made by fingers on the opposite side of the can. These indicators suggest the mark may be most likely to have been made by a right thumb. However, the mark reveals a loop that slopes to the left (very unusual on a right thumb) and more of the area to the left of the core than the right (more of the area to the right would be expected to be seen in a mark made by a right thumb on a surface like this). Also, when examined the marks on the other side of the can all sloped left (which is very unusual for all right digits). As a result, the most likely digit to have made the mark is actually a left thumb. This is because it is more likely that the can has been handled upside down than the mark was made by that area of a right thumb of a donor who has loops that oppose the expected tendency on all right digits.

Considering all indicators to arrive at the most likely area or areas of skin to have made the mark may allow the examiner to quickly identify its donor. However, where prints made by the most likely areas turn out not to have been made by the same area of skin, the examiner must then consider which other areas also need to be compared with the mark.

7.7 Palm Marks

In a study of 480 cases with identifiable marks, Langenburg, Bochet & Ford (2014) report approximately 1 in 7 of them was a palm mark and the UK Home Office (1969) found a similar ratio of 1 in 6 from 3903 cases. In a study of 45 palm marks, Sutton et al. (2013) reported that marks made by right palms were more frequently encountered than those by left palms. Examiners in the study judged 32 to have been made by right palms, and of the 32 that were identified, 25 were identified as having been made by right palms.

A palm mark will usually be larger than a mark made by a distal phalange and may also reveal distinctive ridge flow or creasing. As with marks made by any area, part of determining the most likely area of skin to have made a palm mark involves considering where the mark was found and how that surface could be touched. Additionally, the areas surrounding the mark in the image should be examined to see if there are any other impressions that could have been made by the same hand at the same time and used to indicate which area of palm made the mark in question.

Beyond those general considerations, this section describes three indicators that can be used to determine the most likely area of palm to have made the mark; the shape of the mark, its ridge flow and the creases that may be revealed within it.

Unless otherwise stated, the frequency of occurrence of details used in this section has come from a study of 2525 sets of prints by the MPS. The MPS examined the sets after receiving the 'Demystifying Palm Prints' workshop delivered by Ron Smith & Associates to test the utility of some of the palmar configuration trends observed by Mr Smith against the prints they commonly encountered.

The palm can be divided into three main areas and features three major creases (Figures 7.30 and 7.31).

7.7.1 Triradiate Area

The area of the palm directly below the fingers is known as the triradiate or interdigital area and extends across the width of the hand covering approximately the upper third of the palm.

7.7.1.1 Shape

Marks made by this area often have a rectangular shape in which one of the longer edges undulates significantly (Turnidge, 2010). This edge will usually correspond to the top of the triradiate (Figure 7.32).

Figure 7.30 The ridge flow on the palm can be divided into these three areas.

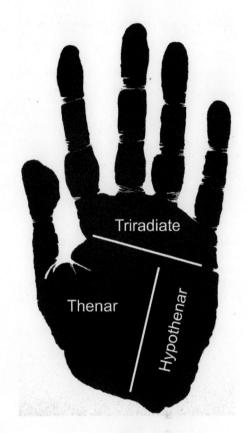

Figure 7.31 The three major creases on the palm are the Distal Transverse Crease, Proximal Transverse Crease and the Thenar (or Radial Longitudinal) Crease. The examiner may find it simpler to refer to these as the Top, Middle and Bottom creases, respectively, and that is how they will be referenced in this chapter. The three phalanges of the fingers are known as the Proximal, Medial and Distal Phalanges and the creases that separate them from each other or the palm are known as the Proximal, Medial and Distal Flexures.

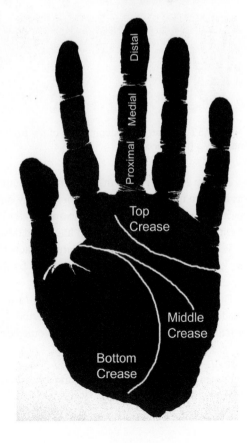

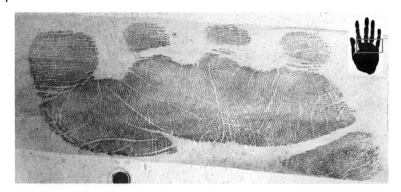

Figure 7.32 This mark was made by the right triradiate on the pole of a coat stand. The mark reveals many indicators of the area that made it, including an undulating top edge and the presence of a series of small, square or rectangular marks made by the proximal phalanges above that edge. The break in the ridge flow on the right side at the bottom of the mark was caused by the top crease that separates the triradiate from the hypothenar area.

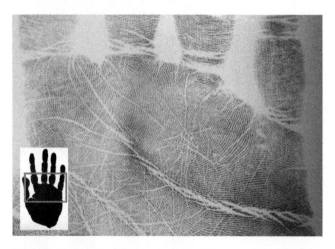

Figure 7.33 The steepest sloping part of the top edge slopes down to the right in this print as it was made by a right hand.

The top edge of a mark made by the triradiate area can also indicate which hand made the mark, as in some marks, it will slope more steeply on one side than the other (Figure 7.33). The steeper part will usually correspond to the area under the ring and little fingers and slope down and to the right in marks made by the right hand and down and to the left in marks made by the left hand.

Where only a part of the triradiate contacts the surface, a small circular or triangular mark may be left. These are most commonly made by the skin below the index or little fingers and often reveal part or all of a delta (Figure 7.34).[1]

Part of the triradiate area is often seen in marks that include the hypothenar area. Such marks made by the left hand may have an inverted 'L' shape with the triradiate area forming the shorter section (marks made by the right hand would be a mirror image of the inverted 'L' shape) (Figure 7.35).

---

1 Though they are commonly referred to as deltas, technically these formations are 'triradiates' as their arms do not typically enclose, or attempt to enclose the core of a pattern.

Figure 7.34 The mark on the left was made by the area under the right index finger and reveals a delta, the end of the top crease and the start of the middle and bottom creases. The print on the right was made by the corresponding area of the same donor.

Figure 7.35 The hypothenar area often contacts a surface with the triradiate area and produces marks that have an inverted 'L' shape if made by the left hand. The undulating edge often seen along the top of triradiate marks is particularly distinct in this mark.

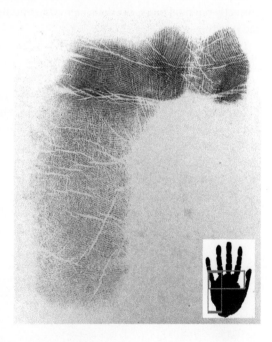

It is also common to see a series of short, often rectangular marks made by the proximal phalanges along the top edge of marks made by the triradiate area (Figure 7.32).

7.7.1.2 Ridge Flow

The ridge flow in the triradiate area is distinctive because it usually features more deltas than are seen in any other area of the palm. The triradiate typically includes one delta below each finger, as seen in Figure 7.33. Other than the carpal delta that is usually located centrally at the base of the palm, these are often the only deltas found on the palm. In a study of 1600 palm prints, Maceo et al. (2013) found that the majority had four deltas in the triradiate area but also recorded instances where only three or as many as seven were seen.

As the angles formed by the 'arms' of the deltas usually differ depending on where they are in the triradiate area, this difference can be used to indicate which area of the triradiate made a mark that reveals a delta (Figure 7.36). For example, the MPS study found a very high level of conformance with Ron Smith's observation that the delta beneath the index finger is usually equiangular – over 99% of the deltas below the index fingers were comprised of three approximately equal angles (Turnidge, 2010). The MPS study found similar conformance with Mr Smith's observation that the deltas below the middle and ring fingers were not usually equiangular, with over 99% being 'Y' shaped with the top of the 'Y' (the most acute angle) opening out directly beneath the digits.

The ridge flow beneath the delta under the middle finger can provide a strong indication of whether a mark revealing this area was made by a left hand or a right hand. Over 97% of the time, these ridges flowed down and to the right in an impression made by the right triradiate (as in Figure 7.33) and down and to the left in an impression made by the left triradiate (the ridges flowing down from the delta associated with the ring finger only conformed to this pattern around 26% of the time). The two deltas under the middle and ring fingers are also different from those under the index and little fingers as they often appear closer to the top edge of a mark as in Figure 7.32. It may be possible to differentiate between the deltas under the middle and ring fingers as, with a correctly oriented mark, the delta below the middle finger will often be higher than the delta below the ring finger.

The delta under the little finger was not found to be equiangular and could also be differentiated from those under the middle and ring fingers as its most acute angle usually points inwards away from the outer edge of the palm. Over 87% of these deltas had their most acute angle opening towards the inner surface of the palm (like a 'Y' on its side). The delta under the little finger is also different from the other deltas in that it does not usually lie directly below the digit – 99% of those seen were found further towards the delta under the ring finger (Figure 7.33).

Beginning between the index and middle fingers, the ridge flow in the triradiate area predominantly flows down and to the right in marks made by the right hand and down and to the left in marks made by the left hand. As they approach the little finger side of the hand, the ridges that remain in the triradiate area (those above the top crease) will often curve upwards slightly and end in an arch formation of ridges beneath the delta that corresponds to the little finger (Figure 7.37). As the ridges beneath the delta under the index finger are commonly relatively straight or only

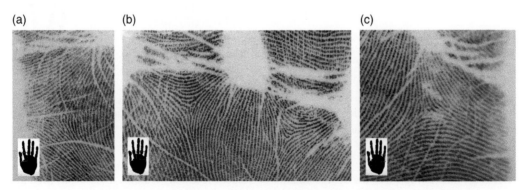

(a)   (b)   (c)

Figure 7.36 These prints show the deltas that correspond to the index (a), middle and ring (b), and little fingers (c) of the right hand.

Figure 7.37 The ridges below the delta that correspond to the little finger often form an arch formation between it and the top crease.

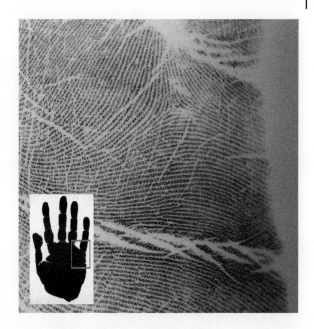

slightly curved, this arching of ridges below the little finger delta is another feature that can be used to differentiate between the deltas in the triradiate area.

By far, the most frequently seen patterns in the triradiate area were loops. The ridges that form each loop usually enter the palm between the digits and turn back on themselves before exiting the palm close to where they entered. As a result, most loops in the triradiate 'point' downwards towards the centre of the palm (forming a 'U' shape) (Figure 7.38).

Triradiate areas were found with multiple loops, but the most commonly encountered formation was a single loop in the area between the ring and little fingers. 54% of loops in the triradiate were found between the ring and little fingers and 45% were found between the middle and ring fingers, meaning loops between the index and middle fingers were rarely seen. After loops, the next most frequently seen pattern was a tented arch type of pattern that was usually found around the delta under the ring finger. Some triradiate areas had no patterns, and very occasionally, a small whorl type pattern was seen (Figure 7.39). One study of twenty thousand palm prints found that only 0.002% featured a whorl in the triradiate area (Leadbetter, 1976). Additionally, Tietze and Witthuhn (2001) and Maceo et al. (2013) both described finding 'column' or 'pillar' type formations in the triradiate, which comprised of several ridges that ended perpendicularly to another ridge and were surrounded by one or more deltas.

7.7.1.3 Creases

The bottom of the triradiate area on the little finger side is marked by the beginning of the top crease at the edge of the palm. At this point, the crease is often comprised of a series of small chevron-shaped creases that can be used as an orientation indicator as they almost always point inwards away from the edge of the palm (Cowger, 1993; Victoria Police, 2006) (Figure 7.40).

From the little finger side, the top crease travels across the hand marking the bottom of the triradiate area. Its path normally curves upwards slightly as it passes across the palm before turning upwards more steeply as it approaches the area under the middle and index fingers. The 82% of top creases were observed to end in this area, often by splitting into several smaller creases (Figure 7.41).

Figure 7.38 Loops were the most frequently seen pattern in the triradiate area and were most commonly seen between the ring and little fingers. *Source:* Courtesy of Mayor's Office for Policing and Crime, London.

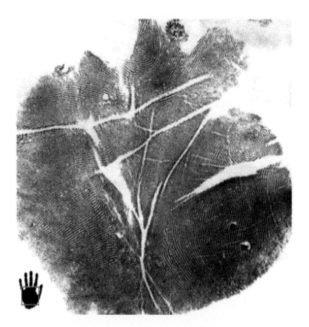

Figure 7.39 This print reveals a whorl type pattern in the triradiate as well as one in the thenar. *Source:* Reprinted with permission from Ray (2012).

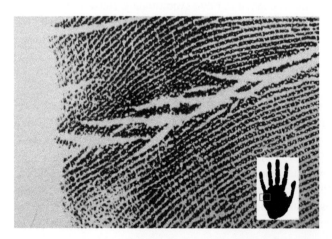

Figure 7.40 The chevron-shaped creases that often form the beginning of the top crease at the edge of the palm point inwards towards the centre. This print was made by a left hand, so the chevrons point to the left.

Figure 7.41 After entering the palm on the little finger side; the top crease usually follows an upwardly curving path across the hand before ending in the area below the index and middle fingers. The crease often splits into several smaller creases at this point, as can be seen in this print.

However, a small number of top creases were observed to reach the edge of the palm between the index and middle fingers.

The bottom of the triradiate area on the thumb side broadly coincides with the point at which the middle crease begins at the edge of the palm. The middle crease travels across the palm approximately parallel to the top crease and usually ends in the hypothenar. If it does not begin separately and below the middle crease, the bottom crease may begin by splitting from the middle crease close to the edge of the palm (Figure 7.42). Major creases do not intersect anywhere else on the friction ridge skin, so where this formation is seen in a mark, it is a strong indicator of the area of palm that made it.

7.7.2 Hypothenar Area

The hypothenar area extends down the little finger side of the palm from the top crease to the carpal flexure (the wrist crease). Sutton et al. (2013) found that marks made by the hypothenar were the most frequently encountered palm marks.

7.7.2.1 Shape
Marks made by this area often have an egg shape which can be used to orient the mark as the widest part will usually have been made by the bottom of the hypothenar (Figure 7.43). Hypothenar marks may also form part of an inverted 'L' shaped mark if some of the triradiate area has also come into contact with the surface (Figure 7.35).

Hypothenar marks are often found on documents due to the position of the hand when writing, and they sometimes reveal the transition between friction ridge skin on the palm of the hand and the non-friction ridge skin on the back of the hand (Figure 7.44). Once such a mark is correctly oriented, the presence of non-friction ridge skin on the right of the mark would indicate the mark was made by a right hand and vice versa.

7.7.2.2 Ridge Flow
The predominant ridge flow in the hypothenar is a strong indicator of the hand that made the mark. Though the ridges at the top and bottom may have a more horizontal flow, in a hypothenar mark made by the right hand, most of the ridges will almost always be flowing down and to the right and in a mark made by the left hand, they will be flowing down and to the left (Figure 7.45).

A particularly distinctive aspect of hypothenar marks is that the predominantly right or left sloping ridge flow is large and usually uninterrupted by other details. Maceo et al. (2013) found that

(a)   (b)

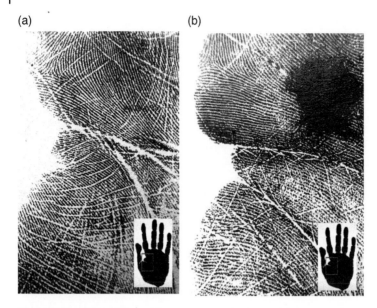

Figure 7.42 The middle and bottom creases both enter the palm on the thumb side and do so as one crease that splits into two as in (a), or separately as in (b). *Source:* Courtesy of Mayor's Office for Policing and Crime, London.

(a)   (b)

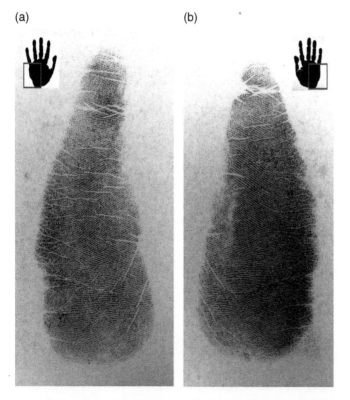

Figure 7.43 Marks made by the hypothenar often have an elongated egg shape with the widest part being made by the bottom of the hand, (a) was made by a left hypothenar and (b) by a right hypothenar.

Figure 7.44 Marks made by the hypothenar sometimes reveal the transition between the friction ridge skin of the palm and the non-friction ridge skin of the back of the hand. This mark reveals non-friction ridge skin (and hair) on its right side, indicating that the mark was made by a right hypothenar.

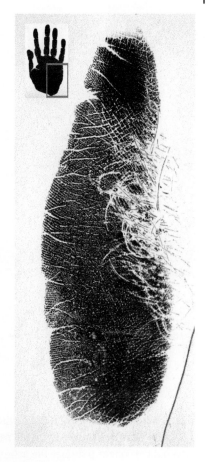

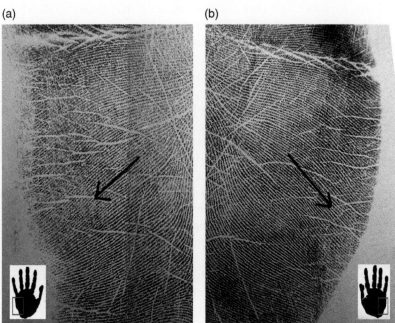

Figure 7.45 The predominant ridge flow in the hypothenar area slopes down and to the left in impressions made by the left hand (a) and down and to the right in impressions made by the right hand (b).

Figure 7.46 Most hypothenar areas did not reveal a pattern but of those that did the most frequently seen pattern was a loop in which the recurve faced the edge of the palm. *Source: Courtesy of Mayor's Office for Policing and Crime, London.*

over 66% of the 1600 palm prints they examined had no pattern present in the hypothenar area. Where patterns were seen in the hypothenar area, they were usually loops in which the recurve faced the edge of the palm, which supported the observation made by Ron Smith (over 81% of loops were found to slope this way) (Figure 7.46).

Both the MPS data and that collected by Maceo et al. (2013) also documented loops with the opposite directional slope (in which the recurve faced inwards towards the centre of the palm) (Figure 7.39). Maceo et al. (2013) reported over 25% of hypothenar areas with a pattern revealed a loop with this slope and also described loops in which the recurve faces the bottom of the palm, which were seen in over 3% of hypothenar areas with a pattern.

Loops in the hypothenar will often appear close to the carpal delta, which is usually found at the base of the palm between the hypothenar and thenar. Loops may also have another delta associated with them that would not otherwise be present. As Ron Smith observed, loops in which the recurve faces inwards towards the centre of the palm usually have the carpal delta below them and another delta above them. This was the case in over 97% of the prints that featured a loop sloping in this direction. A second delta was also seen with prints that featured a loop in which the recurve faced the edge of the palm, though with much less frequency. The additional delta for these loops was normally located near the edge of the palm and in some cases was so far around the side of the hand that it would not contact a surface the palm was placed flat onto.

Other patterns found in the hypothenar area but seen less frequently than loops include whorls, twinned loops, multiple separate loops and arches. Maceo et al. (2013) also observed vertical or horizontal columns of ridges that end perpendicularly to other ridges and are flanked on one or both sides by a delta.

As ridges from the hypothenar approach the centre of the palm, some of them often become finer, and the furrows between them narrower, as they turn upwards to flow around the thenar area (Cowger, 1993). In the same area, some of the ridges usually converge to create a distinctive funnel shape that occurs approximately where the middle crease ends (Figure 7.47). This area is marked by the presence of numerous ridge endings and is also an indicator of the orientation of the mark as the funnel shape is usually not symmetrical, i.e. the side of it nearest the top of the palm is generally flatter than the side nearest the bottom of the palm (Cowger, 1993).

When there is no pattern in the hypothenar, the carpal delta will usually be the only delta in the palm outside the triradiate area. The carpal delta will often be found just above the carpal flexure (wrist crease) along the axis of the metacarpal bone running through the hand from the phalanges of the ring finger, though occasionally it can be found higher up on the hypothenar side of the palm (Cummins and Midlo, 1961).

When a mark is made by the lower half of a palm, the ridge flow on either side of the carpal delta can indicate whether a right or left hand made the mark. The ridges at the bottom of the thenar will usually flow out of the bottom of the mark, whereas the ridges at the bottom of the hypothenar will usually flow out the side (Figure 7.48). So once a mark made by the bottom of the hand is

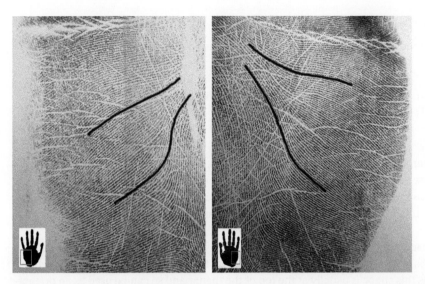

Figure 7.47 These prints show the distinctive funnel formation of ridges seen in the hypothenar. The formation is characterised by multiple ridge endings and is usually flatter at the top than the bottom.

(a) (b)

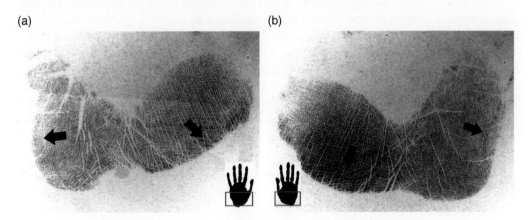

Figure 7.48 As the ridges in the hypothenar generally flow off the side of an impression, and the ridges from the bottom of the thenar generally flow off the bottom, when a mark is made by the lower part of the hand, this difference in ridge flow can indicate which hand made the mark – (a) was made by a left palm and (b) by a right palm.

correctly orientated, if some of the ridges exit out the side on the right, it will probably have been made by a right hand (and vice versa). When marks are made by the lower half of the hand, the upper edge often reveals a 'V' shape as the palms concave nature results in the middle not contacting the surface.

7.7.2.3 Creases

The most prominent crease in the hypothenar area is usually the ending of the middle crease, but others that may be present include short creases that enter the area from the edge of the palm. The presence and abundance of these creases can be used to indicate the proximity of the edge of the hypothenar area in the mark as their frequency increases close to the edge (Figure 7.43).

7.7.3 Thenar Area

The thenar area extends around the base of the thumb and is bounded by the bottom crease.

7.7.3.1 Shape

Like the hypothenar, marks made by the thenar often have an egg shape in which the wider section usually corresponds to the bottom (Figure 7.49).

7.7.3.2 Ridge Flow

Ridges in the thenar often appear indistinct in impressions, and this quality can sometimes be a useful indicator that the mark may have been made by a thenar area.

Impressions made by the thenar area usually reveal a semi-circular or arching ridge flow beginning between the index finger and thumb and curving around the thumb towards the wrist (Figure 7.50).

The bottom crease also follows this path, and whilst the ridges nearest it may have a smooth curve, the ridges nearer the thumb may form more of an 'L' shape (Figure 7.51). The longer section of ridges that forms this shape is usually at the top of the thenar and so can be used as an orientation indicator.

Patterns are rarely seen in the thenar area. In a study of over 34,000 sets of prints, Tietze and Witthuhn (2001) reported that only around 6% of the right thenar and 15% of left thenar areas revealed patterns. Though whorls and other patterns are occasionally seen (Figure 7.39), the most common deviations from the standard semi-circular ridge flow seen in the thenar are known as vestiges. Vestiges are one of the most distinctive formations found on the palm. They are not only a strong indicator that the mark was made by the thenar area but can also indicate the orientation of the mark and so whether it was made by a left or right hand. Vestiges usually feature some ridges that run perpendicular to those surrounding them and often include squared-off loop patterns (as in Figure 7.39) that can appear as a single formation or as two loops that meet core to core. If composed of two loops, the upper loop will usually be shorter and narrower than the lower loop (Victoria Police, 2006).

(a)   (b)

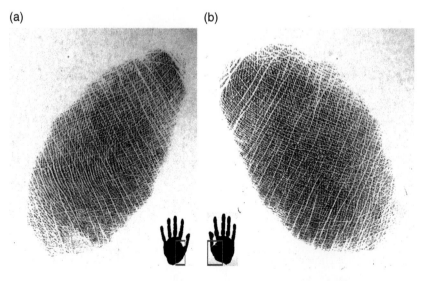

Figure 7.49 Thenar marks often have an egg shape, and the widest point of the shape will usually be at the bottom of the mark – (a) was made by a left thenar, (b) by a right thenar.

(a) (b)

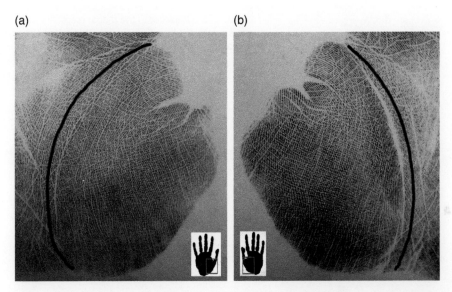

Figure 7.50 The ridges in the thenar broadly follow the path of the bottom crease and flow relatively smoothly around the thumb in a semi-circular formation – (a) was made by a left thenar, (b) by a right thenar.

Figure 7.51 Often the ridge flow in the thenar is not entirely smooth but is composed of a long section of ridge with a slight curvature followed by a sharp turn and then a shorter section of ridge.

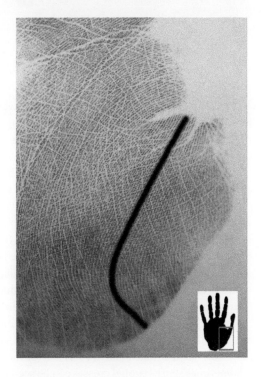

7.7.3.3 Creases

After entering the hand between the thumb and index finger, either as part of or below the middle crease, the bottom crease follows a similar semi-circular path to the ridges in the thenar. Most of the time, the crease either passes the carpal delta on the thumb side before reaching the carpal flexure (48% of thenar creases did this) or splits into several creases that pass either side of the delta (20%). The bottom crease was rarely seen to pass the carpal delta on the hypothenar side (8%).

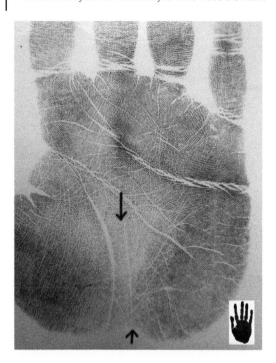

Figure 7.52 Adjacent to the thenar crease in some prints will be a finger crease that follows the line of the middle finger down through the palm.

Some palms feature finger creases that run vertically down the palm following the path of the metacarpal bones connected to each digit. Finger creases that correspond to the index finger are rarely seen, and the most commonly seen finger crease is usually associated with the bones under the middle finger and often appears adjacent to the bottom crease, as in Figure 7.52.

In marks that reveal the area between the thumb and the index finger, there is sometimes the appearance of several creases that radiate from the base of the thumb forming a 'starburst' configuration (Turnidge, 2010; Victoria Police, 2006). The formation is caused by folds in the skin, and the width of the folds and the angles between them will vary depending on the position of the thumb when the impression was made (Figure 7.53). If the thumb is close to the palm when the impression was made, then the 'starbust' will often be larger and more prominent than if the thumb was outstretched from the palm at the moment of deposition.

The thumb crease separates the proximal phalange of the thumb from the thenar area and is often comprised of a series of interlocking creases, though it can also be a solid (single line) crease (Figure 7.54). The wrist crease often has a similar appearance to the thumb crease, though marks made by these areas can usually be differentiated from each other as unlike the thumb crease, there will only be ridge detail on one side of the wrist crease (Figure 7.54).

Sometimes short, parallel creases can be seen entering the thenar on the thumb side between the thumb and the wrist crease (Figure 7.55). The presence of these creases can be used to indicate the proximity of the edge of the friction ridge skin on the thumb side in marks as their frequency increases nearer the edge of the hand.

One of the most notable features of the thenar is the numerous creases that are usually present. These creases will either be a series of roughly parallel creases stretching across the area, or creases that cross the area both horizontally and vertically and intersect each other at right angles to form a 'cross-hatching' effect (Figure 7.56).

(a)   (b)

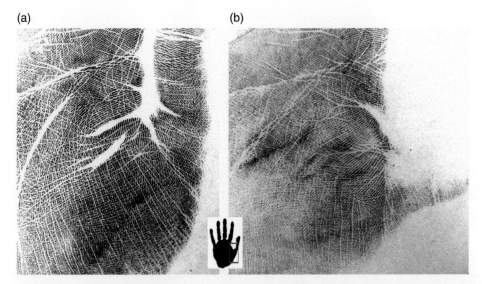

Figure 7.53 The 'starburst' crease is found at the base of the thumb on the index finger side. It is usually more prominent if the impression has been made when the thumb is close to the index finger as in (a) rather than spread apart from it as in (b).

(a)   (b)

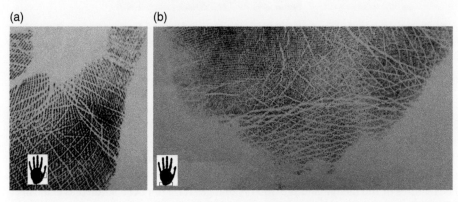

Figure 7.54 The print in (a) shows the thumb crease that separates the proximal phalange of the thumb from the palm. The print in (b) shows the wrist crease (or carpal flexure), which is the only crease that has friction ridges on one side and non-friction ridge skin on the other.

Figure 7.55 Short parallel creases can sometimes be seen entering the thenar from the edge of the palm.

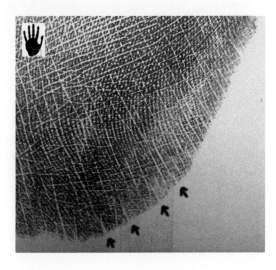

(a)　　　　　　　　　　　　　　　　(b)

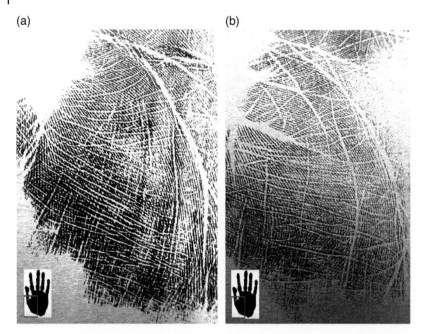

Figure 7.56　One of the most distinctive and ubiquitous features of many thenar impressions is the presence of multiple single creases (a) or creases that cross each other perpendicularly (b). *Source:* Courtesy of Mayor's Office for Policing and Crime, London.

7.8　Medial and Proximal Phalange Marks

The ridge flow and appearance of the flexures in marks made by the medial or proximal phalanges can indicate which phalange was most likely to have made the mark and which digit that phalange is part of. The features of the medial and proximal phalanges have not been studied as extensively as that of the distal phalanges, and unless otherwise stated, the frequency of occurrence of details used in this section comes from an unpublished study of 1000 sets of prints by the MPS (some of which was also conducted to test observations made by Ron Smith).

Marks made by the medial or proximal phalanges are often distinctive because of their rectangular shape and the flexures that interrupt the ridge flow at both ends. Medial phalange marks are commonly accompanied by distal and/or proximal phalange marks, and proximal phalange marks are often seen with triradiate marks (Figure 7.57).

Ploetz-Radman (in Cummins & Midlo, 1961) described twelve distinct configurations of ridge flow that can be found on the medial and proximal phalanges, and some of these can be seen in Figure 7.58.

However, in determining the area of skin that is most likely to have made a mark, whether the ridges reveal a clear left or right slope is a more useful indicator than the configuration. That is because the general trend of the ridge flow across all the phalanges broadly forms an arch shape with its apex in the middle and ring fingers. So, in an impression made by the right hand, it flows down and to the left on the thumb and both phalanges of the index and middle fingers and down and to the right on the middle and little fingers (and vice versa on left hand).

Figure 7.59 illustrates the predominant flow seen on each phalange and shows the ridge flow on the phalanges of both little fingers, and the right index finger conformed very highly with the general trend. This was also the most commonly seen ridge flow on the phalanges of the left index

Figure 7.57 These marks were made by the proximal phalanges of the index, middle, ring and little fingers of a right hand around the neck of a wine bottle.

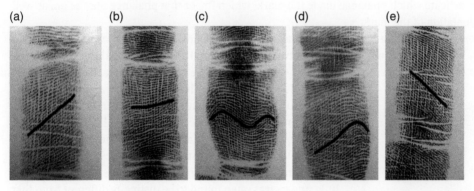

Figure 7.58 Some of the common types of configurations found on the medial and proximal phalanges.

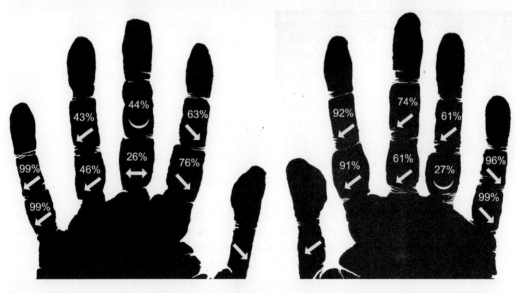

Figure 7.59 To a greater or lesser extent, the ridge flow across all the medial and proximal phalanges broadly forms an arch shape. However, the MPS study found that the predominant ridge flow of the right ring finger and left middle finger phalanges did not conform to this trend. The most common flow in the right ring finger medial phalange opposed it, and concave arch formations were the most common formation in the right ring proximal and left middle medial phalanges. The same proportion (26%) of left middle proximal phalanges conformed to the trend as opposed it.

finger, left ring finger and right middle finger, but there were some other types of flow that were also encountered fairly frequently in those areas – particularly concave and convex arches. The most commonly seen ridge flow on the phalanges of the left middle finger and right ring finger did not conform to the general trend and were much more variable – including ridges sloping the opposite way to the general trend and concave and convex arches.

The ridge flow on the proximal phalange of the thumb echoes that of the index finger by predominantly sloping diagonally downwards towards the thumb side of the hand though the MPS study did not record the extent of the conformance with the trend.

7.8.1 Flexures

The appearance in a mark of the flexures that separate the phalanges from each other and from the palm can indicate which phalange made the mark, which finger that phalange may be on, as well as the orientation of the mark.

The flexures that separate the distal phalanges from the medial phalanges were usually found to be just a single crease, whereas those between the proximal and medial phalanges were often made up of two or more separate creases (Figure 7.60). The MPS study also found very high conformity with Ron Smith's observation about the differences in the widths of flexures in that over 99% of the time, the flexures observed between the proximal and medial phalanges covered an area that was twice the width of those between the medial and distal phalanges.

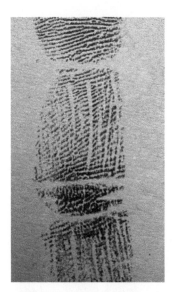

The configuration of the proximal flexures of the index and little fingers was often found to be distinctly different from that of the middle and ring fingers. The majority of proximal flexures of the index and little fingers (86%) were comprised of a series of short chevron-shaped creases that almost always pointed inwards away from the edge of the palm. Whereas over 87% of the proximal flexures of the middle and ring fingers were not made up of short creases but were comprised of two separate continuous parallel creases.

Distinctive vertical creases such as those in Figure 7.61 are often seen in the proximal and medial phalanges.

Figure 7.60 The flexures that separate the distal and medial phalanges were usually revealed in impressions as a single crease, whereas those that separate the medial and proximal phalanges were usually multiple creases that covered a wider area. This difference can be a useful indicator of the orientation of a mark made by a proximal or medial phalange. *Source:* Courtesy of Mayor's Office for Policing and Crime, London.

7.9 Plantar Marks

Marks made by the friction ridge skin on the feet are known as plantar marks, and the plantar surface can be divided into as many as eight areas (Cummins & Midlo, 1961). However, as the examiner is unlikely to be comparing plantar marks frequently and their usual ridge configurations are like those on the palm, the examiner may find it more practical to consider the divisions indicated in Figure 7.62. The toes can be described as the big toe, with the other toes being the second, third, fourth and fifth or little toe.

The plantar area is the largest friction ridge skin area, and so impressions made by it are frequently significantly larger than impressions left by the other areas and are also often distinctive because of their shape. Another crucial indicator that a mark could be made by an area of plantar, irrespective of its size or shape, is that the surface it was found on is something that is likely to have come in contact with bare feet. This could be a bathroom floor, but it could also be something that was found on a floor or, for example, in the foot well of a car.

The information in this section will describe the shapes of marks that different areas of the foot leave, as well as the distinctive ridge flow and creasing that can be found in different areas. Consideration of these factors may allow the examiner to arrive at the most likely area of plantar to have made the mark in question as well as its orientation.

7.9.1 Plantar Triradiate Area

The plantar triradiate area is directly below the toes and extends across the width of the foot, covering approximately the top third of the plantar surface (Figure 7.63). Like marks made by the palm triradiate, marks made by this area often have an oval or rectangular shape in which the top edge may undulate (Figure 7.64).

In addition to being an indicator that the mark may have been made by a plantar triradiate area, the top edge can also indicate which foot made the mark. Usually, the edge is relatively flat under the big and second toes but slopes steeply down and to the left under the fourth and little toes of the left foot and down and to the right under the same toes of the right foot (Figure 7.63).

Figure 7.61 Vertical creasing can be found on the medial and proximal phalanges. *Source:* Courtesy of Mayor's Office for Policing and Crime, London.

Marks that reveal the plantar triradiate area often also reveal part of the plantar hypothenar area and so have an inverted 'L' shape if they are made by a left foot and a mirror image of that shape if they are made by a right foot.

Plantar triradiate marks are often accompanied by marks made by some or all of the toes. Marks made by the big toe will often reveal the proximal phalange and be connected to the triradiate area, whereas marks made by the other toes will not usually reveal the medial or proximal phalanges and will be separated from the triradiate area.

Like the palm triradiate, the ridge flow of the plantar triradiate area is distinctive because it usually features more deltas than are seen in any other area of the foot and like the palm, there is usually a delta beneath each of the toes (Figure 7.63). The deltas associated with the second, third and fourth toes are often very close to the top of the triradiate, whereas those associated with the big and little toes are often very close to the sides of the foot. Due to their proximity to the edges of the triradiate, the deltas beneath the toes are frequently not revealed in marks (particularly those associated with the middle toes) (Figure 7.64).

Fox & Plato (1987) examined 502 plantar prints of American Caucasians and found that approximately 70% of triradiate areas had at least one delta in addition to those associated with the toes. That delta was most commonly located at the bottom of the triradiate, roughly in line with the second toe (Figure 7.65). Other additional deltas were also commonly seen, usually accompanying patterns in the triradiate area.

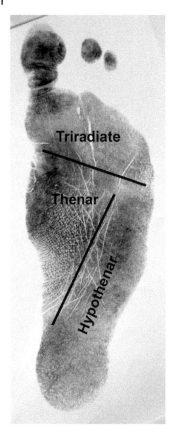

Figure 7.62 Like the palmar surface, the plantar surface can be divided into three main areas.

There have been very few studies of the ridges on the plantar surface, but Fox and Plato (1987) compared their data with that from nine other studies and found that like the triradiate area of the palm, the most common patterns seen in the plantar triradiate area are loops. Patterns were frequently observed in an area known as the 'Hallucal', which is directly under the big toe as well as in the areas between the toes. Fox and Plato found that most studies reported between 50 and 60% of the patterns observed in that area were loops in which the loop opened in the direction of the toes with the core facing the heel (like a 'U' shape) (Figure 7.66). Though less frequently encountered, loops in which the cores face towards the hypothenar and thenar sides of the foot can also be found, as can ones that face the toes.

Most of the studies reviewed by Fox and Plato also found that whorls made up 20–30% of the patterns under the big toe and were the next most frequently seen pattern. Whorls may be accompanied by three deltas when seen in the triradiate, with one often being located below the big toe, the second being lower down on the big toe side with the third opposite it (Cummins & Midlo, 1961).

Plain and tented arch type patterns can also be found in this area, with the plain arches either curving upwards from ridges flowing approximately horizontally through the area or curving towards either side of the foot from ridges flowing vertically through the area (Cummins & Midlo, 1961; Fox and Plato, 1987).

The most frequently encountered configuration in all nine studies between the second and third toes and the fourth and

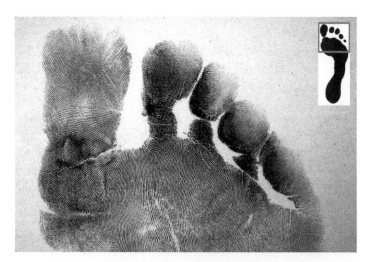

Figure 7.63 Though there are deltas under each toe, due to their positioning on the edges of the plantar triradiate, they are often not revealed in marks. This print was taken by applying powder directly to the skin and then lifting it off with an adhesive label that was able to wrap around the edges of the area and record the deltas associated with each toe (the ones associated with the big and little toes are at the extreme bottom edges of the mark). This method also recorded the proximal and medial phalanges of the smaller toes and their flexion creases which are also not often recorded in marks.

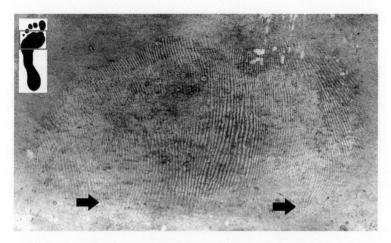

Figure 7.64 This mark was made by the triradiate area of a left foot on the interior windscreen of a car in front of the passenger seat. As is often the case, no deltas can be seen, though two loop type formations can be seen at the bottom of the mark.

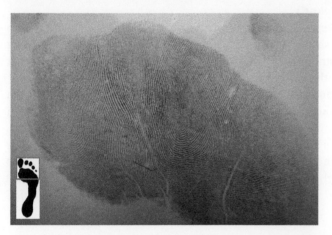

Figure 7.65 As well as the deltas associated with each toe, most plantar triradiate areas have at least one additional delta, which is often found at the bottom of the area roughly in line with the second toe.

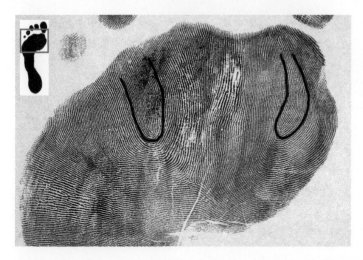

Figure 7.66 The most common patterns found in the plantar triradiate area are loops in which the core faces the bottom of the area like a 'U'. These loops are most often seen below the big toe and between the third and fourth toes.

little toes was a pattern-less area of ridges (Fox and Plato, 1987). After pattern-less areas, loops in which the core faces the top of the foot were the next most commonly observed formation between the second and third toes, as were loops with the opposite direction between the fourth and little toes. Loops in which the core faces the bottom of the foot were the most common configuration between the third and fourth toes in all studies (Fox and Plato, 1987). As with the area under the big toe, whorls and other patterns are occasionally seen between the smaller toes.

Though impressions made by the triradiate area of the palm may often reveal the proximal flexures of the digits at the top, the proximal flexures of most of the toes are rarely seen in impressions made by the plantar triradiate (though they are revealed in Figure 7.63).

7.9.2 Plantar Hypothenar Area

The hypothenar extends from the triradiate area down the outside of the foot and includes the heel area (which is also known as the 'Calcar').

Marks made by this area are often long and thin or, if they include the triradiate area, form inverted or inverted and mirrored 'L' shapes. Alternatively, if the centre part of the plantar hypothenar does not contact the surface, two separate marks may be left – one by the upper part of the hypothenar with the triradiate and the other by the heel (Figure 7.67).

The ridges in the plantar hypothenar area usually cross it at right angles to the length of the foot (Figure 7.68). Like the hypothenar area of the palm, the plantar hypothenar is usually devoid of patterns (Fox and Plato, 1987).

Where patterns do appear in this area, they are most likely to be loops and most commonly appear adjacent to the edge of the foot just below the triradiate area, though they may also occasionally be seen in the heel area (Cummins & Midlo, 1961; Fox and Plato, 1987;

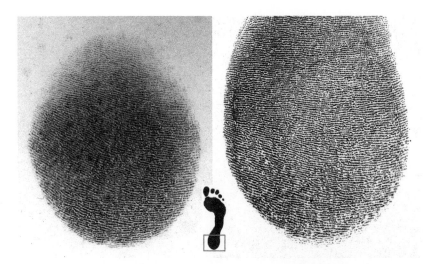

Figure 7.67 The mark on the left was made by the right heel. If the middle part of the hypothenar does not contact the surface, marks that are circular or oval in shape are often left by the heel. The print on the right was made by the same area of skin.

Okajima, 1967). In both cases, the most commonly seen loops were those in which the core faced the outer edge of the foot (Fox and Plato, 1987).

7.9.3 Plantar Thenar Area

The thenar area extends from the triradiate down the inner side of the foot. The majority of this area rarely contacts the floor due to the shape of the foot, so it is not often seen in marks (Figure 7.69).

The ridge flow in the plantar thenar generally continues that of the hypothenar and flows perpendicularly across the foot and rarely features a pattern. Fox and Plato (1987) reported that approximately 90% of the thenar areas they examined did not reveal a pattern but occasionally included a vestige at the bottom of the thenar just above the heel.

When it is revealed in a mark, the plantar thenar is often distinctive because it is usually the area of the foot that has the most creasing (Massey, 2004). Even when the majority of the thenar area is not revealed, significant creasing will often be visible around the edges of the thenar where it meets the triradiate and the hypothenar (Figure 7.70).

Figure 7.68 The ridge flow in the hypothenar area crosses at right angles to the length of the foot and is usually devoid of patterns.

Figure 7.69 The print on the left (a) was taken by rolling the plantar area over a curved surface to record the thenar area. The print on the right (b) was taken on a flat surface. Due to the curvature of the foot, the majority of the thenar area is usually only seen in marks that are made on concave or soft surfaces.

(a)　　　　(b)

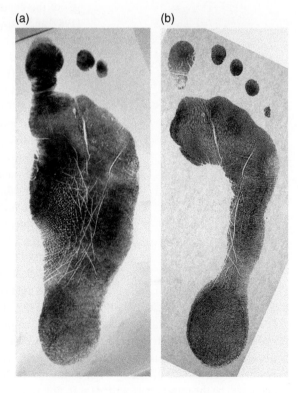

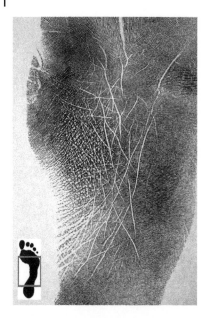

Figure 7.70 Like the palmar thenar area, the plantar thenar area is usually distinctive because of the amount of creasing present.

Figure 7.71 This mark was made by a right big toe.

7.10 Toe Marks

Marks made by the big toe are usually significantly different in shape and size than marks made by the other toes, additionally they also differ in that they usually include a mark made by the proximal phalange (Figure 7.71). However, marks made by the big toe can appear like thumb marks though in a study of 70 toe prints, Bradley (1977) found that as well as being larger, the patterns were lower down in the impressions and the ridges at the tips of big toes sloped the opposite way to fault ridges in thumbs.

Marks made by the other toes are often similar in size and shape to marks made by the tips of the fingers and are rarely accompanied by marks made by their phalanges (Figure 7.72).

All toe marks are usually accompanied by a mark made by the triradiate, and the big toe is often connected to the triradiate by detail from its proximal phalange. Of marks made by the smaller toes, a mark by the little toe is often the closest to the triradiate, and a mark made by the second toe is often the furthest away. This creates a spatial relationship that can be used to indicate whether a right or left foot made the marks (Figure 7.73).

The same pattern types that are found on the distal phalanges of the digits are also found on the toes. Fox and Plato (1987) and Hammer (1988) examined the data from a total of fifteen studies and found that loops that slope to the right on right toes (and vice versa) were the most commonly encountered pattern (Figure 7.74). Most results indicated these were seen on 45–64% of toes, whereas loops sloping the other way were seen on 0–3% and were found more frequently on the big toe than any other digit (Cummins & Midlo, 1961; Fox and Plato, 1987; Hammer, 1988). Whorls were the next most frequently encountered pattern, with most results indicating these were seen on 17–36% of toes (Fox and Plato, 1987; Hammer, 1988). Studies have found there was a very low percentage of whorls on the little toe (Cummins & Midlo, 1961; Fox and Plato, 1987; Hammer, 1988) and a very high percentage on the third toe (Fox and Plato, 1987; Cummins & Midlo, 1961). Arches were the least common, with most results indicating these were seen on 13–29% of toes (Fox and Plato, 1987; Hammer, 1988). Several studies reported a very high percentage of arches being found on the little toes (Cummins & Midlo, 1961; Fox and Plato, 1987; Hammer, 1988).

Because the big toe usually contacts a surface squarely, patterns on that toe may often be clearly revealed in marks, though they are rarely clearly revealed in marks made by any of the other toes because only their tip usually contacts a flat surface. As a result, the cores of the smaller toes, if they can be seen at all, will usually be at or near the bottom of the impression. Often the area above the core will be the only area recorded (as in Figure 7.73), an effect that is usually most pronounced in marks made by the little toe.

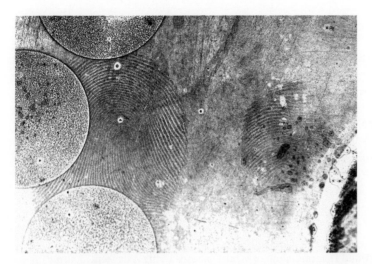

Figure 7.72 These marks were made on the bottom of a bath adjacent to the plughole by the big toe and second toe of a right foot.

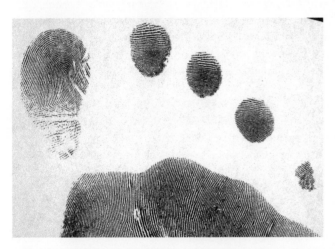

Figure 7.73 Marks made by the big toes often clearly reveal a pattern type, whereas those made by the smaller toes often only reveal the tip of the pattern area. In this image, the top of the core in the mark made by the second toe can be seen at the bottom of the mark, but all the other smaller toe marks only reveal the tips or side of the toes.

(a)　　　　(b)　　　　(c)　　　　(d)　　　　(e)

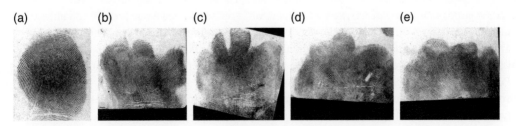

Figure 7.74 These prints were taken from the left toes of the same person by coating the skin in powder and then wrapping adhesive labels around each toe. Print (a) was made by the left big toe, (b) the second toe, (c) the third toe, (d) the fourth toe and (e) was made by the little toe.

References

Ashbaugh, D, R 1999, *Quantitative-Qualitative Friction Ridge Analysis*, CRC Press LLC, Boca Raton, USA.

Babler, W, J 1977, 'The prenatal origins of population differences in human dermatoglyphics', Thesis, University of Michigan.

Bansal, R & Craigen, M, A, C 2007, 'Rotational Alignment of the Finger Nails in a Normal Population', *The Journal of Hand Surgery*, 32E (1), pp. 80–84.

Battley, H 1930, *Single Finger Prints*, His Majesty's Stationery Office, London, UK.

Black, J, P 2006, 'Pilot Study: The Application of ACE-V to Simultaneous (Cluster) Impressions', *Journal of Forensic Identification*, 56 (6), pp. 933–971.

Bradley, P 1977, 'Investigations into Ridge Patterns on Feet', *Fingerprint Whorld*, 3 (10), p. 39.

Brazelle, M 2015, 'Whorl Pattern Analysis: Determining Directional Flow', *Identification News*, 45, pp. 9–11.

Brazelle, M & Brazelle, S 2018, 'Hand Determination of Whorl Patterns Using Axis Slant', *Journal of Forensic Identification*, 68 (1), pp. 77–86.

Brooks, A, J 1975, 'Frequency of Distribution of Crime Scene Latent Prints', *Journal of Police Science and Administration*, 3 (3), pp. 292–293.

Cherrill, F, R 1954, *The Finger Print System at Scotland Yard*, Her Majesty's Stationery Office, London, UK.

Commonwealth vs. Terry L. Patterson 2005, 445 Mass, 626, September 7, 2005 - December 27, 2005. Suffolk County, http://masscases.com/cases/sjc/445/445mass626.html

Cowger, J, F 1993, *Friction Ridge Skin*, CRC Press, Boca Raton, USA.

Cummins, H & Midlo, C 1961, *Finger Prints, Palms and Soles*, Dover Publications, INC. New York, USA.

Doak, R 2004, 'Dominant Deltas – a Concept', *Fingerprint Whorld*, 30 (117), pp. 118–123.

FBI 1972, '*An Analysis of Standards in Fingerprint Identification*', Federal Bureau of Investigation Bulletin, June, http://www.latent-prints.com/images/FBILEB.pdf.

Fox, K, M & Plato, C, C 1987, 'Toe and Plantar Dermatoglyphics in Adult American Caucasians', *American Journal of Physical Anthropology*, 74 (1), pp. 55–64.

Hammer, H, J 1988, 'The Dermatoglyphic Patterns on the Toes and Soles of the Feet in Various Populations and Their Significance for the Forensic Practice', *Fingerprint Whorld*, 14 (54), pp. 44–48.

IEEGFI 2004, Interpol European Expert Group on Fingerprint Identification, Method for Fingerprint Identification II, http://www.latent-prints.com/images/ieegf2.pdf.

Kapoor, N & Badiye, A (2015), 'An Analysis of Whorl Patterns for Determination of Hand', *Journal of Forensic and Legal Medicine*, 32, pp. 32–46.

Langenburg, G, Bochet, F & Ford, S 2014, 'A Report of Statistics from Latent Print Casework', *Forensic Science Policy & Management: An International Journal*, 5 (1–2), pp. 15–37.

Leadbetter, M 1976, 'The 'Simian' Crease', *Fingerprint Whorld*, 2 (1), p. 12.

Maceo, A, Carter, M & Stromback, B 2013, 'Palm Prints', in Siegel, J, A and Saukko P, J (eds.) *Encyclopedia of Forensic Sciences*, Second Edition, vol. 4 pp. 29–36, Academic Press, Waltham.

Mandrah, K & Kanwal, N, K 2015, 'A Preliminary Study on Assertion of Hand from Whorl Pattern on Thumb', *Journal of Medical Toxicology and Clinical Forensic Medicine*, 1 (2), pp. 1–4.

Massey, S, L 2004, 'Persistence of Creases of the Foot and Their Value for Forensic Identification Purposes', *Journal of Forensic Identification*, 54 (3), pp. 296–315.

Murray, M 2008, 'An analysis of the pattern frequencies occurring in each finger, based on the pattern types from 'The Australian Modification of the Henry System'', project submitted in partial fulfillment of the requirements for the Diploma of Public Safety (Forensic Investigation) at the Canberra Institute of Technology.

OIG 2011, A Review of the FBI's Progress in Responding to the Recommendations in the Office of the Inspector General Report on the Fingerprint Misidentification in the Brandon Mayfield Case, U.S. Department of Justice, Office of the Inspector General, Oversight and Review Division, June.

Okajima, M 1967, 'Frequency of Epidermal-Ridge Minutiae in the Calcar Area of Japanese Twins', *American Journal of Human Genetics*, 19 (5), pp. 660–673.

Penrose, L, S & Plomley, N, J, B 1969, 'Structure of Interstitial Epidermal Ridges', *Zeitschrift für Morphologie und Anthropologie*, 61 (1), pp. 81–84.

Peters, M, Mackenzie, K & Bryden, P 2002, 'Finger Length and Distal Finger Extent Patterns in Humans', *American Journal of Physical Anthropology*, 117, pp. 209–217.

Ray, E 2012, 'Frequency of Patterns in Palms', *Journal of Forensic Identification*, 62 (6).

Rife, D, C 1953, 'Finger Prints as Criteria of Ethnic Relationship', *American Journal of Human Genetics*, 5 (4), pp. 389–399.

Scientific Working Group on Friction Ridge Analysis, Study and Technology (SWGFAST) 2008, *Standard for Simultaneous Impression Examination (Latent)*, Document #20, Ver. 2.0. http://www.swgfast.org/Documents.html.

Singh, G, Chattopadhyay, P, K & Garg, R, K 2005, 'Determination of the Hand from Single Digit Fingerprint: A Study of Whorls', *Forensic Science International*, 152, pp. 205–208.

Stoney, D & Thornton, J, I 1987, 'A Systematic Study of Epidermal Ridge Minutiae', *Journal of Forensic Sciences*, 32, pp. 1182–1203.

Stücker, M, Geil, M, Kyeck, S, Hoffman, K, Rochling, A, Memmel, U & Altmeyer, P 2001, 'Interpapillary Lines – The Variable Part of the Human Fingerprint', *Journal of Forensic Science*, 46 (4), pp. 857–861.

Sutton, R, Glazzard, Z, Riley, D & Buckley, K 2013, 'Preliminary Analysis of the Nature and Processing of Palm Marks by a U.K. Fingerprint Bureau', *Journal of Forensic Sciences, November*, 58 (6), pp. 1615–1620.

Swofford, H, J 2005, 'Fingerprint Patterns: A Study on the Finger and Ethnicity Prioritized Order of Occurrence', *Journal of Forensic Identification*, 55 (4), pp. 480–488.

Tan, V, Kinchelow, T & Beredjiklian, P, K 2008, 'Variation in Digital Rotation and Alignment in Normal Subjects', *Journal of Hand Surgery*, 33A, pp. 873–878.

Tietze, S & Witthuhn, K 2001, *Papillarleistenstruktur der menschlichen Handinnenflache*, Luchterhand, Germany.

Turnidge, S 2010, *Understanding Palm Prints and the Phalanges of the Fingers*, workshop for Fingerprint Examiners, Kent Police Training Centre, 13th April.

U.K Home Office, 1969, Police Research and Development Branch, Report No. 1/69, January, J. W. Godsell.

Vanderkolk, J, R 2011, 'Examination Process', in National Institute of Justice, *The Fingerprint Sourcebook*, www.nij.gov.

Victoria Police 2006 '*Palm Print Analysis Workshop*', The Victoria Police Forensic Services Department, Fingerprint Branch, Melbourne, Australia, 11th and 12th May 2006.

Washington, A, J 1993, http://www.dermatoglyphics.com, viewed 23rd July 2020.

8

What Details Does the Print Reveal, and Are They Likely to Be a Reliable Record of the Details on the Skin of the Donor?

Having completed the analysis of the mark, the examiner can begin their analysis of a print that could have been made by the same area.

If a person has been nominated to have their fingerprints compared with the mark, the examiner will likely have access to a set of that person's fingerprints. Each person will usually have been assigned a unique reference number, and before beginning the analysis, the examiner should verify that the number (or other appropriate references) associated with the prints is the same as the number of the person they have been asked to compare the mark with. If more than one person has been nominated to have their fingerprints compared with the mark, the decision about which prints to analyse first can be based on several factors, including whether the result of the comparison with one person is more urgently required than with others. If no one has been nominated to have their fingerprints compared with the mark then, providing it is suitable, the examiner will need to search the mark against a database of prints to locate a print before they can begin their analysis.

8.1 Sets of Fingerprints

Most sets of fingerprints will be recorded electronically or with ink and paper. The most popular method of recording fingerprints electronically is known as 'livescan' and uses a glass platen as the recording surface. The platen is illuminated by a light source and uses a camera to capture light reflected from the platen to produce an image of the skin in contact with it (Lee & Gaensslen, 2001). Electronic capture is advantageous in that it allows the fingerprints to be instantly searched against a database or accessed by examiners in different locations. Additionally, livescan systems automatically perform a quality check of each print in real-time and require the operator to take additional impressions if a print is of poor quality. However, the use of livescan may not always be possible or practical. For example, sets taken by force, from complainants at crime scenes or from the deceased may not always be recorded electronically.

Whether recorded electronically or with ink, a set of fingerprints usually comprises of impressions of the distal phalanges taken by rolling each digit over the recording surface from one nail edge to the other to capture as much detail as possible. These prints are known as the 'rolled' impressions, and each of them will be recorded in a box on the digital or paper fingerprint form endorsed with the name of the digit the print was made by, as in Figure 8.1.

A set of fingerprints will also usually include additional prints of each of the digits taken without rolling them. These prints are known as the 'plain' impressions (or 'slaps'). To record these prints,

The Forensic Analysis, Comparison and Evaluation of Friction Ridge Skin Impressions, First Edition. Dan Perkins.
© 2022 John Wiley & Sons Ltd. Published 2022 by John Wiley & Sons Ltd.

3. Right middle

Figure 8.1 A rolled impression of the right middle finger on a fingerprint form.

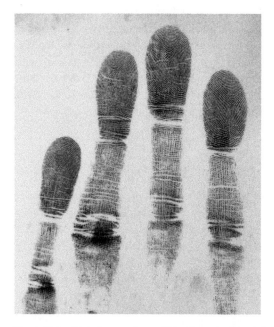

Figure 8.2 Plain impressions of the fingers of the left hand taken simultaneously.

the four fingers of each hand are recorded simultaneously, and prints of both thumbs are also taken together at the same time (Figure 8.2).

The primary purpose of the plain impressions is to provide a means to confirm that the rolled impressions have been recorded in the correct boxes. Impressions that have been taken in the wrong boxes may result in the examiner unknowingly comparing the mark with the wrong area of skin and erroneously concluding it was not made by that person.

One of the differences between a set of fingerprints recorded using livescan and a set taken by any other method is that the livescan set will likely have been through an automatic check to confirm the rolled impressions have been recorded in the correct boxes.[1] This may not have occurred with fingerprints taken by other means, so with those sets, it is good practice to quickly compare the rolled impressions to the plain impressions before beginning the analysis of a print.

Though the plain impressions do not reveal as much detail as the rolled impressions, because the digit is not moving when the print is being made, they often reveal clearer detail than the rolled impressions. Also, the plain impressions may be the only place on a set of fingerprints where impressions of the medial and proximal phalanges are recorded.

A set of fingerprints may also include prints of the palm. Palm prints will usually be taken by placing the palms flat on the recording surface and may also include additional impressions of the fingers.

Occasionally a set of fingerprints will also include prints taken from the feet though usually plantar prints are only taken if plantar marks have been or are likely to be found at the scene of the offence the mark relates to.

A well-taken set of fingerprints will usually record impressions of the areas of skin that leave most marks. However, if the offence the mark relates to is particularly serious, a set of prints may be taken specifically to record additional areas. Such 'major case print' sets may include impressions taken specifically from the tips of the digits and rolled prints of the medial and proximal phalanges (Figures 8.3 and 8.4).

Sets taken in relation to serious offences may also include prints taken from the palm using a curved surface to better record detail from the points the proximal phalanges meet the palm and

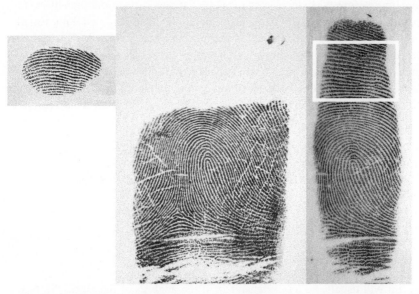

Figure 8.3 The mark on the left was made by the tip of a right middle finger, and the print in the centre is a rolled impression of the same digit. Though rolled impressions record most of the detail on the distal phalanges, they typically do not record much of the detail from the tips. In this example, though the mark and print were made by the same digit, the mark can not be identified using this print as the area that made it has not been recorded. The print on the right was taken by rolling the digit upwards on its end and does record the area seen in the mark.

---

1 However, the system may allow the operator to override the check so it is possible that a set taken with livescan could also include prints recorded in the wrong boxes.

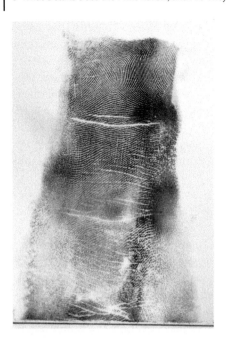

Figure 8.4 Because this impression was taken by rolling the digit, it records more detail from the edges of the medial and proximal phalanges than would normally be seen in a plain impression taken from those areas.

from the centre of the palm, as these areas may not always be recorded on a standard set. The sets may also include palm impressions taken by placing the palm flat on a surface and then lifting the thumb side of the palm to roll the hypothenar edge over the surface to record detail not usually seen in prints taken using the standard technique. If plantar prints are required, these may be taken using a flat recording surface and/or using a curved surface to capture additional areas (particularly the inner arch or thenar area of the foot, which does not usually contact a flat surface).

In addition to electronic capture and ink, there are other methods used to take fingerprints. These may be used in a variety of situations, but they are most frequently employed to record the prints of the deceased. Such methods include:

- applying a light coating of fingerprint powder direct to the ridges and lifting it off using adhesive labels or tape (Figure 8.5)
- applying a casting material direct to the ridges to produce a cast of the details (Figure 8.6)
- photographing the ridges

8.2 The Details in the Print

The analysis of the print will usually be less extensive than the analysis of the mark. Prints are normally taken carefully under controlled conditions for the purpose of comparison with other impressions, and so many considerations relevant for the mark will not usually be relevant for the print. For example, most prints will be suitable for comparison, will not require significant enhancement, will be correctly oriented, and the area of skin that made them will be known. However, the examiner should consider each print on its own merits, as there may be instances where analysis like that performed for the mark will be necessary for the print. When confronted with a print that does require a more extensive analysis, the examiner should first investigate the possibility of using an alternate print of that area of skin. Generally, there are at least two prints of

Figure 8.5 This print was recorded by applying the powder directly to the skin and lifting it off with an adhesive label. The label would be mounted on the underside of an acetate fingerprint form to facilitate comparison.

Figure 8.6 This print was recorded using a casting media applied directly to the skin. The cast may then be photographed to facilitate comparison, or impressions may be taken using ink or other methods from the cast itself.

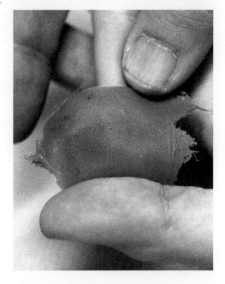

every distal phalange on a set of fingerprints (plain and rolled), and there may be multiple sets available for that person. Even if alternate prints are not available or not any better, the examiner may be able to request that additional prints are taken specifically for comparison with the mark.

In the same way, they did for the mark, the examiner will need to determine what level of documentation is appropriate for the print. Once this has been determined, the examiner should analyse the print to establish the details it reveals in a similar way they did for the mark (though, for example, it is not necessary to locate a target group of characteristics in the print). The entirety of the print does need to be considered, but where it reveals a much larger area of detail than the mark, the examiner should focus their attention on the area that is common to both impressions.

Though there are three distinct stages of ACE, there is overlap between them. For instance, although still in the analysis stage, as the examiner observes the details in the print, they will find it difficult to avoid beginning to compare them with those they are already familiar with in the mark. However, the examiner must resist the temptation to advance to the actual comparison stage until the analysis of the print is complete. This temptation may be particularly great if the examiner is using a computer to search a database of prints for ones that are similar to the mark (rather than using a hardcopy set) as their first exposure to the print maybe when it appears on the screen side by side with the mark. If the examiner advances to the comparison stage prior to completing the analysis of the print, they may miss details that could affect their conclusion. For example, at first glance, the print may appear to reveal a pattern that is different to that in the mark, but with closer inspection, the examiner may observe subtle indicators of distortion that have altered its appearance. Consequently, it is as important to perform a complete analysis of the print as it is of the mark. Vanderkolk (2009) suggests that the acronym for ACE should actually be 'AACE', with the second 'A' representing and signifying the importance of the analysis of the print.

8.3 The Reliability of the Details

Though most prints are taken carefully for the purpose of comparison, all impressions suffer distortion, so the print should be examined in the same way as the mark for indicators of distortion. Most of the distortion factors that can affect the mark can also affect the print, so the indicators and effects of each type of distortion in this chapter will be the same as those described in Chapter 4.

In analysing the print, the examiner should pay particular attention to the following distortion factors:

- **Downward pressure**
  Though it is rarely significant in comparison, rolled impressions of the distal phalanges typically reveal indicators of slightly heavier downward pressure at the bottom of the print in one or both corners as in Figure 8.1, whereas plain impressions will usually reveal consistent pressure throughout.
- **Movement**
  Plain impressions will usually be unaffected by movement, but rolled impressions often reveal slight smearing or blurring along the top edge and occasionally can reveal significant distortion throughout (Figure 8.7). If the examiner observes indicators of movement, they may be able to confirm or refute whether it has occurred by comparing the rolled impression with the plain impression. Occasionally, particularly with prints taken electronically, significant movement can occur with little or no indicators resulting from it as described by Giuliano (2019).
- **Superimposition**
  Superimposition of one print over another on a fingerprint form can occur but is less commonly seen than the superimposition of marks. Generally, when it occurs in an impression taken electronically, it is due to the glass platen not being wiped clean after a print has been taken, resulting in the next print appearing superimposed over the previous one. The examiner may also encounter superimposition in prints taken by other methods. For instance, if a set is taken in ink, the donors inked digits may accidentally contact areas of the form that will later have prints recorded on them, or the person taking the prints may accidentally leave impressions of their ridge detail superimposed with those of the donor.
- **Colour reversal**
  Reverse colour prints are occasionally seen as a result of excess downward pressure during deposition, particularly where the skin has been over-inked, and the ink has subsequently been removed from the ridges but remains in the furrows. However, they are more commonly encountered as a result of techniques used to take fingerprints from the deceased. When a casting

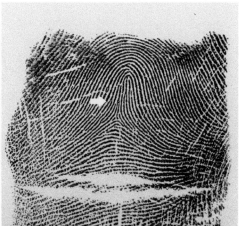

Figure 8.7 These rolled impressions were both taken from the same digit; the one on the right was taken with ink and the one on the left with livescan. The impression in ink reveals minor indicators of movement in the upper corners and slightly heavier downward pressure down the right side. However, the livescan print reveals indicators of significant movement in a couple of areas. The first is marked along the top edge of the print by a sharp change in its contour, which drops suddenly about a third of the way along the impression from the left. Flowing from this point is a line of smeared and misaligned ridges leading down to the top of the core. Comparison with the inked print reveals that movement along this line has created the appearance of characteristics around it that are not on the skin of the donor - known as 'spurious minutiae'. (Note: The author would like to thank Dave Darby for sharing his experiences with spurious minutiae and granting access to his extensive personal collection of examples.) The movement has also obscured characteristics that are on the skin of the donor, for example, in the area just above the core. The second line of movement is present to the left of the delta extending vertically from just above the distal flexure to the middle of the print. This movement is also marked by smeared and misaligned ridges and has created the appearance of a double bifurcation (indicated by the arrow) that is not on the skin of the donor.

technique is used, the examiner may be supplied with a photograph of the cast or a print produced by applying a media (for example, ink) to the cast and using it to make an impression on a fingerprint form. In both a photograph of a cast and an impression made by it with ink, the darker lines will actually be the furrows rather than the ridges (as seen in Figures 8.6 and 8.8).

- **Direction reversal**

   Reverse direction prints will not normally be found on a fingerprint form. An exception to this may be if the fingerprints have been taken from a deceased individual using a casting technique. In that case, the examiner may be supplied with an image produced by applying a media such as ink to the cast and using it to make a print on a fingerprint form. A print made this way will be a mirror image of a mark made by the same area of skin (Figure 8.8). Another exception would be if the examiner is using a photograph of a person's actual skin as the ridge detail in it would also be a mirror image of that in a mark made by the same area.

Because some methods of recording prints can significantly alter the appearance of the details, it is crucial that the examiner is aware of how the print they are analysing was recorded.

Other factors that could cause the details in the print to look different from those on the skin of the donor include the amount of ink used to record the impression or the amount of sweat on the skin if the prints were recorded electronically. A large amount of material on the skin can obscure fine details and alter the appearance of others, for example ridge endings on the skin could appear as bifurcations in the print. A factor that could cause palm prints to look different relates to the

Figure 8.8 The print on the right was recorded in ink using a cast made from the skin of the donor. The impression on the left was made in ink by the same digit of the actual skin of the donor. Using a cast to make a print results in the print being both reverse colour and reverse direction, so where this technique is used, it is crucial the examiner is aware of it. Otherwise, they may erroneously conclude that impressions such as these were not made by the same area of skin.

position of the thumb when the print was taken. If a palm print is taken when the thumb is stretched away from the palm and compared with one from the same hand taken when the thumb is close to the palm, the spatial relationships between some details can appear to be slightly different. Also, creases can appear to be wider and greater in number in one impression than the other. Similar differences can be produced if one palm print is taken with the fingers splayed apart and another with the fingers close together.

Though rarely seen, the examiner may also encounter prints that appear differently as a result of the skin of the donor being intentionally altered to avoid identification. There are numerous documented instances of fingerprint alteration going back to the early 1900s utilising a variety of methods (Cherrill, 1954). One of the simplest was reported by Hastings (2015) and involved a man arrested for driving a stolen vehicle biting off and eating some of the ridge detail from his fingers whilst in a police car on the way to have his fingerprints taken. Generally, prints taken in such cases reveal small voids around the centre of each print as it is difficult and time-consuming to bite off all the ridge detail.

One of the most common ways of altering ridge detail is by cutting or slicing (FBI, 2015). Wertheim (1998b) describes several cases where people made repeated cuts to their skin, resulting in scars or thin voids appearing in the prints. Though these attempts left most of the detail unaltered, cutting or slicing does have the potential to change the appearance of the pattern. In 1995 fingerprints taken from a man in Miami revealed a fine 'N' shape incision on the distal phalanges of each digit (Wertheim (1998a). Close inspection revealed that some of the ridges on either side of the incisions were running perpendicularly to each other, which, as it is not generally a ridge flow that occurs naturally, suggested that those areas were not originally next to each other. To test this theory, fingerprints taken from the man were cut along the lines of the incisions and re-arranged. This resulted in a set of fingerprints that could then be identified as having been made by a person who was wanted for drug offences (Wertheim, 1998a). Wertheim suggests that the mutilation was carried out by cutting the 'N' shape into the skin, taking the two ends of the triangles that form it, and switching them over, then allowing the skin to heal in this new configuration.

Leadbetter (1976) and FBI (1974) describe a method of alteration encountered in Hong Kong in the 1950s and 1960s in which a small section of skin in the core area was cut out, rotated through 180 degrees, and then replaced and allowed to heal. Though some of the prints from skin altered in this way were less obviously mutilated than others, most revealed ridge formations in the core that were abnormal (for example, upside-down loops and ridges running perpendicular to each other). Wertheim (1998b) describes a similar method of alteration encountered in the United States in the 1930s in which the person switched some small sections of skin from one finger to another before inverting them. There have also been instances where much larger areas of ridge detail have been swapped with each other. In 2008 a woman entered Japan illegally after having plastic surgery to graft skin from her thumbs and forefingers from one hand to the other (Fake Fingerprint, 2009). Similarly, Feng et al. (2009) describe a case where a man was caught limping across the US Mexican border in 2005 after having the ridge detail on his fingers swapped with the ridge detail on his toes by a surgeon. Wertheim (1998b) describes an even more unusual case in which a man had the skin from his distal phalanges removed before inserting his fingers into incisions made into the skin of his chest. The digits remained there until non-friction ridge skin from the man's chest had grown onto his distal phalanges.

References

Cherrill, F, R 1954, *The Finger Print System at Scotland Yard*, Her Majesty's Stationery Office, London, UK.

'Fake Fingerprint' 2009, 'Chinese woman fools Japan controls', BBC, 7 December, http://news.bbc.co.uk/1/hi/world/asia-pacific/8400222.stm.

Feng, J, Jain, A, K & Ross, A 2009, 'Fingerprint Alteration', MSU Technical Report, MSU-CSE-09-30, Dec. 2009.

FBI 1974, '"Cutting Up" with Fingerprints', *Identification News*, April 1974, p. 5. Reprinted from *FBI Law Enforcement Bulletin*, vol. 43, no. 4, April 1974.

FBI 2015, 'Altered Fingerprints: A Challenge to Law Enforcement Identification Efforts', *FBI Law Enforcement Bulletin*, https://leb.fbi.gov/2015/may/forensic-spotlight-altered-fingerprints-a-challenge-to-law-enforcement-identification-efforts.

Giuliano, A 2019, 'Artifacts Caused by Livescan Affect a Latent Print Comparison: An Actual Case', *Journal of Forensic Identification*, 69 (1), pp. 20–25.

Hastings, D 2015, 'Fla. suspect chews off fingertips,' *Daily News*, Aug 1, http://www.nydailynews.com/news/national/fla-suspect-eats-fingertips-avoid-identification-article-1.2311613.

Leadbetter, M 1976, 'Transmogrification', *Fingerprint Whorld*, 2 (1), pp.10–11.

Lee, H, C & Gaensslen, R, E 2001, *Advances in Fingerprint Technology*, Second Edition, CRC Press LLC, Boca Raton, Florida.

Vanderkolk, J, R 2009, *Forensic Comparative Science*, Elsevier Academic Press, Burlington, MA.

Wertheim, K 1998a, 'An Extreme Case of Fingerprint Mutilation', *Journal of Forensic Identification*, 48 (4), pp. 466–477.

Wertheim, K 1998b, 'Permanent Intentional Fingerprint Mutilation', Lecture Presentation and Poster, IAI, July 1998.

Part 2

Comparison

Comparison is the examination of the mark and the print to establish agreement or disagreement of their details.

The Forensic Analysis, Comparison and Evaluation of Friction Ridge Skin Impressions, First Edition. Dan Perkins.

9

Are the Details in the Mark in Agreement or Disagreement with Those in the Print?

The comparison will usually take place with the mark and print arranged side-by-side with the same orientation. If the first-level details in both impressions agree, the examiner will compare the second and potentially the third-level details before progressing to the evaluation stage. If any details at any level are not in agreement, then the examiner may end the comparison and progress to the evaluation stage without comparing any further details. Generally, each level of detail is progressively more time consuming to compare, so considering them in level order is the most efficient way to carry out the comparison as it allows the examiner to progress to the evaluation stage at the earliest possible opportunity if any details disagree.

With details that are clearly revealed and have not been subject to significant distortion, determining whether they agree or disagree is usually straightforward – those that appear similarly would be considered to be in agreement and those that appear differently in disagreement.

However, if a detail is unclear or there are indicators that it could have been significantly distorted, the examiner may need to allow for it to look differently but still be considered in agreement (and vice versa). This application of wider tolerances for unclear or distorted areas carries with it a risk that the width of those tolerances may allow details not made by the same area of skin to be considered in agreement.[1] As a result, each such correspondence carries less weight or value during the evaluation (so the examiner needs to find more details in agreement in unclear or distorted areas to support a conclusion than they would in clear, undistorted areas). If the examiner does consider details that appear differently to be in agreement, they should be able to justify that decision by, for example, pointing out indicators of distortion. The justification for considering details that appear differently to be in agreement or those that appear similarly to be in disagreement as a result of distortion must be based on indicators of a type of distortion that is present in the relevant area of the impression, and that is capable of altering the appearance of the detail in the way, and to the extent necessary. The investigations into the erroneous identifications of Shirley McKie and Brandon Mayfield found that there was little support for the examiner's views that differences in the appearance of some details were due to distortion and that those differences should have indicated to the examiners that the impressions were not made by the same area of skin (Campbell, 2011; OIG, 2006). As well as the appearance of the details, the examiner must be aware that their decision about whether two details are in agreement can be influenced by other factors (Campbell, 2011; Dror et al., 2006; Dror & Charlton, 2006). These factors, which are discussed

---

1 Though there is widespread support for varying tolerances according to clarity (Ashbaugh, 1999; SWGFAST, 2013; Vanderkolk, 2011), because of the risks the Interpol European Expert Group on Fingerprint Identification (IEEGFI, 2004) contend that tolerances should not vary.

The Forensic Analysis, Comparison and Evaluation of Friction Ridge Skin Impressions, First Edition. Dan Perkins.
© 2022 John Wiley & Sons Ltd. Published 2022 by John Wiley & Sons Ltd.

more fully in Appendix 3, include beliefs or expectations the examiner may have that the print under comparison is likely (or unlikely) to have been made by the same donor as the mark. For instance, going into the comparison, the examiner could be aware that the donor of the prints is suspected of being involved in the offence the mark came from and was seen to be the last person to touch the surface the mark was found on. They may also not be the first examiner to compare the mark with the print, and they may know that a previous examiner concluded that the mark was made by the same donor as the print. Added to this, the examiner may see a conclusion of identification as being a more desirable or 'positive' outcome to the process that may assist in 'solving' a crime (Charlton et al., 2010). Expectations combined with motivational factors aligned in the same direction may influence the examiner to interpret ambiguous details as being in agreement and to 'explain away' details that do not appear to support their expectation as being products of distortion (Charlton et al., 2010; Darley & Gross, 1983; Nickerson 1998; Lord et al., 1979; Pyszczynski & Greenberg 1987; Snyder & Gangestad 1981). Where the impressions reveal clear and abundant detail, these factors are unlikely to influence the examiners' conclusion, and where that conclusion is more susceptible, such as with complex marks, the organisation they are employed by is likely to utilise measures to minimise the effects of these influences (such as limiting the examiners' exposure to the types of information that may cause expectations).

With impressions that reveal clear and abundant details, the documentation produced during the comparison stage may record little other than which donors prints the mark was compared with, particularly if no details are found to be in agreement. However, if details are found in agreement, the examiner may make a record of those.

9.1 First-Level Detail

As most marks are made by the skin of the distal phalanges, where it is revealed in both impressions, the first detail compared will usually be the pattern family. If the pattern families are in agreement, the examiner should compare the other details that can be used to differentiate between patterns of the same family, i.e. the pattern type, any slope that may be revealed, the ridge count of any loops, the tracing of any whorls and the direction any spiral whorls unwind. If the pattern family is not revealed in either impression, the examiner should work through the other first-level details listed earlier, beginning with pattern type and comparing any that are revealed in both. Even if the entirety of a pattern is not revealed in both impressions, the examiner may still be able to compare distinctive ridge formations such as deltas and cores.

With some impressions, for example, those made by the tips of the fingers, the only first-level detail available for comparison may be ridge flow. In that case, the examiner should compare the straightness or curvature of the ridges plus any convergences or divergences.

If the examiner observes that any of the first-level details are not in agreement, they may progress to the evaluation stage. However, before doing so, they must also consider whether they observed any indicators of distortion during their analysis that could cause any of the first-level details to appear differently. If the examiner thinks the details may have been significantly distorted, then they should progress to comparing second-level details. However, some examiners argue that because it is possible for distortion to occur without leaving obvious indicators, second-level details should always be compared before progressing to the evaluation stage – see Chapter 10 for more information.

If the examiner observes that first-level details agree in both impressions, then they should progress to comparing second-level details.

9.2   Second-Level Detail

To ensure they will be comparing the same area in both impressions, the examiner will start by looking in the print for the fixed point they selected in the mark (around which they located their target group of characteristics). If the area containing the fixed point is not revealed in the print, the examiner should locate a new fixed point (and target group) in the mark in an area that is revealed in the print. If the mark does not reveal a suitable fixed point, then the examiner must start by looking for the characteristics that comprise the target group. A fixed point allows the search for the target group to be restricted to a limited area of the print, so in the absence of one, the examiner will need to look for the group across a larger area of the print that has consistent ridge flow with that found in the mark. If the characteristics in the target group were made by an area of skin that could appear near the edge of a print, the examiner must consider that some or all the characteristics may not be revealed in the print. In that scenario, the examiner should begin by looking for the characteristic most likely to be revealed, which is often the one that is likely to be closest to the centre of the print.

Once the fixed point (or similar characteristic) has been located in the print, the examiner should look for one of the characteristics in the target group around it. The factors the examiner will consider to determine whether characteristics are in agreement in two impressions are:[2]

- The number of intervening ridges between them
- Their spatial relationship
- Their type
- Their orientation

Because skin is flexible, it is the number of intervening ridges between characteristics that is considered rather than the precise distance between them. For example, one study compared measurements between 6 different details in a series of impressions made by the same area of skin and recorded variances of more than plus or minus 20% (Saviers, 1989). Therefore, the examiner should look for a characteristic that is the same number of ridges away from the fixed point in the print as it is in the mark. Whilst the number of intervening ridges must be the same, the examiner should allow that the characteristic could be slightly closer to, further away from, higher or lower than the fixed point in one impression than the other as a result of the flexibility of the skin. They must also be aware that those differences could be much greater if significant distortion has occurred. If a characteristic is found in the print with a similar spatial relationship and the same number of ridges between it and the fixed point, it should be considered to see if it is also of a compatible type and orientation. For instance, a bifurcation that opens along the horizontal plane in the mark could appear in the print to be opening on a slightly more upward or downward plane. Similarly, variances in downward pressure or the amount of material on the skin at the time the impressions were made can result in the same characteristic appearing as a bifurcation in one impression and a ridge ending in the other. However, though they can appear similarly, ridge endings and bifurcations should not simply be considered as interchangeable 'points' of comparison. For instance, in clear impressions that have not been subject to significant distortion and have been recorded with similar pressure, the same characteristic will usually appear similarly in both the mark and print. But where there is

---

2 Where clarity allows, the examiner may also observe some of the more prominent third-level details such as the shapes and widths of the characteristics that may also assist them in determining whether they are in agreement or not.

less clarity or more distortion, greater allowances may need to be made for a characteristic to appear differently. If a characteristic has a different appearance in impressions, but the examiner is considering whether the difference could have been caused by distortion, they should consider whether similar distortion can be seen in other nearby characteristics. Though the extent, presence and type of distortion can vary throughout an impression, it would be unusual to see significant variations in a localised area. If multiple characteristics appear differently and those differences are not consistent with each other (i.e. all affected by the same type of distortion in the same way), the examiner should be very cautious if considering them to be in agreement. As Ashbaugh (1999) explains, 'if each area of friction ridge detail being compared requires justification for why the formation appears slightly different or why it is not spatially correct, be cautious, one may be talking oneself into agreement that is not really there'.

If a characteristic meets the requirements to be considered in agreement in both impressions, the examiner should look for another from their target group that is similar in type, orientation and has a similar spatial relationship with the first characteristic in both impressions with the same number of ridges between them. By relating each newly compared characteristic to the previous one in this way, the examiner can establish whether characteristics across the entire print have the same relationships to each other as they do in the mark. Sometimes a characteristic may be separated from those already compared by an area that, due to its clarity, does not allow the examiner to count the ridges directly between the characteristics. However, the examiner will usually be able to relate such a characteristic to those already compared by following a ridge from them until a point is reached where a count can be made from that ridge to the new characteristic. If there are characteristics that are entirely separated, for example, by voids in the mark, the examiner may only be able to consider their spatial relationship relative to each other in both impressions and, in that case, should also consider whether the areas in question are consistent with being simultaneous impressions (see Chapter 7).

If the second characteristic in the agreement is found, the examiner should repeat the same process to locate further characteristics until either one is found that is not in agreement or all those in the target group have been found to be in agreement. If all those in the target group are found to be in agreement, the examiner should return to the mark to locate further characteristics to compare with those in the print.

If each successive characteristic that is compared is found to be in agreement, the examiner may either compare all the characteristics revealed or may stop when they decide they have compared enough to support a conclusion. Some argue that all characteristics always need to be compared to support an identification; however, this is not always necessary (Champod et al., 2016; IEEGFI, 2006). During a comparison, as each characteristic is deemed in agreement, it will become less and less likely to the examiner that the same configuration could be found in an impression not made by the same area of skin (Neumann, Evett & Skerritt 2012). Marks, for example, those made by palm, may reveal hundreds of characteristics, so whilst they may compare every one with some comparisons, particularly with ones involving complex marks or marks revealing few characteristics, with others the examiner may make a decision that their observations can support a conclusion before they have compared every characteristic. That decision is a subjective one that depends on the clarity and rarity of the characteristics (the clearer and more unusual they are, the fewer are needed) but is also influenced by the examiner's personal experience and training related to how many characteristics they have seen in agreement in impressions that were not made by the same area of skin. For example, the OIG (2006) described the 10 characteristics considered to be in agreement in the mark and print in the Brandon Mayfield erroneous identification as 'an extremely unusual event' and found the similarity contributed substantially to four experienced examiners making the wrong decision.

Several articles have described similar 'close non-matches' with between 7-9 characteristics in agreement (Clark, 2002; Mark & Attias, 1996; Thornton, 2000). Langenburg (2012) and clpex.com (https://web.archive.org/web/20150912071214fw_/http://www.clpex.com/CloseCalls/CloseCalls. htm) have documented other examples (such as those in Figures 9.1 and 9.2) including one where some examiners considered there to be as many as 14 characteristics in agreement (http://www. clpex.com/phpBB/viewtopic.php?f=2&t=2513; Langenburg, 2012).

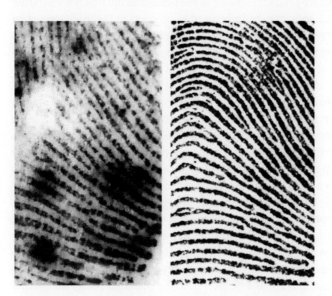

Figure 9.1 These impressions were not made by the same area of skin, but in the areas of them that are shown, examiners have reported finding 9 characteristics that could be considered to be in agreement. The print was found as a result of a search of a computer database. *Source:* Reprinted from Complete Latent Print Examination, http://clpex.com, 'Close Calls', Print 003, https://web.archive.org/web/20150912071214fw_/ http://www.clpex.com/CloseCalls/CloseCalls.htm.

Figure 9.2 These impressions were also not made by the same area of skin, but in the areas shown, examiners have reported finding 12 characteristics that could be considered to be in agreement. *Source:* Reprinted from Complete Latent Print Examination, http://clpex.com, 'Close Calls', Print 009, https://web. archive.org/web/20150912071214fw_/http://www.clpex.com/CloseCalls/CloseCalls.htm.

Because the impressions being compared could be close nonmatching ones, the examiner must be cautious when not comparing all the characteristics in a mark as a further comparison could result in the observation of characteristics that are not in agreement.

Other factors that can, in some circumstances, influence an examiners decision that enough characteristics have been compared include the beliefs or expectations already discussed that the examiner may hold (Charlton et al., 2010; Dror et al., 2012). For example, if the examiner has an expectation that a mark has been made by the same donor as the print, and the first few characteristics they compare are in agreement, they may be influenced to progress to the evaluation stage more quickly, after comparing fewer characteristics than they would if that expectation was not present. Where the examiner is aware of information that suggests one conclusion is more likely than another, it is good practice to consciously seek details that would disprove that conclusion.

If the examiner cannot locate any characteristics in agreement or finds one that is not in agreement, they may progress to the evaluation stage. However, before doing so, they must also consider whether they observed any indicators of distortion during their analysis that could have caused the differences in the characteristics they have observed. As it is possible for distortion to occur without leaving obvious indicators, as with first-level details, some examiners advocate continuing the comparison when a characteristic that is not in agreement is found. For example, the examiner may locate and compare a second or even a third target group in a different area if the characteristics in the first are not considered to be in agreement (Langenburg, 2012). As well as with areas that may have been distorted, this practice is also recommended where the details are unclear, where there is no fixed point, or the examiner is unsure of the area of skin that made the mark. As with making the same decision when characteristics are in agreement, there is no standard dictating how many need to be found in disagreement. It could be one very clear one where the examiner is sure they are comparing the same area, or it could be several, so the examiner makes a subjective decision that they have compared enough to support a conclusion.

During the comparison, the examiner may use dots or symbols on images of the impressions (different from those used in the analysis of the impressions) to indicate the positions of the characteristics they consider to be in agreement. Marking each corresponding characteristic in the impressions creates reference points that make it easier for the examiner to locate further characteristics throughout the print and relate them to each other, as well as allowing the examiner to see which characteristics they have compared. Additionally, the images revealing the annotations made by the examiner may form part of the documentation of the comparison stage and serve as a record of the details compared and considered to be in agreement.

9.3 Third-Level Detail

In most instances, the examiner will progress to the evaluation stage after comparing the first, or the first and second-level details. These details are more reproducible than third-level details and will almost always provide enough information on which a conclusion can be based. Third-level details will usually only be separately compared on the very rare occasions when they are clearly revealed in both impressions, and those impressions reveal some but not enough detail from other levels to support a conclusion. Liddle (in Anthonioz et al., 2011) considered 73 cases in which characteristics were found to be in agreement between two impressions but were not found to be present in sufficient quantity to reach a conclusion. Liddle reported that there was only one case of the 73 in which the comparison of the pores may have facilitated an identification. Nevertheless, third-level details are occasionally relied upon to support identifications, Ashbaugh (1999), Barclay (1997), Clegg (1994), Locard (in Ashbaugh, 1999) and Turner & Weightman (2007) all describe cases where third-level details have been used with characteristics to support identifications.

Though the examiner may observe prominent third-level details whilst comparing other details, to actually compare third-level details significant magnification is required. Once both impressions are enlarged to the same scale, in order to ensure they are comparing the same area in both, the examiner should locate a detail such as a characteristic to use as a fixed point. The examiner should then locate a detail such as a pore in the mark and look for a similar detail with a similar spatial relationship between it and the fixed point in the print. The examiner may do the same with edge shapes, the thickness and thinness of ridges, as well as any misalignments of the individual ridge units that comprise them.

Due to their poor general reproducibility, the examiner should not expect every third-level detail in one impression to look similar or even appear at all in another made by the same area of skin. For example, the FBI advises its examiners to only rely on third-level details where ridge and furrow widths of the two impressions indicates similar downward pressure was used in both (OIG, 2011). The poor reproducibility of third-level details means that it is easy to rely on details that appear to be in agreement and at the same time 'explain away' those that do not as products of distortion. So, it is particularly important that the examiner employ 'fair reasoning' when considering whether details are in agreement or not rather than 'cherry picking' ones that support an expectation or preferred conclusion and discounting the others (IEEGFI, 2004). 'Cherry picking' of third-level details to support an identification was cited as a contributory factor in the erroneous identification of Brandon Mayfield (OIG, 2006). To address examiners susceptibility to cherry-picking third-level details, the Office of the Inspector General recommended that examiners consult all available sets of fingerprints of a person to determine whether third-level details being used to support identification in the print are actually a reliable record of the skin of the donor (OIG, 2011). The examiner may also consider obtaining and examining photos or casts of the skin in order to minimise the distortion caused when making an impression (though they must be aware that no recording technique is perfect, and both of these methods may introduce their own distortions).

If the examiner is comparing third-level details, it would usually be because there are very few details revealed, and so if any are found to be in agreement, the examiner would likely compare all available details. As with other details, if a clear third-level detail in disagreement is found, the examiner could progress to the evaluation stage – but due to their poor reproducibility, the examiner would usually compare all such details before progressing.

Any comparison of third-level detail would likely involve a complex mark and so would require more extensive documentation.

9.4 Creases, Subsidiary Ridges and Scars

The examiner may compare the length and path of any scars or creases as well as the points where any subsidiary ridges begin or end. Such details are usually used in conjunction with characteristics to support a conclusion though there are a few published cases where palmar or plantar creases (Cron, 1997; Hays, 2013; Massey, 2004) or subsidiary ridges (Reneau, 2003) have been used without characteristics to support identifications (Figure 9.3).

As with characteristics, scars, creases and subsidiary ridges can also be related to each other by considering their spatial relationships and the number of ridges between them.

To facilitate and document the comparison of such details, the examiner may draw directly over an image of a crease or scar to record its beginning, end and path. A crease may also cross paths with other creases creating distinctive angles, shapes and configurations which can be compared and often make good fixed points in large areas of similar ridge flow as in Figure 9.4. Where clarity allows the examiner may also compare the finer details such as the shapes and thicknesses of the ridges at the site of scars.

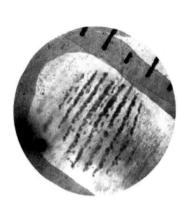

Figure 9.3 The mark on the left was developed on the eyepiece of the scope of an assault rifle and was identified as having been made by the same area of skin as the print on the right – the left thumb (Reneau, 2003). Though there are no characteristics clearly revealed in the mark, the examiners were able to use subsidiary ridges along with third-level details to identify the impressions. *Source:* Reprinted with permission from Reneau, R, D 2003, 'Unusual Latent Print Examinations', Journal of Forensic Identification, 53 (5), pp. 531–537.

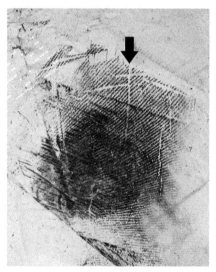
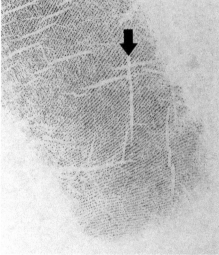

Figure 9.4 Though these impressions were made by the same area of skin, not all the creases in the print are present in the mark. However, the configuration of creases marked by the arrows is in agreement in both impressions. Where they are present in both impressions, creases can sometimes be useful fixed points with which to start a comparison of characteristics in large areas of pattern-less ridge flow such as those in the images.

As with characteristics, if a detail in disagreement is found, the examiner can progress to the evaluation stage. For example, a flexion crease may take a significantly different path or be found to be a different number of ridges away from another detail in one impression than the other. However, the examiner must be aware that some creases may be present in one impression and entirely absent (or less prominent) in another. Additionally, variances in the positions of the digits when the impressions were made or the nature of the surface they were recorded on can also result in differences in appearance in creases from impressions made by the same area of skin. Similarly, as a scar can appear in one impression and be absent in another made by the same area of skin due to the injury not having occurred when the impression was made, the examiner would not usually progress to the evaluation stage based on the presence of a scar in one impression and its absence in the other. The examiner should also be cautious about progressing to the evaluation stage based on a disagreement involving the presence and absence of subsidiary ridges due to possible variances in downward pressure during deposition of the impressions or the age of the donor when the impressions were made.

References

Ashbaugh, D, R. (1999). *Quantitative-Qualitative Friction Ridge Analysis*. Boca Raton, Florida: CRC Press LLC.

Anthonioz, A., Egli, N., Champod, C. et al. (2011). Investigation of the Reproducibility of Third-Level Characteristics. *Journal of Forensic Identification* 61 (2): 171–192.

Barclay, F. (1997). Friction Ridge Identification Based on Pore Structure – A Case Study. *Fingerprint Whorld* 23 (87): 7–11.

Campbell, A 2011, *The Fingerprint Inquiry Report*, Available at: http://www.thefingerprintinquiryscotland.org.uk/inquiry/3127-2.html.

Champod, C., Lennard, C., Margot, P., and Stoilovic, M. (2016). *Fingerprints and Other Ridge Skin Impressions*, Seconde. Boca Raton: CRC Press.

Charlton, D., Fraser-Mackenzie, P, A, F., and Dror, I, E. (2010). Emotional Experiences and Motivating Factors Associated with Fingerprint Analysis. *Journal of Forensic Sciences* 55 (2): 383–393.

Clark, J, D 2002, 'ACE-V – Is it scientifically reliable and accurate?', www.latent-prints.com/ACE-V.htm.

Clegg, D, L, J. (1994). Poroscopy in Practice. *Journal of Forensic Identification* 44 (1): 15–21.

Cron, J, G. (1997). Palmar Flexion Crease Identification Precedent at Trial Testimony. *The Print* 13 (6): 1–3. (reprinted from T.D.I.A.I. April-June 1996 Newsletter).

Darley, J, M. and Gross, P, H. (1983). A Hypothesis-Confirming Bias in Labelling Effects. *Journal of Personality and Social Psychology* 44 (1): 20–33.

Dror, I, E. and Charlton, D. (2006). Why Experts Make Errors. *Journal of Forensic Identification* 56 (4): 600–616.

Dror, I, E., Charlton, D., and Peron, A, E. (2006). Contextual Information Renders Experts Vulnerable to Making Erroneous Identifications. *Forensic Science International* 156 (1): 74–78.

Dror, I, E., Wertheim, K., Fraser-Mackenzie, P., and Walajtys, J. (2012). The Impact of Human-Technology Cooperation and Distributed Cognition in Forensic Science: Biasing Effects of AFIS Contextual Information on Human Experts. *Journal of Forensic Sciences* 57 (2): 343–352.

Hays, M. (2013). An Identification Based on Palmar Flexion Creases. *Journal of Forensic Identification* 63 (6): 633–641.

IEEGFI 2004, Method for Fingerprint Identification II, http://www.latent-prints.com/images/ieegf2.pdf.

IEEGFI 2006, Method for Fingerprint Identification, http://latent-prints.com/images/IEEGFI%201a.pdf.

Langenburg, G 2012, 'A critical analysis and study of the ACE-V process', PhD thesis, University of Lausanne.

Lord, C, G., Ross, L., and Lepper, M, R. (1979). Biased Assimilation and Attitude Polarization: The Effects of Prior Theories on Subsequently Considered Evidence. *Journal of Personality and Social Psychology* 37 (11): 2098–2109.

Mark, Y. and Attias, D. (1996). What Is the Minimum Standard of Characteristics for Fingerprint Identification? *Fingerprint Whorld* 22 (86): 148–150.

Massey, S, L. (2004). Persistence of Creases of the Foot and Their Value for Forensic Identification Purposes. *Journal of Forensic Identification 54* (3): 296–315.

Neumann, C., Evett, I, W., and Skerrett, J. (2012). Quantifying the Weight of Evidence from a Forensic Fingerprint Comparison: A New Paradigm. *Journal of the Royal Statistical Society, A* 175 (Part 2): 371–415.

Nickerson, R, S. (1998). Confirmation Bias: A Ubiquitous Phenomenon in Many Guises. *Review of General Psychology* 2 (2): 175–220.

OIG 2006, 'A Review of the FBI's Handling of the Brandon Mayfield Case', U.S. Department of Justice, Office of the Inspector General, (https://oig.justice.gov/special/s0601/final.pdf).

OIG 2011, 'A Review of the FBI's Progress in Responding to the Recommendations in the Office of the Inspector General Report on the Fingerprint Misidentification in the Brandon Mayfield Case', U.S. Department of Justice, Office of the Inspector General, Oversight and Review Division, June, https://oig.justice.gov/special/s1105.pdf.

Pyszczynski, T. and Greenberg, J. (1987). Toward an Integration of Cognitive and Motivational Perspectives on Social Inference: A Biased Hypothesis-Testing Model. *Advances in Experimental Social Psychology* 20: 297–340.

Reneau, R, D. (2003). Unusual Latent Print Examinations. *Journal of Forensic Identifcation* 53 (5): 531–537.

Saviers, K, D. (1989). The Reliability of Linear Measurement Methods in Friction Ridge Skin Comparisons. *Journal of Forensic Identification* 39 (1): 33–41.

SWGFAST 2013 'Document #10 Standard for Examining Friction Ridge Impressions and Resulting Conclusions (Latent/Tenprint)', Ver. 2.0, 03/13/13.

Snyder, M. and Gangestad, S. (1981). Hypothesis-Testing Processes. In: *New Directions in Attribution Research*, vol. 3 (ed. J, H. Harvey, W. Ickes and R, F. Kidd). Hillsdale, New Jersey: Lawrence Erlbaum Associates.

Thornton, J 2000, 'Setting Standards In The Comparison and Identification', A transcript of the presentation by Dr. John Thornton 84th Annual Training Conference of the California State Division of IAI, Laughlin, Nevada, May 9, www.latent-prints.com/Thornton.htm.

Turner, J, M. and Weightman, A, S. (2007). Focus on Pores. *Journal of Forensic Identification* 57 (6): 874–882.

Vanderkolk, J, R 2011, 'Examination Process', in National Institute of Justice, *The Fingerprint Sourcebook*, www.nij.gov.

Part 3

Evaluation

In the evaluation stage, the examiner decides which conclusion is supported by their observations in the analysis and comparison.

The Forensic Analysis, Comparison and Evaluation of Friction Ridge Skin Impressions, First Edition. Dan Perkins.
© 2022 John Wiley & Sons Ltd. Published 2022 by John Wiley & Sons Ltd.

10

Which Conclusion Is Supported by the Observations in the Analysis and Comparison?

The evaluation will usually result in the examiner reaching one of three conclusions: Exclusion, Identification or Inconclusive.[1,2] Different definitions of the conclusions exist, as do different ideas about the criteria that need to be met to reach them, so the examiner will be guided by the organisation that employs them regarding how they use the conclusions. Some organisations will follow national requirements, such as those set in the United Kingdom by the Forensic Science Regulator (FSR), whilst others will follow requirements determined by professional bodies such as the Scientific Working Group on Friction Ridge Analysis, Study and Technology (SWGFAST) or the Organization of Scientific Area Committees for Forensic Science (OSAC). SWGFAST was a US government funded body of American and international scientists, examiners and researchers established to develop best practices for the fingerprint discipline. SWGFAST operations came to a close in 2014, and their library of documents was transferred to the newly formed OSAC. OSAC is administered by the US government through the National Institute of Standards and Technology (NIST). The OSAC Friction Ridge Subcommittee is responsible for maintaining and updating the existing SWGFAST documents as well as developing new ones.

With many marks the evaluation and comparison stages will be almost indistinguishable in the mind of the examiner. For example, if the mark reveals a clear whorl pattern and the print a clear arch pattern and there are no significant indicators of distortion, then both comparison and evaluation may take place almost simultaneously. With other marks a conclusion will be reached more gradually as a result of the agreement or disagreement of successive details. As such, Egli (2009) and Langenburg (2012) liken the evaluation more to a process of on-going assessment beginning in the comparison stage rather than a separate stage where a conclusion is reached. However, with some marks, particularly complex ones where the examiner extensively documented their analysis and comparison, a distinct and separate evaluation should be performed to allow for a thorough review of the observations the examiner made during the analysis and comparison.

---

1 One scenario in which none of these conclusions is appropriate occurs when the examiner searches the mark through a computer database and does not identify its donor. Although the examiner will have excluded the selection of prints returned by the system as having been made by the same area of skin as the mark, they will not have excluded all the people whose prints are on the database. This is because even if the mark is searched against every print in the database at every orientation, the system will only return a limited selection of prints it considers most similar for the examiner to compare. There are a variety of reasons why that selection may not include the corresponding print even if that print is on the database, including; the mark being poor quality, the print being poor quality or the print not revealing the area that made the mark. As a result, in this scenario rather than reaching one of the three conclusions, the examiner may just document the searches they have carried out and report that the mark remains unidentified.
2 If they do not identify the mark some examiners may be required to report a 'negative' or 'not identified' conclusion rather than specifying exclusion or inconclusive (National Research Council (NRC), 2009).

The Forensic Analysis, Comparison and Evaluation of Friction Ridge Skin Impressions, First Edition. Dan Perkins.
© 2022 John Wiley & Sons Ltd. Published 2022 by John Wiley & Sons Ltd.

In recent years the way conclusions of identification are defined and described has been widely criticised. An identification is an opinion, but critics argue it often sounds like a fact. For example, The Interpol European Expert Group on Fingerprint Identification define identification as meaning the mark 'was made by this donor excluding all other possible donors', and until 2011 the equivalent SWGFAST definition stated that a mark and print 'originated from the same source, to the exclusion of all others' (IEEGFI, 2006; SWGFAST, 2002). Defining identification in these terms draws criticism on the grounds that there is not sufficient justification for the examiner to be able to claim that they have determined the one and only person in the world who could have made the mark (AAAS, 2017; Haber & Haber, 2009; Mnookin, 2008; NRC, 2009; Page, Taylor & Blenkin, 2011; PCAST, 2016; Saks & Koehler, 2008). The only scenario in which that level of certainty may be justified would be a largely hypothetical one in which the mark could only have been made by one of a small number of known individuals. In that case, the mark could be compared with the fingerprints of all those people, which would likely result in details in agreement being found in the print of one person and details in disagreement in the prints of all the others. The examiner could then exclude all those who did not make the mark and identify the mark's donor with absolute certainty – as there is no one else who could have made it. However, in practice, the examiner will almost always be dealing with a mark that was made by one of an unknown number of unknown individuals. In that scenario, the mark cannot be compared with the prints of everyone who could have made it. Instead, it can only be compared with the prints of those nominated for comparison by investigators or returned as the result of a search of a computer database. As a result, unlike the first scenario, if details in agreement are found with any of these prints a further factor must be considered: what is the likelihood that similar agreement could also be found in a print made by someone else? The more detail there is in agreement, the less likely that is, but the examiner has to make a decision on that likelihood based on the details using their training and experience. So rather than an expression of certainty that one person, and one person only could have made the mark, an identification is, in fact, the expression of a probability (Champod & Evett; 2001; Langenburg; 2013; Page, Taylor & Blenkin, 2011; Saks & Koehler, 2008). Specifically, identification is the opinion that the probability the same level of agreement could be found in an impression made by anyone other than the donor of the print is extremely low.

Describing an identification in terms that the person identified is the one and only person who could have made the mark is now widely discouraged (Campbell, 2011; FSR, 2017; Meuwly, 2011; Mnookin, 2008; NRC, 2009; NIST, 2012; SWGFAST, 2012a). Recommendation 3.7 from the Expert Working Group on Human Factors in Latent Print Analysis states: 'Because empirical evidence and statistical reasoning do not support a source attribution to the exclusion of all other individuals in the world, latent print examiners should not report or testify, directly or by implication, to a source attribution to the exclusion of all others in the world' (NIST, 2012). This recommendation also applies to other ways that have been used by examiners to express the certainty of their conclusions in addition to phrases such as 'to the exclusion of all others'. For example, in US v. Hugh, 2005 (NO. 03-829, E.D. Pa. Jan. 28, 2009), an examiner testified that he was '100 per cent, without a doubt' certain that the mark was made by the defendant, and similar language was also encountered by those who investigated the erroneous identifications of Shirley McKie, Marion Ross and Brandon Mayfield (Campbell, 2011; OIG, 2006). Additionally, when giving evidence in court, examiners have stated that the process used to arrive at their conclusions is 'infallible' or has a 'zero error rate' (McCartney, 2006; US v Mitchell, 2004, 365 F.3d 215, 3d Cir.). As Eldridge (2012) explains, when examiners say '"I am 100% certain of my conclusion," we might mean that we have conducted a careful examination, reached the best conclusion possible with the data available, and that we would not have reported that conclusion unless we were confident that we had done our work well. But what does the jury hear? They hear, "I'm an expert, and I'm telling you that this conclusion is fact and cannot possibly be wrong."' Courts do not require conclusions of absolute certainty from the examiner; they require the examiners opinion and so as Campbell (2011) states: 'In testifying on the basis of '100% certainty' fingerprint examiners are, at one and the same time, addressing a standard not set by law and overstating their evidence in order to meet it.'

In response to the criticisms, SWGFAST (2011) changed its definition of identification, removing the words 'to the exclusion of all others' and adding language acknowledging the mark could have been made by another donor but stating the 'likelihood' of that to be 'so remote that it is considered as a practical impossibility.'

However, some think the changes have not gone far enough (Cole, 2014; Swofford, 2016). For example, the American Association for the Advancement of Science recognises that the new SWGFAST definition acknowledges 'an element' of uncertainty but consider that it does not deal forthrightly with the level of uncertainty that exists (AAAS, 2017).[3] Various alternative ways of articulating what identification means have been suggested (AAAS, 2017; NIST 2012) and in 2018 following on from work began by SWGFAST, OSAC published a new draft 'Standard for Friction Ridge Examination Conclusions'.[4] The new OSAC identification conclusion is defined as meaning that the examiners observations 'provide extremely strong support for the proposition that the impressions originated from the same source and extremely weak support for the proposition that the impressions originated from different sources' (OSAC, 2018). So, under this definition, when reporting a conclusion of identification, the examiner is no longer saying that it is their opinion that the mark and print 'were made by the same person' but instead is saying that it is their opinion there is extremely strong support for the proposition that they were. The definition explains the strength of support there is for the conclusion (the degree to which the support for one proposition outweighs the other) and articulating it in this way clearly communicates that there is some uncertainty involved but also quantifies the amount of support there is for the conclusion the examiner has reached.

The OSAC document also offers two new conclusions, 'Support for Same Source (SSS)' and 'Support for Different Sources (SDS)', in addition to the traditional Identification, Inconclusive and Exclusion conclusions to form a scale of five (Figure 10.1).

Using a range of conclusions provides additional options to allow the examiner greater precision in communicating the strength of support their observations provide. For example, a mark revealing a small amount of clear details may just barely meet one examiners threshold for identification, whereas another examiner may consider the agreement is below that threshold. Under the traditional categoric conclusions, the first examiner would report a conclusion of identification. This would place the conclusion under the same broad category as another identification in which there was an overwhelming amount of clear details in agreement, even though there may be much more uncertainty and risk of error with the former. Conversely, the second examiner would report a conclusion of inconclusive, and this may deprive a judge, jury or investigator of potentially useful

---

3 Irrespective of how it may be defined, some argue that the discipline should stop using the term 'Identification' entirely as they consider that it now carries with it an inappropriate implied certainty because of the way it has historically been defined and described (Champod, 2015; Cole, 2014; FSR, 2020; NIST, 2012). One organisation, the US Army's Defence Forensic Science Centre stopped using the term in 2015 and replaced it with 'association' which is defined as 'The conclusion that the two impressions have corresponding ridge detail and, in the opinion of the examiner, the likelihood of observing this amount of correspondence when made by different sources is considered extremely low (DFSC, 2015).'

4 At the time of writing the OSAC conclusions had been submitted for development as an American National Standard but were at the 'Draft Document' stage so the conclusions themselves may be subject to change during the standards development process.

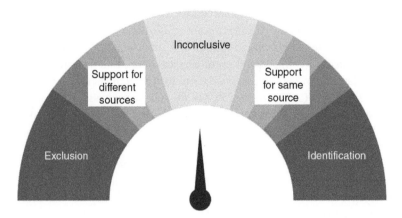

Figure 10.1   A representation of the OSAC conclusions (Based on OSAC, 2018).

information, i.e. that the examiner did observe agreement in the impressions (Champod & Evett, 2001; Neumann, Evett & Skerrett, 2012).[5] The scale recognises that fingerprint evidence exists along a continuum, at one end of which is a finger-shaped smudge revealing no detail and at the other is a mark revealing an abundance of clear detail. In between the two are marks revealing progressively increasing levels of support and so the practice of only reporting conclusions once that support reaches an arbitrary point designated as the beginning of a category will result in potentially useful evidence being lost (Champod & Evett, 2001). Recognising that the SSS and SDS conclusions are broad, OSAC also allows the examiner to further indicate the extent of support there is within them by grading it using terms such as limited, moderate or strong if they wish. This would mean that the inconclusive conclusion would only be used when the balance of the details does not favour one side of the scale or the other.

Whichever conclusions examiners use, Eldridge (2011) offers the following advice: 'Remember, you are not stating an unequivocal fact; you are offering an opinion based upon knowledge (your experience and training) and observed data (the two impressions). As long as you are transparent about how you reached your conclusion and the limitations of your conclusions, that is all you need to do.'

10.1   Exclusion

A conclusion of exclusion describes the opinion that there is an extremely low probability the donor of the mark was the donor of the print.

Current requirements for exclusion call for the examiner to find 'sufficient' details in disagreement before excluding (FSR, 2017; SWGFAST, 2013a). Whilst the FSR's requirement does not specify what is, or is not sufficient, SWGFAST does set out conditions under which details at different levels can be used to exclude. SWGFAST allows exclusion solely on disagreeing first level details, but only with simple comparisons (for example, where there are clear indicators of the pattern type and orientation of the mark, and no indicators of significant distortion). With more

---

5 Whilst some organisations will qualify such conclusions to explain that there is some support for an alternative conclusion, others may simply report that the conclusion was inconclusive and offer no further information.

complex comparisons, for example, where the details may be distorted or the examiner may be less sure of the area of skin that made the mark, SWGFAST requires the examiner must also observe disagreement of second-level details before excluding. SWGFAST also specifies that third level details can only be used in conjunction with first or second level details to exclude (due to their lack of reproducibility).

However, some examiners argue that irrespective of the apparent simplicity of the comparison, details in disagreement at both first and second levels should be observed before excluding. For example, the impressions in Figures 10.2 and 10.3 appear to reveal first-level details that support simple conclusions of exclusion.

Figure 10.2 These prints were both taken from the right index finger of the same person. The one on the right was taken 12 years after the other and appears to have three ridges missing between the core and the delta. The examiner who published the prints suggest this may have been caused by an injury that resulted in a portion of the skin being sliced out or folded over, which unusually has not resulted in an obvious scar. *Source:* Reprinted with permission from Carrick (1982).

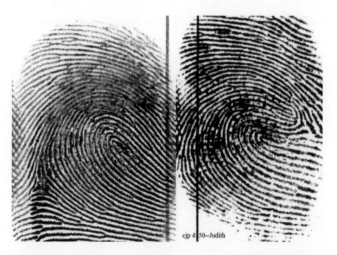

Figure 10.3 These prints were both made by the same area of skin but appear to reveal a different pattern as a result of distortion believed to have been caused by scarring, which left little or no indicators of the injury. *Source:* Reprinted from Complete Latent Print Examination, http://clpex.com, Judith Miller, Austin, TX, FIG 22.

However, they were actually identified as having been made by the same area of skin and only appear to have first-level details in disagreement as a result of scarring, which unusually has not resulted in the presence of obvious indicators of significant distortion. Whilst significant distortion that leaves little or no indicators of its occurrence may be extremely rare, Triplett (personal communication, 24 October 2014) argues the time required for the examiner to compare the additional second level details to reach the correct conclusion is worth it when balanced against the potential consequences of an erroneous exclusion. Such an approach may also provide a safeguard with impressions such as those in Figure 10.4, in which, though obvious indicators of distortion are revealed, it would be easy for the examiner to fail to properly consider the effect the distortion may have had on the first level details. However, other examiners would argue that no approach is entirely risk free and in the absence of data indicating how many erroneous exclusions such an approach would prevent, it does not make sense to, for example, compare the characteristics in an arch with those in a whorl before excluding because of the remote possibility they could have been made by the same area of skin.

Ray & Dechant (2013), citing the fact that erroneous exclusions are the most common technical errors made by examiners, also advocate an approach that requires both first and second level detail to be in disagreement to support exclusion. Their approach (which is in use in several organisations in the United States) requires the examiner to find two or more groups of clearly disagreeing characteristics that can be related to first level details (such as a core or delta) before exclusion can occur. The function of the first level details in this approach is to act as a fixed point to ensure that the examiner is comparing the same area at the correct orientation in both impressions. Ray (DoubleLoop Podcast, 2017) uses marks such as the one in Figure 10.5 to show that in the absence of a core or delta, it can be easy to compare a mark with the wrong area of a print. Ray reports this mark being erroneously excluded by approximately 25% of the examiners he tasked with comparing it, which may be because the mark reveals indicators that led them to compare it with the print at the incorrect orientation shown in the figure, as opposed to the correct orientation shown in Figure 10.6.

Some support for the Ray & Dechant approach can be found in the experiments Ulery et al. (2017) conducted, which showed that exclusion conclusions were more likely to be correct when a delta or core was present. However, critics of the approach may argue that it is possible to be confident that the correct area is being compared in both impressions without it revealing a core or delta

Figure 10.4 Though these impressions were identified as having been made by the same area of skin, SWGFAST report that multiple examiners erroneously excluded them due to distortion giving the mark on the left the appearance of a right sloping pattern. *Source:* Reprinted with permission from SWGFAST (2013a).

Figure 10.5 Just looking at the appearance of the mark on the left, it would be easy to conclude that the most likely orientation for it is the one shown. A comparison with the print on the right at this orientation would almost certainly result in a conclusion of exclusion. However, the mark was found on the trigger of a firearm, and careful consideration of how that surface is likely to be touched would lead to a possible alternative orientation of the mark, as shown in Figure 10.4. *Source:* Reprinted from Complete Latent Print Examination, http:// clpex.com, FIG 24.

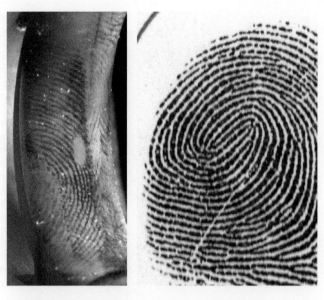

Figure 10.6 When the same mark in 10.5 is oriented correctly, it can be seen why it was identified as having been made by the same area of skin as the print. *Source:* Reprinted from Complete Latent Print Examination, http://clpex.com, FIG 24.

(or other fixed points).[6] For example, though a mark may not reveal a fixed point, it may be made on part of a surface that is only likely to have been contacted by a particular area of skin, and it may reveal ridge flow typically seen in that area. In that scenario, following the Ray & Dechant approach would mean that even if all the details were in disagreement, the examiner would have to reach an inconclusive conclusion rather than one of exclusion. Ray (DoubleLoop Podcast, 2017) suggests this may be a scenario in which the new OSAC conclusion of SDS could be used as a compromise.

Another approach to address common causes of erroneous exclusions, such as the examiner getting the orientation of the mark wrong or not comparing it with the right area of the prints (Busey et al., 2021), is suggested by Eldridge, de Donno & Champod (2021). The authors recommend that 'examiners need to make re-considering their initial analysis before rendering an exclusion decision part of their comparison routine'. Eldridge, de Donno & Champod (2021) suggest the review could

---

6 Ray & Dechant's approach does allow some flexibility in that certain other details, like creases or even large distinctive areas of ridge flow (such as that of the hypothenar) can be used instead of a core or delta.

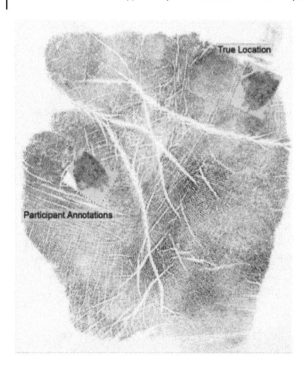

Figure 10.7 This image shows a mark and print that were erroneously excluded as having been made by the same donor during a study described in Eldridge, De Donno & Champod (2021). Two images of the mark are visible, one adjacent to the corresponding area of the print (in the triradiate area) and one adjacent to the area where the examiner found some characteristics that were potentially in agreement (in the thenar). The comparison was conducted using software (PiAnoS) that tracked the amount of time the examiner spent comparing the mark with different areas, which showed that they spent only 3 seconds comparing it with the correct area and 480 seconds comparing it with the area in the thenar before wrongly excluding it. *Source:* Reprinted with permission from Eldridge, De Donno & Champod (2021).

help to mitigate the effects of a 'mind-set' bias in which the examiner becomes so sure that decisions like those made about orientation or digit/area determination during the analysis stage are correct that they ignore or fail to consider alternative interpretations (Figures 10.7 and 10.8).

If the mark is complex and, for example, the only details revealed lack clarity, the examiner may need to review the documentation they produced during the analysis of the mark to consider the confidence level they had in each of those details before reaching a conclusion. For instance, the absence of a detail in the print the examiner saw in the mark and had high confidence in may add considerable weight to a conclusion of exclusion, whereas much less weight may be added if the examiner had low confidence in the detail. With comparisons such as those in Figures 10.2, 10.3 and 10.4 where there are details that appear to support identification and exclusion, the examiner should weigh the value of those details against each other (SWGFAST, 2013a). For instance, in Figure 10.2, the apparent difference in the ridge count, which could support an exclusion, would be weighed against the many characteristics in agreement and result in a conclusion of identification.

An exclusion can relate to a single print or all the prints of a donor. Generally, if the examiner is working with a set of fingerprints and excludes one print, they would then carry out further comparisons with prints from all the other areas of that donors skin that are likely to have made the mark (completing an analysis of each print before they compare it).[7] The examiner must consider whether the set of fingerprints reveals comparable impressions made by all the areas of the donor's skin that

---

7 The examiner may not always compare each mark with a print made by every area of skin that *could* have made it. For example, unless it is on a surface where reverse direction marks are commonly encountered, a mark revealing a left loop would not usually be compared with prints revealing a right loop – though it is possible, but very unlikely they were made by the same area of skin. Similarly, as Champod (2016) points out, marks made by thumbs and big toes can look alike but if such a mark was found on a door handle it is unlikely the examiner would compare it with toe prints of the donor as it is much more likely to have been made by a thumb and sets of prints do not usually include toe prints.

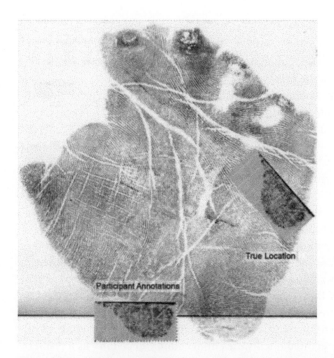

Figure 10.8 Like Figure 10.7, this image also shows a mark and print that were erroneously excluded as having been made by the same donor during a study described in Eldridge, De Donno & Champod (2021). The mark is shown next to the area of the print that was made by the same donor in the mid hypothenar, and another copy is shown next to the bottom of the hypothenar, which the examiner wrongly believed was the corresponding location. From the examiner's documentation, the authors report that the curve along the bottom of the mark led the examiner to believe that it was made by the bottom of the hypothenar, and since they could not find agreement in that area and could not see the curved shape anywhere else, they excluded the mark. However, the curved shape of the mark in this instance was created as a result of the donor holding a piece of paper that did not contact all the palm around the area. *Source:* Reprinted with permission from Eldridge, De Donno & Champod (2021).

are likely to have made the mark, and where necessary obtain other sets if possible. If the only set(s) do not reveal comparable impressions of all those areas and other sets are not available, then a conclusion of inconclusive for the donor (or for particular areas of skin of the donor) should be reported.

In contrast with conclusions of identification, there are few generally accepted requirements for the documentation of exclusions. Other than the conclusion itself, SWGFAST (2013a) only mandates that the documentation should clearly record whether the exclusion relates to some or all the prints of a donor. With comparisons where there are clear and abundant details in disagreement, and the excluded prints are retained in the case file for review, some organisations may only require their examiners to record that information; however, some may also require the examiner to document the justification for their exclusion (particularly with complex comparisons). This may include recording details such as the orientation the mark was compared at, which prints, or areas of prints were compared and the specific details that were used to exclude.

A conclusion of exclusion does not establish that the donor of the prints did not touch the surface the mark was found on. They may not have touched it, but it is also possible to touch a surface without leaving a mark (for example, by wearing gloves or as a result of only a very small amount of material transferring from the skin to the surface). It is also possible that a mark was left but was not found, or that one was not developed by the media used or was no longer present when the surface was examined (for instance as a result of the surface being wiped down or the action of environmental factors).

10.2 Identification

A conclusion of identification describes the opinion that there is an extremely high probability the donor of the mark was also the donor of the print.

To support the conclusion the examiner may use details from all three levels but generally only require details from the first and second levels.

First level details must be used in conjunction with details from other levels because different people share them, and third-level details must also only be used with other details due to their lack of reproducibility.

Like those for exclusion, requirements for identification mandate there must be 'sufficient' details in agreement (FSR, 2017; SWGFAST, 2013a). Whilst the examiner may occasionally need to consider the sufficiency of other details such as creases, most identifications will rely on the agreement of multiple characteristics, so considerations of sufficiency principally apply to them.

Some examiners will be required by national or local standards to find a pre-determined number of characteristics in agreement before they can report a conclusion of identification. Such 'numeric', 'empirical' or 'point' standards specify an amount that it is considered two different areas of skin would not share in agreement – so if that number of characteristics is found, then it would be extremely unlikely that the impressions were made by different donors. Since the first numeric standard was proposed in 1911, numerous countries have instituted one (Champod et al., 2016). The actual number of characteristics required often varies from country to country as a result of some standards being based on different statistical models and others on anecdotal evidence (Champod et al., 2016; Evett & Williams, 1995).

Whilst the numeric approach is attractive because it offers a clear standard that must be satisfied for identification, in practice, given the same mark and print, different examiners may report different numbers of characteristics in agreement. For example, Evett & Williams (1995) tasked 130 examiners with counting how many characteristics were in agreement in nine pairs of marks and prints. Though it had little impact on the conclusion that was reported, the authors found that there was a wide variation in the reported amounts; for instance, with three of the pairs the number of characteristics considered to be in agreement ranged from 8 to 26, 10 to 40 and 14 to 56. Langenburg (2004), Langenburg, Champod & Wertheim (2009) and Ulery et al. (2014) also observed considerable variability between examiners in similar experiments (Figure 10.9).

Figure 10.9 Langenburg, Champod & Wertheim (2009) conducted an experiment in which they provided instructions as to how each type of characteristic should be counted to 15 examiners and asked them to count the number they considered to be in agreement in a series of impressions. The authors report that the count for these impressions ranged from 17 to 35. *Source:* Reprinted with permission from Langenburg, Champod & Wertheim (2009).

Critics of the numeric approach also argue that it does not allow the examiner to take into account other details (such as creases or open fields) or the varying rarity and clarity of characteristics and instead requires they all be given the same weight in the evaluation (Champod et al., 2016; NIST, 2012; Stoney, 2001). However, proponents of the approach argue that they do consider and use other details to support identifications and some countries with a numeric standard have rules in place that allow identifications below that standard in particular circumstances (Champod et al., 2016; Zeelenberg, 2011). Also, whilst the numeric approach may not be as objective as it appears due to variations in the number of characteristics examiners perceive to be in agreement, the non-numeric or 'holistic' approach relies entirely on the individual examiner to decide what is or is not sufficient.

In 1973 after a three-year review of the numeric approach, the International Association of Identification (IAI) issued a resolution stating, 'no valid basis exists at this time for requiring that a pre-determined minimum number of friction ridge characteristics must be present in two impressions in order to establish positive identification' (IAI, 1973). The IAI position was agreed upon by examiners and scientists from eleven countries at an International Symposium on Fingerprint Detection and Identification in Ne'urim, Israel in 1995 and endorsed by SWGFAST in 2003 (Margot & German, 1995; SWGFAST, 2003). However, whilst the IAI resolution and the Ne'urim 'declaration' influenced some countries to adopt a non-numeric approach, many chose to retain their numeric standard. For example, a 2011 survey of 73 countries by Interpol found 44 of them used a numeric standard (Farelo, 2011 in Ulery et al., 2013). One reason for this may be that examiners in those countries consider the standard contributes to the strength of their conclusions as their experience and statistical modelling indicate it is very unlikely that different people would have that amount of details in agreement (Langenburg, 2012; Zeelenberg, 2011).

For determinations of sufficiency, examiners working to a numeric standard will be guided by that threshold. Polski, Smith, Garrett et al. (2011) and Farelo (2012 in Ulery et al. 2013) reported standards of between 4 and 17 characteristics being used, though Farelo (2012 in Ulery et al. 2013) found 24 of 44 countries who reported using a standard all required a minimum of 12 characteristics in agreement for identification.

Examiners not working to a numeric standard will reach a conclusion of identification when they decide the amount of details in agreement is at a level they would not expect to find in a print made by another donor. The examiner will make this decision using their training and experience, but there are a number of statistical models that provide an idea of how infrequent and discriminating configurations of characteristics are.

Since the first one was published in 1892, there have been over 20 models proposed, and detailed reviews of them can be found in Stoney (2001), Langenburg (2011), Abraham et al. (2013) and Neumann (2013). Table 10.1 shows data from Langenburg (2011) demonstrating the probability of matching a specific configuration of 12 characteristics according to fifteen of the published models.

Whilst the probabilities are all very small, due to differing approaches by the modellers and levels of sophistication of the models, the numbers do vary considerably. For example, Galton's (1892) model indicates the probability of 12 matching characteristics is one in eleven thousand, whereas Kingston's (1964) model estimated it as one in three hundred octillion (Langenburg, 2011).

In addition to providing examiners with a general idea of the discriminating power of configurations of characteristics, some of the most recent models are beginning to be used in casework to provide an estimate of the strength of the support that the details in agreement between a specific mark and print provide. For example, Neumann, Evett & Skerrett (2012) developed a model that measures the strength of the support the examiners observations provide using a likelihood ratio (LR). An LR represents the probability of observing the configuration of characteristics in the mark relative to two propositions: the donor of the print was the donor of the mark, or the donor of the

Table 10.1 The probability of finding a specific configuration of 12 characteristics in agreement in impressions from different people according to different statistical models. Originally published by the National Institute of Justice, U.S. Department of Justice/Public Domain.

Probability of matching a specific configuration of 12 characteristics	Author and year
0.00009%	Galton (1892)
0.000005%	Balthazard (1911)
0.000005%	Bose (1917)
0.0000003%	Henry (1900)
0.0000002%	Trauring (1963)
0.000000000003%	Amy (1946–1948)
0.000000000001%	Gupta (1968)
0.0000000000003%	Osterburg et al. (1977–1980)
0.000000000000008%	Pearson (1930)
0.0000000000000003%	Roxburgh (1933)
0.00000000000000000001%	Pankanti et al. (2001)
0.000000000000000000004%	Wilder and Wentworth (1918)
0.000000000000000000001%	Cummins and Midlo (1943)
0.0000000000000000000000003%	Stoney and Thornton (1985–1989)
0.00000000000000000000000000000003%	Kingston (1964)

print was not the donor of the mark. For instance, an LR of 10 000 would indicate that it is 10 000 times more likely to observe the agreement in the impressions if the donor of the print was the donor of the mark than if they were not. The Neumann, Evett & Skerrett model can generate LRs for comparisons of between 3 and 12 characteristics and Figure 10.10 shows the average LRs calculated by the model were between 1000 (for 3 characteristics) and 1 trillion (for 12 characteristics).

To make it easier to appreciate what an LR value means, the UK Association of Forensic Science Providers (AFSP, 2009) recommend using the scale shown in Table 10.2 to equate the numerical value with a verbal description of the strength of support it provides.

Using the scale, it can be seen that even the lowest average LRs generated by the Neuman, Evett & Skerrett model provide strong support for the proposition that the mark and print were made by the same person with the highest average value being significantly in excess of that for extremely strong support.

Langenburg (2013) also suggests analogies and comparisons to everyday events can assist in understanding the significance of LRs. Langenburg explains tossing a coin 13 times and it landing heads up on every occasion is an example of a probability that is relatively similar to a likelihood ratio of 10 000, whereas a likelihood ratio of greater than a billion is on par with tossing the coin 30 times and observing 30 heads in a row. So, for LRs of those values to be achieved by chance, rather than as a result of the same donor making the mark and the print, would be akin to something that is extremely unlikely to occur. For example, the average LR generated for 12 characteristics by the Neuman, Evett & Skerrett model was approximately one thousand times greater than the probability of tossing 30 heads in a row.

The Neumann, Evett & Skerrett model has been validated against a large database and has also been tested on real casework as part of a field study by Neumann et al. (2011). During the study, corresponding characteristics were labelled in the mark and print using the Universal Latent Workstation, that data was then exported into the model to calculate the likelihood ratios for each comparison, as shown in Figure 10.11.

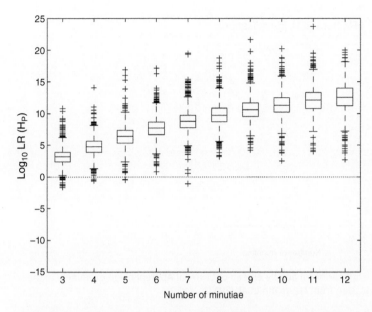

Figure 10.10 Distributions of LRs for configurations of between 3 and 12 characteristics when impressions were made by the same area of skin. *Source:* Reprinted with permission from Neumann, Evett & Skerrett (2012).

Table 10.2 Verbal equivalents of likelihood ratios recommended by the AFSP (AFSP, 2009/Elsevier).

Proposition supported	Value of LR	Verbal equivalent
The donor of the mark was the donor of the print	>1 000 000	Extremely strong support
	10 000–1 000 000	Very strong support
	1000–10 000	Strong support
	100–1000	Moderately strong support
	10–100	Moderate support
	>1–10	Weak support
1 = Inconclusive		
	0.1–< 1	Weak support
	0.01–0.1	Moderate support
The donor of the mark was not the donor of the print	0.001–0.01	Moderately strong support
	0.0001–0.001	Strong support
	0.000001–0.0001	Very strong support
	<0.000001	Extremely strong support

As was the case with all other models, generally the more characteristics there were, the higher the LR was. However, the model also shows that there can be a wide variation in the LRs for marks with the same number of characteristics and that these values can overlap. For example, the mark with 7 characteristics in agreement in Figure 10.11 had an LR of around ten billion but there are numerous other marks with more characteristics that had smaller LRs.

Whilst most models are not available for use by examiners, one based on a model proposed by Egli (2009) that also uses LRs to measure the strength of evidence has been integrated into the

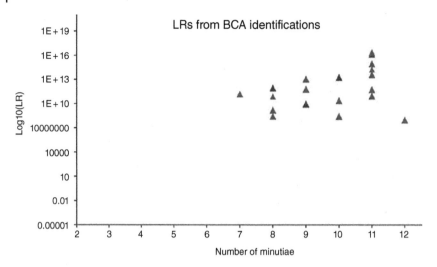

Figure 10.11 LRs assigned to identifications made by Bureau of Criminal Apprehension examiners in the study. *Source:* Reprinted with permission from Neumann et al. (2011).

Picture Annotation System (PiAnoS) software (M. DeDonno, personal communication, 1 May 2020). PiAnoS is open-source software and was created by the School of Criminal Justice at the University of Lausanne and is available at https://ips-labs.unil.ch/doc/. Once the images of the mark and print are captured into the software, to calculate the LR the examiner simply labels the corresponding characteristics and presses a button to perform the calculation and generate the LR.

A different type of model that is also available has been developed and used to provide evidence in military courts by the US Army Defence Forensic Science Centre (DFSC, 2017).[8] FRStat is a stand-alone software application that provides a numerical value to represent the strength of the evidence (Swofford et al., 2018). To use the application, the examiner first records the characteristics they see in the mark using the application as part of their analysis (DoubleLoop Podcast, 2018).[9] The examiner would then carry out their work as normal, either using a computer to search a database of prints to find one to compare with the mark or by comparing the mark directly with a set of prints from a nominated person. If a comparison resulted in a conclusion of identification, after that conclusion had been verified, the examiner would use FRStat to record between 5 and a maximum of 15 characteristics they consider to be in agreement in the impressions. The software would then measure the similarity of the configuration of characteristics designated by the examiner in the mark with the configuration in the print and calculate a score that represents the statistical strength of the correspondence. The score is then compared with a database of scores produced in the same way from impressions that were known to have been made by the same source and those that were known to have been made by different sources. An estimate is then produced of how likely that amount of correspondence is to be found in impressions made by the same area of skin as opposed to those made by different areas (DoubleLoop Podcast, 2018). For example, the FRStat similarity score for the fifteen characteristics that were considered to correspond in the impressions in Figure 10.12 was 70.0149, and the probability of observing that value was calculated

---

8 FRStat is available at https://doi.org/10.5281/zenodo.4426484
9 This is necessary as FRstat compares these characteristics with those the examiner plots during the comparison and only uses ones that the examiner marked in both analysis and comparison as part of the score. This is done to minimise the potential for the details in the print to influence the examiners perception of details in the mark.

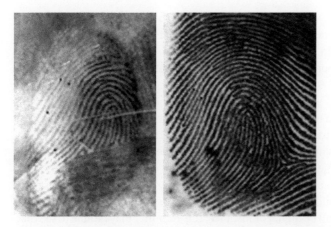

Figure 10.12 A mark and print used to generate a score using FRStat. *Source:* Reprinted with permission from Swofford et al. (2018), Supplemental Appendix V.

as approximately 662 838 times greater when the impressions were made by the same area of skin rather than by different areas of skin.

The use of a statistical model could provide an important tool to underpin the examiner's conclusion by providing a more objective assessment of the strength of their observations to allow a judge or jury to easily understand the significance of the agreement the examiner has observed. However, only a few models are available for use, and all have limitations. For example, they don't take into account the clarity of the details and none considers all the details an examiner could (for example, pattern type, creases, edge shapes etc.). Furthermore, though some have been through validation experiments, none has yet been tested, refined and developed to the point where the model could be adopted as the standard way of estimating the strength of support that exists for examiners observations.

Though examiners not working under a numeric standard will not have to find a pre-determined number of characteristics for identification, the number they find is still significant, and there is evidence that some may have their own personal threshold for identification. Ulery et al. (2014) found a close relationship between the number of characteristics in agreement and the identification conclusion, specifically that when there were 7 or more characteristics in agreement, most resulting conclusions were identifications. Swofford et al. (2013) also reported in their experiment that when 7 or more characteristics were found in agreement, examiners reached a conclusion of identification more than 80% of the time. Similarly, Langenburg, Champod & Genessay (2012) found that under the conditions of the study they conducted, examiners appeared to be working to an operational decision threshold for identification of around 8 or 9 characteristics in agreement. These sorts of numbers are also broadly in line with those that have been put forward by examiners and researchers, either from their experience or experiments. For example, the UK Home Office (1969) reported that experienced examiners are satisfied that an identification has been established when 7 clear characteristics are found in the agreement. Similarly, Stoney (2001) suggests in the United States 7–8 are generally accepted as enough if they satisfy an experienced examiner and Kirk (1985) also referenced the same numbers – though suggests more than 7–8 is customary. Moenssens (1971) indicates that 10–12 are needed but states that it is possible to identify with 7–8 if some of the characteristics, or combinations of characteristics are particularly rare, whereas Gupta (in Cowger 1993) suggested the number was fixed at 11 (based on a statistical model). Locard (in Champod et al., 2016) suggested that 12 were sufficient but that between 8 and 12 could be used dependent on their clarity, rarity and the presence of other details. Balthazaard (in Langenburg, 2011) suggested 17 were necessary, but 10–12 could be used if it could be established that the donor of the mark was one of a limited population in a geographic region, rather than potentially being anyone in the world.

It can be seen from the use of the model proposed by Neumann, Evett & Skerrett (2012) that the marks they examined with around 8-9 characteristics produced LRs in the range of 10 billion, which according to the AFSP (2009) scale would equate to 'extremely strong' support for the proposition that the impressions were made by the same person. However, it is also clear looking at the same model that some comparisons with more characteristics in agreement resulted in lower LRs and configurations of the same number can have quite different LRs, so the number of characteristics cannot be the only sufficiency factor that the examiner considers.

For example, clarity is also crucial as characteristics that are clearly revealed in impressions made by the same area of skin should appear very similarly, so the weight that the examiner attaches to each corresponding pair will be high. For instance, the examiner may be able to see that two ridge endings both have a bulbous tip, and both end closer to the ridge on one side rather than the other. The additional correspondence of these smaller, third level details would add greater support to the conclusion that the impressions were made by the same area of skin than if, as a result of poor clarity, the examiner could only be confident that an ending in the print corresponded to an ending or bifurcation in the same place in the mark. Therefore a few clearly corresponding characteristics may have the same weight as a larger number of less clearly corresponding ones. SWGFAST (2013a) produced the Sufficiency Graph in Figure 10.13 based on the consensus of a number of experienced examiners to show the relationship between the amount of characteristics needed and their clarity (or 'quality'). For example, in the area marked 'A' an identification conclusion may not be warranted, whereas in 'B' the comparison would be considered complex, but identification may be warranted, whilst in 'C' the comparison would be considered noncomplex, and identification may well be warranted.

The examiner may also attach more weight to a correspondence of characteristics as a result of their rarity. Though rarity is not usually a significant consideration in comparisons with many

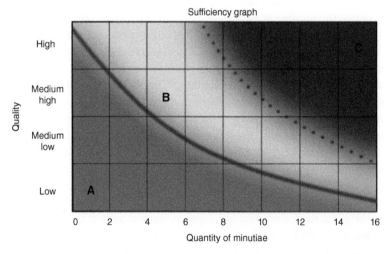

Figure 10.13 The SWGFAST sufficiency graph showing the relationship between the number of characteristics and their clarity ('quality') as it relates to the sufficiency for identification and complexity of comparisons. The dotted line represents an approximation of the boundary between complex and non-complex comparisons. *Source:* Reprinted with permission from SWGFAST (2013a).

characteristics in agreement,[10] a correspondence of rare characteristics (either because of their type, location, orientation or configuration) may be given more weight than a correspondence that is routinely encountered. Rarity of other details in agreement should also be considered, for example, a rare pattern, or unusually high ridge count or rare crease can also add more weight to the conclusion.

When considering whether an amount of agreement is sufficient, the examiner must be aware of the potential for factors other than the details revealed in the impressions to influence them, particularly with complex marks or comparisons which are close to a decision threshold. For example, the Interpol European Expert Group on Fingerprint Identification (IEEGFI, 2004) describe a tendency to lower the threshold for identification when the mark relates to a serious offence. This is known as a 'gravity standard' as it results in the threshold being lowered due to the potentially serious consequences of being unable to identify someone who may be the perpetrator. Similarly, the circumstances of how a donor's prints came to be compared with the mark also have the potential to influence the examiners assessment of what is sufficient. For instance, if the examiner searches a mark against a large database of prints, the computer system will return ones for comparison that an algorithm determines are similar. As a result, those prints are likely to reveal similar configurations of characteristics to one in the mark, even if none of them was made by the same donor. However, it is much less likely that a print from a person nominated by investigators for comparison with the mark as a result of their suspected involvement in the offence or because they are a complainant, would also have a configuration of characteristics like that in the mark if they were not the donor. Consequently, in the latter scenario, when the examiner does observe agreement between the mark and the print, they may be influenced to reach a conclusion of identification sooner, based on fewer characteristics than they would in the former scenario. In the scenario where the examiner is comparing a print that has been returned as a result of the search of a database, some examiners and researchers advocate consciously raising the threshold for sufficiency in order to mitigate against the increased risk of an erroneous identification posed by a coincidentally similar print – as occurred in the erroneous identification of Brandon Mayfield (Dror & Mnoonkin, 2010; Neumann et al., 2015; NIST, 2012; OIG, 2006). In relation to the use of computers to search databases and return respondent prints, Dror et al. (2012) reported results that may suggest the amount of time and effort examiners spend looking at the prints depends on where they are ranked in the list of respondents. Generally, the most similar will be at the top of the list and the least similar at the bottom. The authors found that when the prints of the donor who made the mark were in the list, the examiner was more likely to identify them when they were at the top than the bottom, and they were also more likely to wrongly identify marks with prints made by different donors when they were at the top than at the bottom. A lowering of the sufficiency threshold as a result of an expectation that if a print made by the same donor is in the list, it is most likely to be at the top, could be a contributory factor to the effect seen in this study.

Langenburg (2012) describes the possibility that observing similarities in a mark and a number of prints that were not made by the same donor as a result of doing many comparisons may 'prime' an examiner to employ a higher decision threshold for identification with that mark than a different examiner who only compares that mark with one print and identifies it.

Other factors including knowledge of another examiner's conclusion regarding the same comparison also have the potential to influence considerations of sufficiency and are discussed further in the Appendices on Bias and Verification.

---

10 This is because the rarity of there being so many characteristics in agreement outweighs the need to assess the rarity of any of those characteristics individually.

Because of the subjective nature of the decision, what is sufficient for identification for one examiner will not always be sufficient for another. A study conducted by Ulery et al. (2012) found that 13% of the time, a second examiner would disagree with the initial examiner whether information content was sufficient for identification with the same comparison. The authors reported that much of the variability was due to the examiners having to make categorical decisions in borderline cases, i.e. both examiners may agree that a comparison is on the border between identification and inconclusive but are forced to choose one of those two options and select different ones (Ulery et al. 2014).

With most comparisons the examiner may only need to consider whether the overall amount of details in agreement is enough to warrant a conclusion of identification, however, with some, particularly complex ones that are near a decision threshold, the examiner may need to consider the weight that each detail should contribute to the conclusion. This should involve reviewing the documentation produced during the analysis and comparing it to that produced during the comparison. This will allow the examiner to assign weight to details based on the confidence level they had in those details during the analysis. For instance, one high confidence detail in agreement would contribute more weight to a conclusion of identification than one low confidence detail.

Such a review may draw the examiners attention to instances where the way they view the details in the mark has changed since the analysis. Ulery et al. (2015) found that in 90% of identifications they studied, examiners had revised the way they had seen characteristics in the analysis (for example, by adding, deleting or moving characteristics). Ulery et al. (2015) attribute most of the moved characteristics and some of the deleted and added ones to examiner's being certain of the presence of characteristics but uncertain of location, for example, they may be able to see three ridges becoming two but the precise point they do lacks clarity. The authors also suggest that the high rate of revision seen with identifications may reflect a motivation on the part of the examiner to ensure the annotation of the mark accurately shows their final interpretation of the correspondence of the characteristics. However, though the authors of this study found that the mere occurrence of changes was not associated with error, they did observe that in the only erroneous identification made during the study, the examiner involved based the conclusion almost entirely on characteristics that had not been detected during analysis (Ulery et al., 2015).

One example of a change that a review may draw attention to would be the examiner failing to notice a characteristic during the analysis that they considered to be in agreement with a characteristic in the print during the comparison. This could be because the presence of a characteristic in the print has led the examiner to more closely scrutinise an area of the mark that has resulted in them seeing a characteristic they missed during their analysis. However, it could also be that exposure to the print in conjunction with a form of bias has influenced the examiner to see a characteristic in the mark that is not actually there. As The Expert Working Group on Human Factors in Latent Print Analysis (NIST, 2012) point out, 'The difficulty is that the comparison print may lead both to accurate and inaccurate judgments about previously unnoticed features of the [mark], and there is no fool proof way to tell the two apart.' Employing an entirely 'linear' approach to the ACE process would mean the examiner would not be able to use such a characteristic to support a conclusion because its perception may have been driven by the print. However, most examiners and commentators accept that, on occasion, it is acceptable for the examiner to return to the mark and reanalyse it after seeing the print (Vanderkolk, 2011). As Fraser-Mackenzie, Dror & Wertheim (2013) state: 'It is generally understood that the process cannot, practically, be absolutely linear, if only to allow for correction of missed data during Analysis.' Therefore, rather than discounting such details, as a compromise, Langenburg & Champod (2010) propose the examiner should only rely on previously unseen details in the mark if those details are specifically documented as having first been observed during the comparison and given less weight during the evaluation than details that were observed during the analysis. For example, in Langenburg & Champod's (2010) 'GYRO' system, any such characteristics are marked in orange and can be given higher weight than red characteristics (used to indicate very low confidence details) but should be given lower weight than yellow characteristics (medium level of confidence details) (Langenburg & Champod, 2010).

A review of the documentation produced during the process may also draw the examiners attention to a characteristic they did observe in the mark, but which is not in the print. In this instance, the examiner would need to consider the confidence level they had in the characteristic during the analysis. If it was a low confidence characteristic then it may be that they were simply mistaken in thinking that they had seen it, perhaps as a result of distortion or a lack of clarity. If, however, it was a high confidence characteristic, then it may be that that detail actually adds significant weight to a conclusion of exclusion rather than identification.

The review will also allow the examiner to identify and reflect upon instances where their initial interpretation of details has changed after seeing the print. For example, they may have seen a ridge as forming a bifurcation with the ridge on its left during the analysis but now see that characteristic as being in agreement with one in the print in which the same ridge is bifurcating with the ridge on its right. Where they have changed their view on details, the examiner should again consider the confidence levels they had in the appearance of those details during the analysis and whether that change is appropriate. For instance, if they had high confidence in the exact location of the bifurcation when they looked at the mark, that should cause them to question how they can now justify it being one ridge over, which may lead to the re-evaluation of that change or the attaching of lower weight to the agreement. Dror et al. (2015) propose that whilst the examiner may be able to justify changing their initial view of low confidence details, they should ether not be permitted to rely on high confidence characteristics they now see differently, or where they are permitted to rely on them, the implementation of additional quality control measures (such as blind verification) for that conclusion may be appropriate.

A review may also draw attention to instances where the examiner has not used 'fair reasoning' when considering whether details are in agreement or not (Interpol European Expert Group on Fingerprint Identification (IEEGFI, 2004). For example, there could be an area in the mark that the examiner has designated as having low clarity during the analysis in which, during the comparison, they observe a detail that appears to be in agreement with one in the print and another detail that does not. The examiner may have been influenced to see the detail in agreement as support for a conclusion that the impressions were made by the same donor and to 'explain away' the other one as being too unclear to be relied upon. However, they must consider both details fairly and objectively, rather than 'cherry picking' one that supports an expectation or preferred conclusion and discounting the other one.

Because details in impressions made by the same area of skin will never be identical, requirements for identification stipulate that their agreement is 'within tolerance' with respect to the effects of distortion, rather than that they are the same (SWGFAST, 2013a). In order to deem details that appear differently to be within tolerance, the examiner must consider what type of distortion could have caused the difference and how it could have occurred. Any explanations must be capable of accounting for the differences observed and should be supported by the presence of indicators in the impression of the type and extent of distortion that the examiner believes has occurred.[11] On most occasions,

---

11 Indicators of potential distortion may also be found in information gathered during the analysis rather than in the impression itself, for example the surface the mark is on may be a surface on which reverse direction marks are commonly encountered.

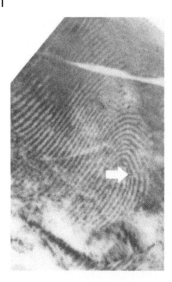 

Figure 10.14 These impressions were identified as having been made by the same area of skin though there is the appearance of a bifurcation to the right of the core in the mark that is not in the print. The mark was found on the door handle of a vehicle, and the examiner who published it suggested the distortion could be due to movement of the digit along the edge of the curved surface. *Source:* Reprinted from Complete Latent Print Examination, http://clpex.com, Siegel, TX, FIG 84.

with impressions that were made by the same area of skin, the examiner will be able to point to clear indicators of a particular type of distortion in or around the areas of the impressions where the differences occur. However, the examiner may occasionally encounter impressions like those in Figure 10.14 in which, though distortion significant enough to create the appearance of a characteristic has occurred; there are little or no indicators of it in the relevant areas. Whilst such impressions are unusual, similar examples have been documented on several occasions (Cowger, 1993; Galton 1892; Hancox, 1979; Langenburg, 2012; Puri, 1967; Thornton, 1977; Triplett, 2012).

The evaluation of impressions like those in Figure 10.14 can sometimes be problematic as some requirements for identification, such as the one set by the UK FSR, state there must be no 'unexplainable differences' for an examiner to reach a conclusion of identification (FSR, 2017). Some examiners consider that requirement means that a conclusion of identification cannot be reported unless the specific cause of the difference can be established. For instance, Bunter (2017) describes a case where he was tasked with reviewing an identification that was being used as evidence in a prosecution. Bunter found 45 characteristics in agreement (24 of which he had high confidence in) but also found 1 in disagreement that he had no explanation for. Bunter produced a report stating that whilst he considered it very unlikely the mark was made by anyone else, according to the FSR requirement, as he could not explain the cause of the difference in appearance, he could not reach a conclusion of identification. Instead, Bunter reported a conclusion of inconclusive, and as a result, the prosecution was dropped.[12] However, other examiners, when faced with a comparison like that in Figure 10.14 would argue that whilst the specific cause of the difference may be unexplained; it is not unexplainable. There are several factors that could account for differences the examiner may see in impressions

---

12 Bunter was surprised by the decision to drop the prosecution and considers that the case may indicate the need for a broader scale of conclusions that provides something between inconclusive and a categorical identification (like for example the OSAC SSS conclusion).

made by the same area of skin. These include movement, superimposition and damage to the skin and whilst it may not always be possible to definitively state which one of them caused a specific difference; they are capable of it. Thornton (1977) argued that in the examples he had seen, it would be absurd for the examiner to reach any conclusion other than identification because of the overwhelming amount of details in agreement – even though the precise cause of the differences in appearance was unknown. Other examiners agree that it is not always possible to ascertain the cause of a difference in appearance, but that does not mean a conclusion of identification cannot be reached (Cowger, 1993; Ray & Dechant, 2013). Many examiners argue that rather than basing or ruling out a conclusion on a single detail, a more balanced and rational approach would be to weigh the cumulative agreement and disagreement of details in each comparison (Baumgartner, 2010; Ray & Dechant, 2013; Triplett, 2011). This approach of each detail being taken into account without any one 'counting as an infinite weight on the scales' is also the one endorsed by SWGFAST (2013a) (Ray & Dechant, 2013). The SWGFAST conclusion also does specifically allow an identification to be reported where the abundance of details supporting it outweighs a single difference even if the examiner does not know the cause of the difference (SWGFAST, 2013a).

There is no standard to define whether a detail that differs in appearance in two impressions is within or without of tolerance, so that decision is based on the clarity of the detail and the presence and extent of indicators of distortion and relies on the examiners training and experience. When evaluating such details, the IEEGFI (2004) suggest that if a difference in appearance is attributed to distortion, those details should not then contribute the same weight to a conclusion as details that clearly are in agreement – in effect, an explained difference in appearance should not come to be seen as a similarity. Ultimately the use of details that appear significantly differently to support a conclusion of identification must be thoroughly justified and documented. As Maceo (2009) explains: "Distortion' is not a wild card to be played to dismiss unexplainable regions of concern. . .' Examiners inappropriate and unjustified allowances for details to appear differently were cited as causes of the erroneous identifications of Shirley McKie, Marion Ross and Brandon Mayfield (Campbell, 2011; OIG, 2006).

With a conclusion of identification, the examiner must carefully document which mark was identified and for which area of skin of which person (the prints of whom may be retained with the case file). The examiner will likely also be required to document the basis for their conclusion, which may include recording the details they consider to be in agreement. The justification for relying on details that differ significantly in appearance, or details that were not seen during the analysis or were seen significantly differently, should also be recorded.

When reporting a conclusion of identification, the examiner must be conscious of the limits of that conclusion, namely that the identification does not establish the donor of the print's involvement in the offence. Rather it is simply extremely strong evidence that the donor contacted the surface the mark was on at some point in time.

Though the examiner may gain satisfaction or appreciation as a result of conclusions of identification, they must avoid seeing identifications as 'positive' outcomes of the process and other conclusions as 'negative' outcomes. The examiner's aim is to reach an objective, balanced and reliable conclusion that is supported by their observations irrespective of the nature of that conclusion.

10.3 Inconclusive

When used with Identification and Exclusion as part of the traditional three options, a conclusion of inconclusive describes the opinion that it is not possible to say whether there is an extremely high or extremely low probability that the donor of the mark was the donor of the print.

The examiner may reach this conclusion for several reasons, including:

- The impressions revealing details in disagreement but not enough to support a conclusion of exclusion.
- The impressions revealing details in agreement but not enough to support a conclusion of identification.
- Prints of an area of skin that may have made the mark not being available for comparison.

Inconclusive is a very broad conclusion encompassing opinions ranging all the way from just below the threshold for exclusion to just below the threshold for identification. Whilst some organisations will qualify the conclusion by saying, for example, that there is some agreement between the mark and a print of a particular person but not enough for a conclusion of identification, some may simply report the result of the comparison as inconclusive and provide no other details. Many argue this way of reporting prevents a judge, jury or investigator from access to potentially meaningful information (Champod & Evett, 2001; Dror & Langenberg, 2019; Neumann, 2013; Swofford, 2015). For example, Attias, Hefetz & Ben-Shimon (2015) describe a case in which the mark in question was found on a cigarette packet at the scene of an assault. A search of the mark through a computer system resulted in comparison with a print that the examiners concluded was inconclusive as there was some but not enough detail in agreement to support identification. However, the name of the donor of the print was on a police list of people who may have been involved in the offence, and that person's DNA was recovered from the scene. As a result, though the comparison was inconclusive the authors argue that communicating that there was some agreement with a particular individual to investigators in an appropriate manner could be a useful means of assisting an investigative process – particularly one where other evidence was lacking. Champod (2016) calculated the LR for that comparison was 8333, and on the AFSP scale, that would mean the observations provide 'strong support' for the proposition that the mark was made by the donor of the print. Neumann et al. (2011) also reported high LRs for comparisons examiners made inconclusive, with some of them providing similar support to that seen for some identifications. Similarly, Figure 10.10 shows that some comparisons that have few characteristics in agreement, and so would likely be considered inconclusive, actually have LRs that would indicate they have an extremely high evidential value which, as Lennard (2013) explains, scientists in other disciplines would not hesitate to present in evidence (Neumann, Evett & Skerrett, 2012). Champod & Evett (2001) question the logic of a reporting practice that allows the examiner to voice an opinion that the donor of the mark 'was the donor of the print' the second the threshold for identification is met yet would not allow them to voice any opinion at all if the process had been interrupted a second before that. Langenburg (2012) agrees, stating that such a practice 'is inconsistent with the view that evidence is a continuum and can be expressed in varying degrees of strength. Essentially, out of convention and tradition, the profession has taken a continuum of evidence and parsed it into three large bins: 'identification', 'exclusion' and 'inconclusive'.'

To alleviate its ambiguity, The Las Vegas Metropolitan Police Department utilise three categories within the inconclusive conclusion; 'Cannot exclude' (to describe where there is detail consistent with, but insufficient to identify), 'Incomplete' (to describe where the prints are inadequate and further prints are required) and 'Not Compared' (to describe where the mark is made by an area not revealed in the prints or there are no prints available) (Maceo, 2011). Alternatively, SWGFAST (2013b) put forward a draft document that also separates the inconclusive conclusion into three categories; 'Lack of Comparable Areas' (to describe where a comparison has taken place as far as possible but additional prints are needed), 'Lack of Sufficiency for Individualisation' (to describe where corresponding details are observed but are not sufficient to support an identification) and

'Lack of Sufficiency for Exclusion' (to describe where non-corresponding details are observed but are not sufficient to support an exclusion).

Where an inconclusive conclusion is qualified in some way, care must be taken to ensure those in receipt of the conclusion understand its meaning and its limitations – specifically that the impressions have not been excluded or identified. Some consider that the ways of qualifying the inconclusive conclusion that are in use could result in the evidence being given more weight than is justified as they do not quantify the strength of the support provided (Champod et al., 2016). The new OSAC options of SSS and SDS may provide some form of compromise in this regard as they do communicate a subjective level of the amount of support the examiner considers there to be, and ultimately, they could also be used in conjunction with a statistical model to provide a more objective reporting option.

Inconclusive conclusions are different from exclusion or identification conclusions in that if one examiner identifies a mark for a particular donor and another examiner excludes the same donor with that mark, one of those examiners would likely be considered to have made a mistake. However, if one examiner identifies or excludes impressions and another examiner reaches an inconclusive conclusion with the same impressions, the latter conclusion is typically seen as being correct or as representing a 'difference of opinion' rather than being a mistake. This attitude is reflected in the SWGFAST (2012b) requirement that only recognises two types of error: erroneous identifications and erroneous exclusions. According to SWGFAST, an erroneous inconclusive conclusion would be regarded as a 'non-consensus decision' that could not be considered an error as it 'cannot be proved correct or incorrect'. Because the examiner knows that they will likely not be considered as errors, there is a danger that inconclusive conclusions can be seen as a way out of making a difficult decision that could result in an error – in effect, 'deciding not to decide' (Dror & Langenberg, 2019). Adding to this temptation may be the fact that, unlike identification conclusions, inconclusive conclusions are less likely to be verified (and so any error discovered), and the examiner is also much less likely ever to need to justify them in court. However, inconclusive conclusions can be erroneous where the information available clearly supports an exclusion or identification. For example, Langenburg, Hall & Rosemarie (2015) reported an inconclusive error in their experiment that was made as a result of the examiner incorrectly orienting the mark. To lessen the potential appeal of 'deciding not to decide', Dror & Langenberg (2019) propose several solutions that would continue to allow examiners to make inconclusive conclusions where appropriate but would minimise their use as a way to avoid making a decision when one is warranted. For example, the authors suggest a requirement that examiners clearly document and justify inconclusive conclusions. Dror & Langenberg (2019) observe that in comparison with the other conclusions, transparency and accountability is missing with inconclusive conclusions and suggest that introducing this requirement may reduce the attractiveness of the decision as an easy way out.

Like all conclusions, the extent of the documentation for an inconclusive conclusion should be in proportion to the complexity of the comparison but the name/reference number of the donor of the prints must be recorded in all cases, and the documentation should clearly indicate why the comparison was inconclusive with this person including which impression(s) it is inconclusive with (SWGFAST, 2012c). This may include annotating images of the mark and print to show areas that the examiner considers to be in agreement/disagreement or are unsure of, for example, as a result of distortion.

Before reaching a conclusion of inconclusive, the examiner should ensure they are satisfied that the consideration they gave earlier in the process to whether or not the image of the mark (or print) needs to be enhanced was appropriate.

References

AAAS 2017, A Quality and Gap Analysis - Latent Fingerprint Examination, Report prepared by Thompson, W, Black, J, Jain, A & Kadane, J, doi: https://doi.org/10.1126/srhrl.aag2874.

Abraham, J., Champod, C., Lennard, C., and Roux, C. (2013). Modern Statistical Models for Forensic Fingerprint Examinations: A Critical Review. *Forensic Science International* 232: 131–150.

AFSP (2009). Standards for the Formulation of Evaluative Forensic Science Expert Opinion. *Science and Justice* 49: 161–164.

Attias, D., Hefetz, I., and Ben-Shimon, E. (2015). Latent Fingerprints of Insufficient Value Can be Used as an Investigative Lead. *Journal of Forensic Science & Criminology* 3 (3): 1–4.

Baumgartner, B 2010, 'Weekly Detail – FBI document – discrepancies', Public CLPEX Message Board, Tuesday, Jan 19th, 2:52pm, http://www.clpex.com/phpBB/.

Bunter, S 2017, *Cognitive Bias and ACE-V: What DO you think?*, presentation to The Chartered Society of Forensic Sciences Fingerprint Division Conference, Manchester 2017, viewed 27th May 2020, available at: https://player.vimeo.com/video/216719324.

Busey, T, A, Heise, N, Hicklin, R, A, Ulery, B, T & Buscaglia, J 2021, 'Characterizing Missed Identifications and Errors in Latent Fingerprint Comparisons Using Eye-Tracking Data', *PLoS ONE*, 16 (5), e0251674. https://doi.org/10.1371/journal.pone.0251674.

Carrick, M, F 1982, 'Ridge Count Reduction', *Fingerprint World*, 7 (27), pp. 71–72.

Campbell, A 2011, *The Fingerprint Inquiry Report*, available at: http://www.thefingerprintinquiryscotland.org.uk/inquiry/3127-2.html.

Champod, C. and Evett, I, W. (2001). A Probabilistic Approach to Fingerprint Evidence. *Journal of Forensic Identification* 51 (2): 101–122.

Champod, C. (2015). Fingerprint identification: advances since the 2009 National Research Council report. *Philosophical Transactions, The Royal Society, B* 370: 20140259. http://dx.doi.org/10.1098/rstb.2014.0259.

Champod, C., Lennard, C., Margot, P., and Stoilovic, M. (2016). *Fingerprints and Other Ridge Skin Impressions*, Seconde. Boca Raton: CRC Press.

Cole, S, A 2014, 'Individualization Is Dead, Long Live Individualization! Reforms of Reporting Practices for Fingerprint Analysis in the United States', *Law, Probability and Risk*, 13, pp. 117–150, doi: https://doi.org/10.1093/lpr/mgt014.

Cowger, J, F. (1993). *Friction Ridge Skin - Comparison and Identification of Fingerprints*. Boca Raton, Florida: CRC Press.

DFSC 2015, Information Paper, Department of the Army, CIFS-FSL-LP, 3rd November.

DFSC 2017, Information Paper, Department of the Army, CFIS-FSL-LP, 9th March, https://osf.io/8kajs/.

DoubleLoop Podcast 2017, White Box Exclusion Paper - Episode 149, May 25th, http://doublelooppodcast.com/2017/05/white-box-exclusion-paper-episode-149/.

DoubleLoop Podcast 2018, Henry Swofford Interview, Episode 189, December 23.

Dror, I, E & Langenberg, G 2019, '"Cannot Decide": The Fine Line Between Appropriate Inconclusive Determinations Versus Unjustifiably Deciding Not To Decide', *Journal of Forensic Sciences, January*, 64 (1), pp. 10–15, doi: https://doi.org/10.1111/1556-4029.13854.

Dror, I, E. and Mnoonkin, J, L. (2010). The Use of Technology in Human Expert Domians: Challenges and Risks Arising from the Use of Automated Fingerprint Identification Systems in Forensics. *Law, Probability and Risk* 9: 47–67.

Dror, I, E., Thompson, W, C., Meissner, C, A. et al. (2015). Context Management Toolbox: A Linear Sequential Unmasking (LSU) Approach for Minimizing Cognitive Bias in Forensic Decision Making. *Journal of Forensic Sciences* 60 (4): 1111–1112.

Dror, I, E., Wertheim, K., Fraser-Mackenzie, P., and Walajtys, J. (2012). The Impact of Human-Technology Cooperation and Distributed Cognition in Forensic Science: Biasing Effects of AFIS Contextual Information on Human Experts. *Journal of Forensic Sciences* 57 (2): 343–352.

Egli, N, M 2009, Interpretation of partial fingermarks using an automated fingerprint identification system, PhD Thesis, Faculty of Law and Criminal Sciences, University of Lausanne.

Eldridge, H. (2011). Meeting the Fingerprint Admissibility Challenge in a Post-NAS Environment. *Journal of Forensic Identification* 61 (5): 430–446.

Eldridge, H 2012, 'I am 100% Certain of my Conclusion. (But Should the Jury Be Certain?)', *Evidence Technology Magazine – The Friction Ridge*, 10 (2). evidencemagazine.com

Eldridge, H., De Donno, M., and Champod, C. (2021). Mind-Set – How Bias Leads to Errors in Friction Ridge Comparisons. *Forensic Science International* 318: 110545.

Evett, I, W. and Williams, R, L. (1995). A Review of the Sixteen Points Fingerprint Standard in England and Wales. *Fingerprint Whorld* 21 (82): 125–141.

Fraser-Mackenzie, P, Dror, I & Wertheim, K 2013, 'Cognitive and Contextual Influences in Determination of Latent Fingerprint Suitability for Identification Judgments, Supported by a research grant from the National Institute of Justice (NIJ)', Document 241289, February, 2010-DN-BX-K270.

FSR 2017, Codes of Practice and Conduct, Fingerprint Comparison FSR-C-128, Issue 2.

FSR 2020, Annual Report, 17 November 2018 – 16 November 2019, 25th February, Dr Gillian Tully.

Galton, F. (1892). *Fingerprints*. London: MacMillan and Co.

Haber, L. and Haber, R, N. (2009). *Challenges to Fingerprints*. Tucson, Arizona: Lawyers & Judges Publishing Company, Inc.

Hancox, P. (1979). Ridge Detail Imperfections. *Fingerprint Whorld* 4 (15): 96.

Home Office 1969, The Action Taken When Comparing Finger Impressions, Police Research and Development Branch, Report no. 1/69, January.

IAI 1973, Resolutions & Legislative Committee, Resolution 1973-5, *Identification News*, August, www.theiai.org.

IEEGFI 2004, Method for Fingerprint Identification II, http://www.latent-prints.com/images/ieegf2.pdf.

IEEGFI 2006, Method for Fingerprint Identification, http://latent-prints.com/images/IEEGFI%201a.pdf

Kirk, P, L. (1985). Fingerprints. In: *Crime Investigation*, Seconde (ed. J, I. Thornton). Malabar, Florida: Krieger Publishing Company.

Langenburg, G. (2004). Pilot Study: A Statistical Analysis of the ACE-V Methodology – Analysis Stage. *Journal of Forensic Identification* 54 (1): 64–79.

Langenburg, G. (2009). A Performance Study of the ACE-V Process: A Pilot Study to Measure the Accuracy, Precision, Reproducibility, Repeatability, and Biasability of Conclusions Resulting from the ACE-V Process. *Journal of Forensic Identification* 59 (2): 219–257.

Langenburg, G 2011, 'Scientific Research Supporting the Foundations of Friction Ridge Examinations', in National Institute of Justice, *The Fingerprint Sourcebook*, www.nij.gov.

Langenburg, G 2012, 'A critical analysis and study of the ACE-V process', PhD thesis, University of Lausanne.

Langenburg, G 2013, 'The Consideration of Fingerprint Probabilities in the Courtroom', *Australian Journal of Forensic Sciences*, 45 (3), pp. 296–304, doi: https://doi.org/10.1080/00450618.2013.784360.

Langenburg, G. and Champod, C. (2010). The GYRO System – A Recommended Approach to More Transparent Documentation. *Journal of Forensic Identification* 61 (4): 373–384.

Langenburg, G., Champod, C., and Genessay, T. (2012). Informing the Judgments of Fingerprint Analysts Using Quality Metric and Statistical Assessment Tools. *Forensic Science International* 219 (1–3): 183–198.

Langenburg, G, Champod, C & Wertheim, P 2009, 'Testing for Potential Contextual Bias Effects During the Verification Stage of the ACE-V Methodology when Conducting Fingerprint Comparisons', *Journal of Forensic Science, 54* (3), pp. 571–582, doi: https://doi.org/10.1111/j.1556-4029.2009.01025.x.

Langenburg, G., Hall, C., and Rosemarie, Q. (2015). Utilizing AFIS Searching Tools to Reduce Errors in Fingerprint Casework. *Forensic Science International* 257: 123–133.

Lennard, C. (2013). Fingerprint Identification: How Far Have We Come? *Australian Journal of Forensic Sciences* 45 (4): 356–367.

Maceo, A, V. (2009). Qualitative Assessment of Skin Deformation: A Pilot Study. *Journal of Forensic Identification* 59 (4): 390–440.

Maceo, A. (2011). Documenting and Reporting Inconclusive Results. *Journal of Forensic Identification* 61 (3): 226–231.

Margot, P & German, E 1995, Fingerprint Identification Breakout Meeting, "Ne'uirm Declaration", International Symposium on Fingerprint Detection and Identification, http://www.latent-prints.com/images/1995%20Ne' Urim%20declaration.pdf.

McCartney, C. (2006). *Forensic Identification and Criminal Justice*. Devon, U.K: Willan Publishing.

Meuwly, D. (2011). Position of the European Fingerprint Working Group (EFPWG) of the European Network of Forensic Science Institutes (ENFSI) Regarding the NRC report. *Journal of Forensic Identification* 61 (6): 677–679.

Mnookin, J, L. (2008). The Validity of Latent Fingerprint Identification: Confessions of a Fingerprinting Moderate. *Law, Probability and Risk* 7 (2): 127–141.

Moenssens, A, A. (1971). *Fingerprint Techniques*. Radnor, PA: Chilton Book Company.

Neumann, C. (2013). Statistics and Probabilities as a Means to Support Fingerprint Identification. In: *Lee and Gaensslen's Advances in Fingerprint Technology*, Thirde (ed. R, S. Ramotowski). Boca Raton, Florida: CRC Press.

Neumann, C., Champod, C., Yoo, M. et al. (2015). Quantifying the Weight of Fingerprint Evidence Through the Spatial Relationship, Directions and Types of Minutiae Observed on Fingermarks. *Forensic Science International* 248: 154–171.

Neumann, C., Mateos-Garcia, I., Langenburg, G., and Kostrowski, J. (2011). Operational Benefits and Challenges of the Use of Fingerprint Statistical Models: A Field Study. *Forensic Science International* 212: 32–46.

Neumann, C., Evett, I, W., and Skerrett, J. (2012). Quantifying the Weight of Evidence from a Forensic Fingerprint Comparison: A New Paradigm. *Journal of the Royal Statistical Society, A* 175 (Part 2): 371–415.

NIST 2012, *Latent Print Examination and Human Factors: Improving the Practice through a Systems Approach*, Report of the Expert Working Group on Human Factors in Latent Print Analysis, US Department of Commerce.

NRC (2009). *Strengthening Forensic Science in the United States: A Path Forward*. Washington, DC: National Academy Press.

OIG (2006). *A Review of the FBI's handling of the Brandon Mayfield Case*. U.S. Department of Justice https://oig.justice.gov/special/s0601/final.pdf.

OSAC, 2018, 'Standard for Friction Ridge Examination Conclusions', Version 1.0, June, https://www.nist.gov/topics/organization-scientific-area-committees-forensic-science/friction-ridge-subcommittee.

Page, M., Taylor, J., and Blenkin, M. (2011). Uniqueness in the Forensic Identification Sciences – Fact or Fiction? *Forensic Science International* 206: 12–18.

PCAST 2016, *Forensic Science in Criminal Courts: Ensuring Scientific Validity of Feature-Comparison Methods*, September.

Polski, J, Smith, R, Garrett, R 2011, The Report of the International Association for Identification, Standardization II Committee, 233980, March.

Puri, D, K, S. (1967). Thought on Fingerprinting. *International Criminal Policing Review* 178 (130): 225–227.

Ray, E. and Dechant, P, J. (2013). Sufficiency and Standards for Exclusion Decisions. *Journal of Forensic Identification* 63 (6): 675–697.

Saks, M, J. and Koehler, J, J. (2008). The Individualization Fallacy in Forensic Science Evidence. *Vanderbilt Law Review* 61 (1): 199–219.

Stoney, D, A. (2001). Measurement of Fingerprint Individuality. In: *Advances in Fingerprint Technology*, Seconde (ed. H, C. Lee and R, E. Gaensslen). Boca Raton, Florida: CRC Press.

SWGFAST, 2002, Friction Ridge Examination Methodology for Latent Print Examiners, ver. 1.01, 8/22/02.

SWGFAST 2003, Standards for Conclusions, 9/11, ver. 1.0.

SWGFAST 2011, Standards for examining friction ridge impressions and resulting conclusions (Latent/Tenprint), ver. 1.0, 09/13/11.

SWGFAST 2012a, Individualization/Identification Position Statement, 3/06, ver. 1.0.

SWGFAST 2012b, Document #15, Standard for the Definition and Measurement of Rates of Errors and Non- Consensus Decisions in Friction Ridge Examination (Latent/Tenprint), Ver. 2.0, 15 November.

SWGFAST, 2012c, Document #8, Standard for the Documentation of Analysis, Comparison, Evaluation and Verification (ACE-V) (Latent), Ver. 2.0, 09/11/12.

SWGFAST, 2013a 'Document #10 Standard for Examining Friction Ridge Impressions and Resulting Conclusions (Latent/Tenprint), Ver. 2.0, 03/13/13.

SWGFAST, 2013b 'Document #10 Standard for Examining Friction Ridge Impressions and Resulting Conclusions (Latent/Tenprint), Draft for comment, Ver. 2.1.

Swofford, H., Steffan, S., Warner, G. et al. (2013). Impact of Minutiae Quantity on the Behaviour and Performance of Latent Print Examiners. *Journal of Forensic Identification* 63 (5): 571–591.

Swofford, H, J. (2015). The Emerging Paradigm Shift in the Epistemology of Fingerprint Conclusions. *Journal of Forensic Identification* 65 (3): 201–213.

Swofford, H, J 2016, 'A New Paradigm for Fingerprint Reporting. . .Without Individualization', Defence Forensic Science Center, Forensic Technology Center of Excellence Webinar, July, available at https://onin.com/fp/RTI_Webinar_2016_Reporting_Swofford_Final.pdf.

Swofford, H.J., Koertner, A.J., Zemp, F. et al. (2018). A Method for the Statistical Interpretation of Friction Ridge Skin Impression Evidence: Method Development and Validation. *Forensic Science International* 287: 113–126.

Thornton, J.I. (1977). The One-Dissimilarity Doctrine in Fingerrprint Identification. *International Criminal Police Review* 306: 89–95.

Triplett, M. (2011). Standards for Friction Ridge Identifications. *Fingerprint Whorld* 37 (143): 57–62.

Triplett, M. (2012). The Need to Validate Principles and the Value of Reproducible Results. *Identification News* 42 (4): 10–12.

Ulery, B, T., Hicklin, R, A., Buscaglia, J., and Roberts, M, A. (2012). Repeatability and Reproducibility of Decisions by Latent Fingerprint Examiners. *PLoS ONE* 7 (3): e32800. https://doi.org/10.1371/journal.pone.0032800.

Ulery, B, T, Hicklin, R, A, Roberts, M, A & Buscaglia, J 2014, 'Measuring What Latent Fingerprint Examiners Consider Sufficient Information for Individualization Determinations', *PLoS ONE*, 9 (11), e110179. doi: https://doi.org/10.1371/journal.pone.0110179.

Ulery, B, T., Hicklin, R, A., Roberts, M, A., and Buscaglia, J. (2015). Changes in Latent Fingerprint Examiners' Markup Between Analysis and Comparison. *Forensic Science International* 247: 54–61.

Ulery, B, T., Hicklin, R, A., Kiebuzinski, G, I. et al. (2013). Understanding the Sufficiency of Information for Latent Fingerprint Value Determinations. *Forensic Science International* 230: 99–106.

Ulery, B, T., Hicklin, R, A., Roberts, M, A., and Buscaglia, J. (2017). Factors Associated with Latent Fingerprint Exclusion Determinations. *Forensic Science International* 275: 65–75.

Vanderkolk, J, R 2011, 'Examination Process', in National Institute of Justice, *The Fingerprint Sourcebook*, www.nij.gov.

Zeelenberg, A 2011, 'A Matter of Standards', http://www.latent-prints.com/images/A%20Matter%20 of%20Standards2.pdf.

Appendix 1

Fabrication, Transplantation and Forgery

If a mark has been fabricated, forged or transplanted, then it will not have any evidential value, so the examiner must be aware of indicators that this may have occurred. If carried out by someone with the appropriate knowledge, skill and access, such marks may be undetectable by the examiner. However, by understanding the methods that could be used and being familiar with the indicators that may result, the examiner can increase the chances they will detect such marks.

A.1   Fabrication

In general terms, a fabricated mark is one that did not come from the surface it is reported to have come from. This can occur by accident, as happened in a case that led to the arrest of a woman in the United Kingdom in 2008 after a mark made by her was found at the scene of a burglary ('Signing a wedding card', 2009). Charges against the woman were dropped after it was discovered that the mark had been mistakenly documented as having been found on a games console box when it had actually come from a greeting card. The accused was a colleague of the person whose house was burgled and had touched the card when she signed it at their workplace.

In a case in South Africa in 2007, the judge in a homicide trial acknowledged that the origin of a mark that was a crucial part of the prosecution's case may also have been incorrectly recorded (Altbeker, 2012). In 2005 Fred van der Vyver was charged with the murder of his partner, and one aspect of the evidence against him was that a mark made by him was reported to have been found on a DVD case at the crime scene (Figure A1.1).

van der Vyver told the police that he had been at the victim's apartment on the morning of the offence but had left to go to work, leaving his partner alive and well. Numerous marks made by him were found, but the one on the DVD case was significant as the victim had rented the film after the accused claimed he left the apartment, so if he was telling the truth, it would be difficult to explain the presence of his mark on the case. van der Vyver's defence team hired Pat Wertheim who gave evidence at the trial that, in his opinion, the mark had not come from a DVD case but had, in fact, come from a drinking glass. Wertheim highlighted several factors that were inconsistent with the lift having come from a DVD case, including (Wertheim, 2008a):

- The presence of curved, parallel lines at the top and bottom of the lift that were consistent with having been made by the curved top and bottom edges of a glass.
- The area of skin that made the mark (the thumb side of the left index finger) being an area that commonly contacts a curved surface but is less frequently seen on a flat surface.

The Forensic Analysis, Comparison and Evaluation of Friction Ridge Skin Impressions, First Edition. Dan Perkins.
© 2022 John Wiley & Sons Ltd. Published 2022 by John Wiley & Sons Ltd.

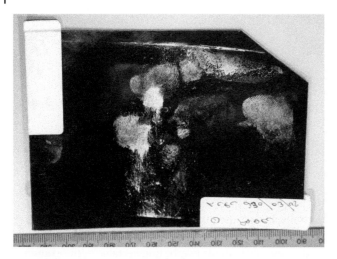

Figure A1.1   The lift reported to have been taken from the DVD case. *Source:* Printed with permission from Pat Wertheim.

- The edge of the mark being curved as would normally be seen on a curved surface but would not normally be seen on a flat surface.
- The presence of another mark along with one of the curved parallel edges that were consistent with being made by a lip.

Fred van der Vyver was found not guilty at the trial, and the judge stated that the description of the mark as having come from the DVD case may have been the result of negligence or incompetence (Altbeker, 2012).

As well as accidental fabrication, there are instances of marks being deliberately fabricated in order to implicate their donors in the commission of an offence. Wertheim (1994) describes a case in which a prison inmate deposited several of his marks on pieces of broken glass and passed them to a visitor. The visitor then committed a burglary, broke some glass in the process and left the pieces with the marks on them at the scene to divert the attention of the police from himself – knowing the inmate had an alibi. The attempted fabrication was detected when a police officer noticed that the pieces of glass with the marks on them were of a different thickness to the glass that was broken at the scene.

In 1967 a mark reported to have come from the counter of a bank that had been robbed was identified as belonging to William DePalma (Sherrer, 2011). DePalma claimed he had never been in the bank, but as a result of the mark and eyewitness testimony, he was convicted of robbery and sentenced to 15 years in prison. After the trial, a private investigator hired by DePalma discovered that one of the examiners involved in the identification had lied about his background and qualifications during the trial. As a result of this deceit, the fingerprint evidence was microscopically reviewed by an examiner who concluded the mark had been fabricated (Foley, 1975). The examiner found particles in the mark that were consistent with the toner used in photocopiers rather than the powder used to develop marks. The examiners' view was that the mark was lifted from a photocopy of a set of fingerprints taken from DePalma in 1957. Figures A1.2 and A1.3 are examples of marks created using this method of fabrication. DePalma served over two years in prison before his conviction was overturned in 1974.

Holik (1979) describes a similar case in which a mark that was reported to have been found on radio and used to implicate a suspect, had also come from a fingerprint form. The appearance of

(a)   (b)

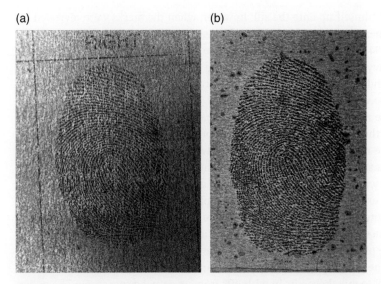

Figure A1.2 These impressions are examples lifted from a fingerprint form, (a) was taken from a photocopy of a form after it had been powdered and (b) from an inked form after it had been powdered.

(a)   (b)

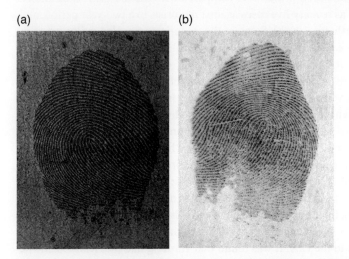

Figure A1.3 These impressions were both lifted from sets of fingerprints that had been taken electronically, (a) was lifted after the form had been powdered and (b) was lifted without powder.

the mark aroused suspicion and when viewed using a microscope, tiny fibres could be seen running through it. A set of the suspect's fingerprints were then examined, and the mark was found to be an exact copy of one of the prints. This inked print was slightly lighter in tone than the other prints on the form, and it was discovered that a police officer had used tape to lift the print from the form and then presented it as having come from the radio. Under magnification, one of the paper fibres above the mark, which had a red colouring, was found to fit exactly into a void in the red lettering of a printed word on the fingerprint form above the print.

Wertheim (1994) describes another case in which prints on a fingerprint form were presented as marks from a crime scene. In 1928 a man was convicted of burglary based on his fingermarks being identified on the wooden windowsill of a second-floor window. It was alleged that the marks were made when the man pulled himself up onto the sill; however, when the marks were reviewed by another examiner, their appearance was not considered consistent with marks left as a result of this activity (Wertheim, 1994). The examiner found that the marks did not reveal indicators of being deposited under heavy pressure (such as a widening of the ridges and a corresponding narrowing of the furrows) and that their relative positions were different to those that would be left by someone pulling themselves up onto a windowsill. The examiner also noted there was no appearance in or around the marks of wood grain, dirt or weathering – some or all of which may be expected to be seen had the marks been left on an exterior wooden windowsill. The examiner found the appearance of the marks was actually consistent with that of prints taken on a finger-print form of all four fingers simultaneously. This included the presence of faint lines framing the marks that were consistent with being made by the edges of a fingerprint form. It was concluded that the original fingerprint examiner in the case had simply photographed prints on the convicted man's fingerprint form and presented them as having come from the windowsill (Wertheim, 1994).

Another type of fabrication was perpetrated in the 1970s by a San Diego police officer who would stop drivers and ask them to place their hands on his car while he spoke to them (Wertheim, 1994). When the driver had left, the police officer would powder and lift their marks from his car and then wait for an opportunity to mix those lifts in with lifts from an actual crime scene. The police officer would then pretend that he had information that linked the driver to that offence and nominate them as a suspect to ensure their prints were compared with the marks. It is estimated that the officer fabricated as many as seventy-five cases against motorists in a two-year period, and most of them were convicted (Wertheim, 2008b). Some of the motorists even pled guilty, presum-ably as they felt they had no chance of convincing a court of their innocence and would receive a lighter penalty if they admitted guilt.

A similar case came to light in 1992 when New York State police officer David Harding admitted during a job interview with the CIA that he had fabricated fingerprint evidence in order to secure several convictions (Cole, 2001). The subsequent investigation led to the officer and five others from the same unit pleading guilty to fabricating evidence in approximately forty cases over an eight-year period (Cole, 2001). One of the cases involved a multiple homicide in which the victims had been tied up in their home, shot, doused in petrol and set on fire (Possley, 2012). The suspect in the case was shot by police and died; however his mother, Shirley Kinge, was charged with helping her son burn down the house. The charge came about as a result of Harding's evidence that marks made by Kinge had been identified on gasoline can which was found in the victim's house (Possley, 2012). Kinge was convicted and served two years in prison until a re-trial took place as a result of Harding's admission to the CIA. It was found that Harding had met Kinge while he was undercover at a restaurant, lifted her marks from a drinking glass and then presented them as having come from the gasoline can (O'Brien, 2015). The charges against Shirley Kinge were dismissed, and Harding served four years in prison for fabricating evidence against her and other defendants (Possley, 2012).

There are many more documented instances of marks being fabricated than there are of them being forged or transplanted, but it is difficult to say how often fabricated marks are encountered. The FBI report receiving at least 15 cases involving fabricated marks between 1930 and 1950 (Bonebreak, 1976) and Wertheim (2008b), who conducted a study of fabricated marks, estimates that just the cases that have been discovered may number in the hundreds or even thousands over the last 100 years.

For a fabricated mark to be successfully used to implicate its donor, the person doing the fabrica-tion would need access to an identifiable impression made by the donor. They would also need to have the knowledge and skill to lift or photograph that impression and know that the police have access to the donor's fingerprints – or have a means of ensuring the police obtain those fingerprints. However, perhaps the biggest obstacle to overcome for the fabricator would be that they would need to be able to seamlessly insert the fabricated mark into the evidence recovered from a crime scene. Therefore, generally, the person doing the fabrication would either need to be the person who examines the scene or would need their cooperation, which explains why in most cases where fabrication is found to have occurred, it was carried out by someone employed by the police.

The indicators that a mark may have been fabricated will vary according to how the fabrication was carried out, but in general, the examiner should consider that fabricated marks may reveal some or all of the following indicators:

- The appearance of the mark and the surface around it may not be consistent with the surface it was reported to have come from.
- The area of skin that made the mark and the marks orientation may not be consistent with that which would typically result from normal handling of that surface.
- The mark may have the appearance of a print (if it came from a fingerprint form).
- Due to the difficulties involved, there may be only one identified mark.

A.2 Transplantation

A transplanted mark is one that was made on one surface and then transferred to another. This can happen accidentally as a result of the two surfaces being in contact with each other. This type of transfer would produce a 'reverse direction' mark that was a mirror image of the original deposit and is discussed in Chapter 4. This section addresses deliberate transference from one surface to another with the intention of implicating the donor of the mark in the commission of an offence.

Though the way this kind of deliberate transfer could be accomplished has been in the public domain since the early 1900s there are no proven instances of it actually occurring in a real case. The most likely explanation for this relates to the practical difficulties involved in being able to successfully implicate the donor of the transferred mark.

Figure A1.4 shows two marks that were transplanted by the author for demonstration purposes. The marks were made on different drinking mugs, and adhesive tape was used to lift them from those surfaces and transfer them to the screen of a mobile phone.[1] Once the tape was removed from the phone, the screen was powdered to develop the marks, and they were lifted from the surface. Though the technique was successful on this occasion, it is not always – for instance, Beaudoin (2004) attempted unsuccessfully to transfer marks in the same way from a metal surface to paper.

One of the factors that contributes to the success or failure of an attempt to transplant a mark is the amount of material on the skin of the donor when the mark is made. The tape used in the transfer generally will not lift all the material off the original surface, nor will it transfer all the material it does lift to the new surface. Therefore, even when transplantation is successful, the process usually results in the production of a faint or 'weak' mark on the new surface. Other factors that may contribute to success or failure include the condition, type and nature of the two surfaces used. For instance, though both mugs used as the original surfaces in Figure A1.4 were clean and dry, the mug mark (a) was deposited on appears to have afforded a better surface for transplantation than the

---

1 One way to improve the chances of a successful transfer would be to develop the mark with powder on the first surface before attempting to lift it. However this would result in an already developed mark appearing on the second surface which may arouse the suspicions of a crime scene examiner tasked with developing marks on that surface.

(a)   (b)

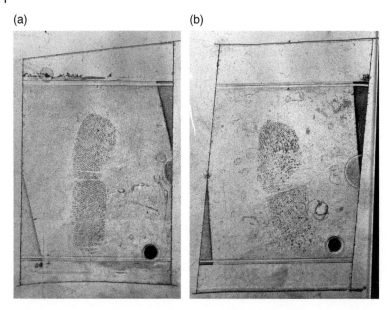

Figure A1.4   These marks were deposited on different mugs and then transferred to the screen of a phone, where they were developed with powder and lifted. The edges of the phone are revealed in the lifts and can be seen running perpendicularly to the tape at the top and bottom. Part of the circular 'Home' button can also be seen midway along both right edges. The parallel diagonal lines in the top right and bottom left of the images were caused by the tape that was used to transfer the mark to the phone, lifting off residue that was present on the screen (even though the screens were 'clean'). This residue was still in place on the rest of the screen, and the powder used to develop the marks has adhered to it, leaving a clean area where the tape was. The same type of tape that was used to transfer the marks was also used to lift them from the phone, so the distance between the diagonal lines is the same as the width of the actual tape in the images. Except for the button, the background to mark (a) reveals no obvious indicators of the surface the mark was originally on, though the lower part of mark (b) reveals some writing running vertically through it which was printed on the mug it came from. There is a slight curvature to both marks that is consistent with what is seen in marks made on curved rather than flat surfaces (most obvious along the left edge of mark (b)). The black circle in the bottom right of both marks was caused by a reflection during the photography of the lifts.

other mug. This may be because of differences in the paint, finish or age of the mugs. The transfer medium can also be a factor as better results can be achieved with some types of lifting tape than others. The amount of time between the mark being deposited and transferred can also affect the likelihood of an identifiable mark resulting on the second surface. It seems logical that the best results are likely to be produced if the transfer takes place as soon as possible, though experiments by the London Metropolitan Police Service (MPS) resulted in identifiable marks being produced despite not being transferred for several hours. One successful test involved the transfer of a mark from a piece of glass to a plastic bottle with a gap of 16 hours in-between deposition and transplantation.[2] The media used to develop the transplanted mark can also be a factor, as some may be more effective than others. For example, Jabbal, Boseley & Lewis (2018) found that the marks they transferred from glass to paper were obscured because the media also reacted strongly with the background of the mark – likely because of the presence of components transferred from the adhesive on the tape.

The transplantation of a mark would only be considered successful if it was not detected, so the appearance of traces of the original surface or the transfer medium around the mark are also

---

2 This test was conducted under controlled conditions and in a real scenario the environmental conditions and whether the mark was protected before being transferred would also likely affect how long it remained suitable for transfer.

factors that can contribute to success or failure. The tape may pick up debris, surface decoration, damage or the traces of the edges of the original surface and transfer some or all those features to the second surface (Figure A1.4). The tape itself may also leave traces on the second surface or remove material from it – the absence of which may be obvious when the second surface is powdered, as in Figure A1.4. Any traces left may not be clearly visible until the second surface has been powdered so the person carrying out the transfer may have no way to be sure that they are not present. Whilst it is difficult to transfer a mark without leaving traces of the tape on the new surface, that difficulty is somewhat dependent on the surfaces involved and the skill of the person carrying out the transfer – transfers can be carried out without leaving obvious traces.

Another factor that could lead to the detection of a transplanted mark would be a failure to consider the position and orientation of the mark on the second surface. Whilst marks can be found in positions and orientations that are not consistent with the normal handling of a surface, where this is the case, it may arouse the suspicions of the examiner and cause them to investigate further. The area of skin that made the mark may also be factor as some areas commonly contact some surfaces but are less likely to be seen on other surfaces. For example, the index finger side of thumbs often contacts a surface the hand grasps between the thumb and fingers but does not often contact surfaces that are not grasped. Similarly, some surfaces will result in marks that may have a particular appearance, so if such marks are transplanted to a different surface, the examiner may begin to consider the possibility that the mark was not made on that surface. For instance, the curvature seen in the marks in Figure A1.4 is commonly seen on curved surfaces rather than flat ones.

Whilst it is possible to transfer marks from one surface to another, there are many practical difficulties that would need to be overcome for the donor of that mark to be identifiable. Whilst transplantation could be successful without some of the following conditions being met, to maximise the likelihood of it being successful, the person carrying out the transfer would need:

a) access to a suitable surface that the donor has touched;
b) to be able to locate a mark without using a development medium on that surface that:
  – was made by the specific donor
  – reveals enough clear detail for its donor to be identified
  – can be transferred;
c) to have the knowledge, materials and skill to lift the mark off the surface;
d) to have access to the crime scene before it is examined;
e) to be able to locate a suitable second surface that:
  – the appearance of the mark will be consistent with
  – the digit or area of skin that made the mark will be consistent with being found on;
f) to place the mark in a position and orientation on the surface that is consistent with the natural handling of that surface;
g) to know the surface the mark is transferred onto will be examined;
h) to be able to complete the transfer leaving a mark that is capable of being identified;
i) to complete the transfer without leaving signs of the tape on the new surface;
j) to know that the police have access to the donor's fingerprints or to be able to ensure the donor is fingerprinted by the police.

When observing the mark for signs that it may have been transplanted, the examiner should consider that such marks may reveal some or all the following indicators:

- They may be faint.
- They may be surrounded by an area that has been 'cleaned' as a result of the removal of material by the tape used to deposit the mark.

- They may be bordered by tape marks.
- They may be close to traces that are inconsistent with the surface they are on (for example, decoration or writing from the original surface).
- They may not be in positions, orientations or left by areas of skin that are consistent with normal handling of the surface.
- Due to the difficulties involved, there may be only one of them.

A.3 Forgery

A forged mark is one that was made by a copy of a person's ridge detail rather than their actual skin. Such a mark could be left at the scene of an offence by the real offender in order to implicate someone else, though there are very few documented instances of this occurring.

Bonebreak (1976) describes a plan devised by two prison inmates to leave forged marks on cheques. When one inmate was released, he took with him a plastic cast of the other inmate's thumb with the intention of using it to leave marks on forged checks to implicate the other inmate, who would have a perfect alibi as he was still in prison. However, before the plan could be put into action, the freed inmate was caught in possession of the forged cheques and confessed to the scheme.

Forged marks were used to implicate innocent people by a burglar in Bulgaria in the 1940s (Wertheim, 2008b). The burglar posed as a fortuneteller and claimed to be able to predict his clients' futures by looking at impressions of the ridge detail they made in clay. However, once the clients had left, the burglar would pour liquid rubber onto the clay to create casts of the ridge detail that he would use to leave marks at his burglaries.

McGuire (in Genge, 2003) describes a case where around 60 forged marks were left at the scene of a multiple homicide to divert attention away from the perpetrator. The marks were immediately observed to be unusual for several reasons, including their size (they appeared to be made by fingers but were much larger) and their position on only one side of objects and surfaces that would normally be grasped. McGuire also observed the scene itself to be unusual because of the state of the premises. For example, though the homicides appeared to have taken place in or around the bedrooms, cupboards and drawers were open in other rooms, and their contents had been smashed or thrown on the floor. Additionally, the refrigerator had been yanked out of position, and a ceiling fan was now dangling from its fixture in the living room. McGuire reports that the perpetrator's plan was to make the scene look like a gorilla (which have ridge detail on their fingers and palms) had committed the homicides. To accomplish this, the perpetrator had hired a gorilla costume on the day of the offence and used the ridge detail on the rubber hands to leave the marks.

Wertheim (2008a) describes a case in which he believes forged fingermarks were used as proof that a painting by an unknown artist was actually by Jackson Pollock. In 2004 the owner of the painting, suspecting it may be by Pollock, took it to an art restorer to see if it could be authenticated (Grann, 2010). The restorer had previously authenticated another painting by an unknown artist as having been made by Pollock using a fingermark he found on it. A fingerprint examiner had compared that mark with a mark on a can of paint that is on display in Pollock's former studio (now a museum) and identified them both as having been made by the same person (Grann, 2010). When the restorer examined the painting in question, he reported finding a mark, and when this mark was compared with the mark on the paint can, it was also found to have been made by the same person (Grann, 2010). The restorer said that the mark proved the artist was Pollock and that as a result, the estimate of the value of the painting may have risen from around $50 000 to

Figure A1.5 One of the marks on the wooden frame of the painting. *Source:* Printed with permission from Pat Wertheim.

approximately $100 million. The owner of the painting began to investigate its provenance and spoke to its first owner, who told him that she had bought it from a young artist who was skilled at imitating the work of famous artists and that she was certain it was not by Pollock (Grann, 2010). As a result, the owner contacted Global Fine Art Registry, who reviewed the painting and arranged for a fingerprint examiner (Pat Wertheim) to look at the mark.

Wertheim found four marks on the painting, three on the frame and one on the canvas itself – all of which he described as having the appearance of inked fingerprints (Figure A1.5).

Wertheim also observed that all the marks were the same shape and that the outlines of each of them appeared to be exact overlays of each other. The shape of the marks was also unusual in that in places, there were significant concave indentations that are not commonly seen in marks (where the edges are usually straight or convex). Wertheim also noted that each mark revealed the same area of detail and was oriented in the same way. He also observed large uneven voids in the ridges that were not consistent with having been made by the sweat pores. After examining the marks, Wertheim's opinion was that their appearance was not consistent with them having been made by friction ridge skin but was consistent with them having been made by a rubber stamp. Wertheim tested his theory by making a cast of the impression on the paint can in the artist's former studio. When Wertheim made impressions in ink using part of the cast as a stamp, he found that they revealed the same area of skin seen in the marks on the painting. Considering all his observations collectively, Wertheim concluded that, in his opinion, the four marks on the painting had been forged with a rubber stamp made from the impression on the paint can (Wertheim, 2008a). The restorer denies forging any of the marks, and by 2008 the painting had been sold to investors who were satisfied with his authentication. The paintings former owner also takes the view that there is no reason to believe it was not painted by Pollock (Grann, 2010).

Sellenraad (2018) reports the use of forged marks in a money laundering offence in 2013. The case involved hundreds of marks made in ink on cheques, and Sellenraad observed that three of them appeared to be very similar in several ways. First, these marks stood out from all the others as they were significantly larger. Second, all three were the same size and shape. Third, the details they revealed did not display the normal variation that would be expected to occur in marks made by the same area of skin. Specifically, there were voids of the same size and shape present in the same locations in all three marks. Also, there were small sections of ridge along the edges of the marks that were re-produced exactly in all three. Though there were also some differences in the appearance of the impressions, such as different shaped voids in the core areas and some distortion caused by movement in the lower right sides of the marks, Sellenraad concluded that the overall similarities of the marks were not consistent with them having been made by the skin of a digit. The suspect later admitted to using a stamp to make the marks on some of the cheques. It may be that the differences observed were due to variations in the amount of ink on the stamp, variations in the surface under the cheque or the amount of pressure or action used to make the marks.

There are various ways to create a copy of an area of ridge detail that can be used to leave a forged mark. For example, marks (a) and (c) in Figure A1.6 were made by a cast of the author's right index finger (the mark in the middle was made by the authors' actual right index finger).

The cast was made by immersing the hand in a casting material (alginate) to create a mold and then filling the mold with silicon to create a replica of the entire hand. This method re-produces all the ridge detail on the hand and so allows marks to be left from any area. The detail is very accurately recorded on the cast and includes the undulations along the edges of the ridges and the pore openings along the summits. However, as it is difficult to eliminate all the air bubbles in the casting material, small indentations of various sizes often affect some areas and appear as voids when marks are made by those areas. The major downside of this method for forgery purposes is that it would be very difficult to produce a cast in this way without the donor being aware of it.

Barton & Matthias (2019) used an alternative method to produce forged marks as part of a study to determine whether examiners could differentiate between real marks and forged, fabricated or

<table>
<tr><td>(a)</td><td>(b)</td><td>(c)</td></tr>
</table>

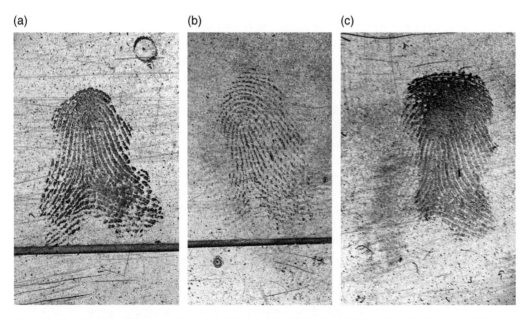

Figure A1.6 All three of these marks were made on glass. Marks (a) and (c) were made by a cast of the author's right index finger, whereas mark (b) was made by the author's actual right index finger. A silicon cast is not as flexible as skin, and so to produce a large enough mark, considerable pressure had to be used, which is why the ridges are wider in the marks in (a) and (c). Having a cast of an entire hand or digit does mean that multiple marks can be left that reveal different areas of skin, but obviously, it would be difficult to create a cast of this type without the donor's knowledge.

transplanted ones.[3] The donor placed their finger into a wet casting material and allowed it to dry. The digit was then removed to reveal an indented mark in the material. The excess casting material was cut off around the mark before sweat was applied to the cast, and it was used to leave marks on a surface. Whilst this technique can produce what appear to be natural looking marks, there are several drawbacks to it that may limit its success.

First, because this technique requires that the digit be pressed into the cast, and the cast then be turned over and used to leave the mark, the details will be reversed in the mark. Whilst reverse direction marks can be identified; unless the examiner anticipates a mark may be reversed, it is much less likely that it will be. For instance, only 46% of examiners participating in the Barton & Matthias study (2019) correctly identified such marks, and that number would likely have been much lower if the examiners were not aware that some of the marks they were looking at had been forged. Because the ridges of the donor's skin were pressed into the casting material leaving an impression of the furrows on the surface of the cast, marks made with a cast like this will also be in reverse colour, so some of the details will have the opposite appearance (for example a ridge that ends will instead appear to be a ridge that splits into two). Both the reverse colour and direction aspects of marks created using this technique could be overcome by making a cast of the cast and using that to leave the marks, but this adds a further complication and can also be problematic as it may be difficult to separate the two casts. A further drawback of casts made in this way is that as the furrows of the cast do not protrude above the surface of the surrounding casting material, it is difficult to leave a mark without the casting material contacting the surface. Where this occurs, the edges of the material can leave a distinctive outline around the mark that can appear very similarly around all marks left by the same cast. Lastly, like the previous casting technique, this method also requires the forger to obtain an impression of the donor's skin in a casting material, which may be difficult to accomplish without the donor's knowledge.

However, there are ways of producing a forged mark without the donor's knowledge. For instance, if an indented mark made by the donor can be found in a suitable soft substance, then a casting material can be applied to the mark to create a cast that can then be used to leave further marks. Samischenko (2001) produced a realistic-looking mark in this way using playdough as the soft substance. Marks produced in this way, unlike those produced by the previous technique, would be the correct colour and direction and would be easier to leave without other parts of the casting material contacting the surface. Alternatively, if a suitable impression of the donor's skin can be located and photographed, then a stamp can be made from it by a manufacturer (Figure A1.7).

Unlike the previous casting methods discussed, which can produce more natural-looking marks, stamps tend to be flat and inflexible and, as a result, leave marks that are all the same shape and reveal the same area of detail. Because of the skin's flexibility and the way it can contact surfaces at different angles, multiple marks made by the same area of the actual skin of the donor would be very unlikely to be the same shape and reveal exactly the same details. So, if multiple marks were left by a stamp of ridge detail, such similarities may be an indicator that they have been forged – as was the case with Sellenraad (2018). However, if only one mark is left, or the forger can locate

---

3 The study found that none of the 26 examiners correctly determined whether all 18 marks were real or forged/fabricated/transplanted. Under the conditions of the study examiners detected marks that were forged 83% of the time, ones that were fabricated 45% of the time and transplanted 56% of the time. The examiners also thought that real marks were forged, fabricated or transplanted 45% of the time. Barton & Matthias point out this may be because examiners were told there were forged, fabricated or transplanted marks among those they were looking at and so they were in a different state of awareness that may have made them doubt the real marks more than they usually would have.

(a)   (b)

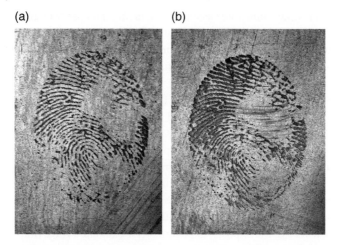

Figure A1.7 These impressions were made by a plastic stamp of a donor's ridge detail. Sweat was applied to the stamp, and then the marks were made on glass, developed with powder and lifted. They are both the same size and reveal the same area of ridge detail and include areas where detail is missing in the same places. The thicker ridges in (b) are likely caused by more sweat being on the stamp than there was in mark (a) and/or slight movement of the cast on the surface during deposition.

multiple impressions and produce stamps of different areas, then the forgery may be much more difficult to detect.

Champod & Espinoza (2014) describe an alternative method to produce a forged mark without the donor's knowledge that could be used to leave a more natural-looking mark than that left by a stamp. The method involves locating a suitably clear impression made by the donor, photographing it and then inverting the image so that the furrows are black and the ridges are white (this step is necessary to avoid the forged mark being in reverse colour). Once inverted, the image is then printed onto an acetate sheet to form a mould onto which a casting material is applied to create a cast that can be used to leave marks. Whilst this method overcomes some of the practical difficulties of the casting method described by Barton & Matthias (2019) in that the marks would not be reverse colour or reverse direction, it would still be necessary to cut off the excess casting material around the cast and as the ridges in the cast would be at the same level as the surrounding material it may still be difficult to leave a mark without the edges of the material contacting the surface and leaving traces. Champod & Espinzoa (2014) also describe a similar but more sophisticated method that uses ultraviolet light to etch metal to create the mould for the cast.

Kiltz et al. (2011) tested an alternative method that relies on capturing an image of a mark and printing it directly onto different surfaces using an inkjet printer. After marks were printed onto the surfaces, standard chemical and physical treatments were used to develop the marks. Examining the ridges in the marks under magnification and comparing them with those produced by actual skin, Kiltz et al. were able to identify differences (such as the regularity and frequency of the dots that made up each ridge) that meant they could differentiate between the forged and real marks in most cases. However, there were marks, for example, those that had been printed onto paper and then immediately powdered, that were very similar to real marks. As a result, the method at the moment is impractical as the mark would need to be printed immediately before it was examined, and the forger may have no control over when the examination would begin – particularly with an item like paper as typically paper exhibits would be retrieved from the scene and subjected to chemical treatments in a laboratory rather than being examined with powder at the scene.

Whilst it is possible to forge marks, there are many practical difficulties that must be overcome for the person whose skin was copied to be identified. Whilst forgery could be successful without some of the following conditions being met, to maximise the likelihood of it being successful, the person carrying it out would need:

a) access to the donor's skin or an identifiable impression they have made;
b) to create an exact copy of the skin or impression that can be used to leave marks at the scene;
c) to have access to the crime scene before it is examined;
d) to place the mark on a surface that the digit or area of skin that made it would be consistent with being found on;
e) to place the mark in a position and orientation on the surface that is consistent with the natural handling of that surface;
f) to know the surface the mark is left on will be examined;
g) to complete the process by leaving a mark that is capable of being identified;
h) to know that the police have access to the donor's fingerprints or to provide details of the donor to the police.

The indicators that a mark may have been forged will vary according to how the forgery was carried out, but in general, the examiner should consider that forged marks may reveal some or all the following indicators:

- Due to the difficulties involved, there may be only one mark.
- If there is more than one, the marks may all be the same shape and reveal the same area.
- They may not be in positions, orientations or left by areas of skin that are consistent with normal handling of the surface.
- They may be bordered by an outline.

References

Altbeker, A 2012, *Fruit of a Poisoned Tree*, Jonathan Ball Publishers, Johannesburg & Cape Town.

Barton, K & Matthias, G 2019, 'Distinguishing Forged and Fabricated Prints', *Journal of Forensic Identification*, 69 (2), pp. 195–206.

Beaudoin, A 2004, 'Research on Transferring a Fingerprint to a Ninhydrin-Treated Document', *Journal of Forensic Identification*, 54 (2), pp. 178–184.

Bonebreak, G, C 1976, 'Fabricating Fingerprint Evidence', *Identification News*, October, pp. 3–13.

Champod, C & Espinoza, M 2014, 'Chapter 2: Forgeries of Fingerprints in Forensic Science', in Marcel, S, Nixon, M, S & Li, S, Z (eds.), *Handbook of Biometric Anti-Spoofing*, Springer, London.

Cole, S, A 2001, *Suspect Identities: A History of Fingerprinting and Criminal Identification*, Harvard University Press, United States of America.

Foley, C 1975, 'Case of the Faked Fingerprint', *Observer Magazine*, 9 February, pp. 20–23.

Genge, N, E 2003, *The Forensic Casebook: The Science of Crime Scene Investigation*, Second Edition, Ebury Press, Great Britain.

Grann, D 2010, 'The mark of a masterpiece', *The New Yorker*, July 12. http://www.newyorker.com/magazine/2010/07/12/the-mark-of-a-masterpiece.

Holik, W 1979, 'Latent Print Forgery', *Identification News*, September, pp. 8–9

Jabbal, R, S, Boseley, R, E & Lewis, S, W 2018, 'Preliminary Studies into the Secondary Transfer of Undeveloped Latent Fingermarks Between Surfaces', *Journal of Forensic Identification*, 68 (3), pp. 421–437.

Kiltz, S, Hildebrant, M, Dittmann, J, Vielhauer, C & Kraetzer, C 2011, 'Printed Fingerprints: A Framework and First Results Towards Detection of Artificially Printed Latent Fingerprints for Forensics', in Farnand, S, P & Gaykema, F (eds.) Image Quality and System Performance VIII, Proceedings of SPIE-IS&T Electronic Imaging, SPIE, vol. 7867, 78670U.

O'Brien, J 2015, 'Shirley Kinge Dies, 25 Years After State Troopers Railroaded Her With Fake Evidence', Syracuse.com, September 21, http://www.syracuse.com/crime/index.ssf/2015/09/shirley_kinge_dies_25_years_after_she_was_railroaded_with_state_polices_doctored.html.

Possley, M (2012), 'Shirley Kinge', The National Registry of Exonerations, Posting date: before June 2012, Last updated: 21st September 2015, https://www.law.umich.edu/special/exoneration/Pages/casedetail.aspx?caseid=3352.

Samischenko, S, S 2001, *Atlas of the unusual papilla patterns*, IUrisprudentsiia, Moskva.

Sellenraad, A 2018, 'A Case Report: The Analysis of Patent Prints Identified as Forgeries', *Journal of Forensic Identification*, 68 (1), pp. 3–9.

Sherrer, H 2011, 'William DePalma Was Framed For Bank Robbery By A Policeman Faking His Fingerprint', Justice Denied, May 7, http://justicedenied.org/wordpress/archives/1117.

'Signing a wedding card landed me in court' 2009, *Metro*, 16th July http://metro.co.uk/2009/07/16/signing-a-wedding-card-landed-me-in-court-278109/.

Wertheim, P, A 1994, 'Detection of Forged and Fabricated Latent Prints: Historical Review and Ethical Implications of the Falsification of Latent Fingerprint Evidence', *Journal of Forensic Identification*, 44 (6), pp. 652–681.

Wertheim, P, A 2008a, 'Fingerprint Forgery: A Case Study', *Identification News*, 38 (6), pp. 12–17.

Wertheim, P, A 2008b, 'Latent Fingerprint Evidence: Fabrication, Not Error', *The Champion*, National Association of Criminal Defense Lawyers, November/December, p.16, https://www.nacdl.org/Champion.aspx?id=1608&terms=fingerprint+fabrication.

Appendix 2

Verification

ACE is also known as 'ACE-V' because some or all the examiner's conclusions will be subject to verification by another examiner. Verification is a quality control measure employed to establish whether an examiner's conclusion is supported by the details in the impressions.

Verification is defined as the process of finding out if something is real or true. With a mark from a crime scene, the 'truth' of the original examiner's conclusion may never be known, but having another examiner reproduce that conclusion gives confidence in its validity. Organisations routinely use verification to detect erroneous conclusions, and there are studies showing that verified conclusions are more likely to be correct than unverified conclusions (Pacheco, Cerchiai & Stoiloff, 2014; Ulery et al., 2011; Ulery et al., 2012; Wertheim, Langenburg & Moenssens, 2006).

Though the word verification is synonymous with confirmation and that may be the most frequent outcome, it is not the goal of the verifier to confirm the original examiner's conclusion. If the verifier reaches the same conclusion, then that conclusion has been verified, but verification may also result in the conclusion not being verified. In that scenario, the organisations' conflict resolution procedure would be employed.

There are many ways in which verification can be carried out, but broadly it is either 'blind' or 'open'.

A2.1   Open Verification

Open verification is the most common type and occurs when the verifier is aware of the original examiner's conclusion.

Fingerprint verification is usually defined as an independent process, but if the verifier is aware of the original examiner's conclusion, critics argue it is akin to ratification and is unlikely to detect errors (Haber & Haber, 2009). As evidence, critics point to erroneous conclusions that have been open verified multiple times. For instance, the FBI's erroneous identification of Brandon Mayfield was verified twice internally and once by a court appointed external examiner before it was discovered by the Spanish Police (OIG, 2006). The Scottish Criminal Records Office (SCRO) erroneous identifications of Shirley McKie and Marion Ross were verified three times internally and twice externally (Campbell, 2011). In addition to those errors, studies have also recorded instances of erroneous conclusions being verified (Langenburg, 2009; Langenburg, Champod & Wertheim, 2009; Pacheco, Cerchiai & Stoiloff, 2014; Wertheim, Langenburg & Moenssens, 2006). One study provided verifiers with 60 conclusions which included six erroneous exclusions – all of which were verified as if they were correct (Langenburg, 2009). The author suggests the reason none of the erroneous exclusions was detected could be related to the fact that the verifiers knew the skill level and identity of the examiners whose work they were verifying. Langenburg argues this knowledge may have led to the verifiers not looking as thoroughly as they normally would because they believed the original examiners would not have 'missed' any identifications. Interestingly, contrary to the verifier's beliefs, when the original examiners knew their work was going to be verified, they actually looked less thoroughly than they did when they thought it was not going to be verified (they made double the number of erroneous exclusions when they knew their work was to be verified than when they knew it was not going to be). Though it was not a verification study, Dror, Charlton & Péron (2006) also showed that examiners could be influenced when they thought impressions had been excluded by other examiners. The authors presented the examiners with impressions they had identified as having been made by the same area of skin several years previously but informed them that the impressions were those wrongly identified by the FBI in connection with the Madrid train bombings in 2004. Under those conditions, only one of five examiners reached the same conclusion of identification (one reached a conclusion of inconclusive, and the remaining three reached conclusions of exclusion). Whilst the conditions of the study are extreme, Dror, Charlton & Péron (2006) suggest it shows that the information provided resulted in the examiners finding what they expected to find, rather than evaluating what was there. One study also found verifiers who were aware of a previous examiner's conclusion of inconclusive were more likely to reach that conclusion themselves (Langenburg, Champod & Wertheim, 2009). The same study also reported that verifiers who were aware of a previous examiner's conclusion (irrespective of what it was) produced more inconclusive conclusions than those who were not. Another study looked at suitability decisions and found that verifiers were more likely to conclude a mark was unsuitable for comparison if the original examiner had done so, but knowledge that the original examiner concluded a mark was suitable did not make it more likely the verifier would also reach that conclusion (Fraser-Mackenzie, Dror & Wertheim, 2013).

Though the studies showed verifiers being influenced by knowledge of previous examiner's conclusion of exclusion, inconclusive and unsuitable for comparison, they also showed verifiers were much less susceptible when the original conclusion was identification. For instance, all the studies found the knowledge that one examiner had reached a conclusion of identification did not result in experienced verifiers reaching the same conclusion when that conclusion was erroneous (Langenburg, 2009; Langenburg, Champod & Wertheim, 2009; Pacheco, Cerchiai & Stoiloff, 2014; Wertheim, Langenburg & Moenssens, 2006).[1] Part of the reason may be that examiners view an erroneous identification as the most serious mistake they can make and so naturally may be more resistant to influence with that conclusion. The susceptibility to confirming inconclusive or not suitable for comparison conclusions also indicates that examiners adopt a more conservative approach when they assume the role of a verifier (Fraser-Mackenzie, Dror & Wertheim, 2013; Langenburg, Champod & Wertheim, 2009). Further support for this notion can be found in two of the studies in which verifiers did not verify all the correct identifications (Pacheco, Cerchiai & Stoiloff, 2014; Wertheim, Langenburg & Moenssens, 2006).

However, one study did find that knowledge that an examiner *and* another verifier had reached a conclusion of identification did influence verifiers. Pacheco, Cerchiai & Stoiloff (2014) asked

---

1 Wertheim, Langenburg & Moenssens (2006) did report that a verifier in their experiment agreed with one erroneous identification. However this experiment was part of a training course that was open to any examiner and the one in question was a trainee with less than a year of experience who was not performing unsupervised case work in their organisation.

examiners to act as a second verifier for 85 comparisons where the impressions were made by different areas of skin but were presented as if one examiner had identified them as having been made by the same area of skin and a verifier had agreed. Under those conditions, there were three instances of verifiers agreeing with erroneous identifications. However, the findings of the studies indicate that open verification is generally an effective means of detecting erroneous identifications. They also indicate the verifier is more susceptible to influence with other conclusions and that the extent of influence (irrespective of the conclusion) is also crucially dependent on the impressions themselves - verifiers will be less susceptible to influence with impressions revealing clear and abundant detail than they will with complex ones (Langenburg, Champod & Wertheim, 2009). The results of the study where verifiers did verify three erroneous identifications may also indicate the extent of the influence on the verifier depends on its strength and that the knowledge that multiple examiners had reached the same conclusion may be a greater influence than the knowledge that just a single examiner had (Pacheco, Cerchiai & Stoiloff, 2014). The studies indicate that open verification does not amount to ratification as critics have suggested, but as the verifier can be improperly influenced, Langenburg (2012) suggests rather than independent, a better word to describe that type of verification may be separate or subsequent. To address the influence, when performing open verification some verifiers, rather than approaching the comparison with an open mind as the original examiner would have done, consciously attempt to prove that examiner's conclusion wrong.

A2.2   Blind Verification

To address the influence that the original examiner's conclusion can have on the verifier, organisations may utilise 'blind' verification. This involves the verifier having no knowledge or expectation of the original examiner's conclusion.

Blind verification should be more independent than open verification, and though there are no studies that directly compare the two, there is some research to indicate blind verification would detect more errors. Two studies tasked examiners with comparing impressions that, unbeknownst to them, other examiners had already reached a conclusion on (Ulery et al., 2011; Ulery et al., 2012). As in all the open studies involving one verifier, no two examiners both erroneously identified the same impressions. However, in contrast to the open Langenburg (2009) study in which none of the erroneous exclusions was detected, conducting verification in this way would have resulted in the discovery of most, but not all, erroneous exclusions (Ulery et al., 2012).

Though it may detect more errors than open verification, organisations will likely need to balance the use of blind verification with the increased time it requires. Historically many organisations have only routinely verified identifications, so conducting blind verification may double their caseload as rather than verifying only the identified mark, the blind verifier would be required to repeat all the work of the original examiner. Due to the increase in time blind verification requires and the effectiveness of open verification in most scenarios, organisations will typically use open verification in most instances and apply blind verification in circumstances where there is a greater risk of error. For example, OSAC (2019) recommend blind verification should be used in several scenarios including; where only one mark has been identified, and that identification resulted from a search of a computer database, in high profile cases or where the mark or comparison is complex. Rather than only using blind verification for single identifications arising from the search of a database, some organisations such as the FBI will subject all single mark identifications to blind verification (OIG, 2011).

A2.3 The Use of Verification

As well as specifying the type of verification to be used, the organisation that employs the examiner will have procedures detailing which conclusions should be verified. For example, a 2012 survey of 56 organisations (almost all in the United States) found that only 36% verify all conclusions (Black, 2012). OSAC's recommendations for verification state that it should apply to all decisions, including those for suitability, but as a minimum, it shall apply to conclusions of identification, support for same source and exclusion (OSAC, 2019). The survey conducted by Black (2012) indicated that whilst almost every organisation that responded was verifying 100% of conclusions of identification, only 55% were verifying all conclusions of exclusion, only 52% were verifying all inconclusive conclusions, and only 36% were verifying all no value conclusions. Black (2012) and others argue that if all conclusions were verified, then a greater proportion of errors would be detected, but some of the organisations that responded to his survey cited backlogs and a lack of resources as reasons this was not currently being done (Montooth, 2019). Whilst verifying 100% of conclusions may not be practical, many argue that because the most common error is erroneous exclusion, a greater proportion of those conclusions should be verified than are being currently.

As well as applying blind verification in scenario's where there is a greater risk of error, an organisation may also apply more verifications. Whilst there have not been any studies to determine whether multiple verifiers are more likely to detect errors than single verifiers, it seems reasonable that whilst one verifier might repeat the mistake of the original examiner, a second or third verifier would be less likely to repeat it. However, the results of Pacheco, Cerchiai & Stoiloff (2014) suggest that with open verification, if a second verifier is already aware that an examiner and another verifier have already reached the same conclusion, they are more likely to also reach the same conclusion than if they were the first verifier.

In addition to having procedures in place to specify what will be verified, how it will be verified and how many verifications there will be, the organisation may also specify how verifiers are engaged to avoid examiners seeking opinions from potential verifiers and being able to select one who agrees with their conclusion. Such a procedure should also direct that where an examiner has sought advice on a comparison from another examiner, that examiner should not then become a verifier for that comparison (Campbell, 2011). The organisation must also ensure a culture exists that enables verifiers to freely oppose and challenge the conclusion of the original examiner and each other. The inquiry into the erroneous identifications of Shirley McKie and Marion Ross made a number of recommendations relating to procedural measures to ensure the independence of verifiers, including that; 'No discussions should take place between verifiers and preceding examiners until they have completed their work and reached their conclusions' (Campbell, 2011).

Verification procedures should also specify how and to what extent verifiers should document their work and how conflicting conclusions should be documented, resolved and disclosed to the criminal justice system.

A2.4 Technical Review

In addition to verification, many organisations may also employ technical reviews of some or all examiner conclusions. Verification is an effective means of establishing whether a conclusion is reproducible – whether different examiners will arrive at the same conclusion with the same impressions. However, the fact that two or more examiners arrive at the same conclusion independently does not mean that conclusion is correct. For example, Ulery et al. (2012) estimate from their data that whilst blind verification would detect most erroneous exclusions, 19% would be agreed by the verifier. A technical review provides an additional quality control measure as it allows another examiner to scrutinise how the original examiner arrived at their conclusion. It also allows them to consider whether processes were followed, whether all the examiner's interpretations, opinions and decisions that led to their conclusion are appropriate and also affords them the opportunity to attempt to prove the conclusion wrong (Black, 2012; Triplett & Cooney, 2006).

Triplett & Cooney (2006) explain that having a verifier just reproduce a conclusion rather than taking the step of scrutinising how it was arrived at it can be acceptable with many comparisons where the details in the impressions are clear and abundant because the process used to arrive at a conclusion will be self-evident. However, with other comparisons, for example, complex ones, a complete review of how the examiner came to a conclusion may be necessary. To do this requires the reviewer to scrutinise the documentation the original examiner produced to support their conclusion, which could improperly influence them. As a result, organisations may conduct technical reviews separately from verifications or where the verifier is also the technical reviewer, conduct them only after the verifier has reached and documented their own conclusion.

References

Black, J, P 2012, 'Is There a Need for 100% Verification (Review) of Latent Print Examination Conclusions?', *Journal of Forensic Identification*, 62 (1), pp. 80–100.

Campbell, A 2011, *The Fingerprint Inquiry Report*, Available at: http://www. thefingerprintinquiryscotland.org.uk/inquiry/3127-2.html.

Dror, I, E, Charlton, C, Péron, A, E 2006, 'Contextual Information Renders Experts Vulnerable to Making Erroneous Identifications', *Forensic Science International*, 156, pp. 74–78.

Fraser-Mackenzie, P, A, Dror, I, E & Wertheim, K 2013, 'Cognitive and Contextual Influences in Determination of Latent Fingerprint Suitability for Identification Judgments', *Science & Justice*, 53 (2), pp. 144–153.

Haber, L & Haber, R, N 2009, *Challenges to Fingerprints*, Lawyers & Judges Publishing Company, Inc., Tucson, Arizona.

Langenburg, G 2009, 'A Performance Study of the ACE-V Process: A Pilot Study to Measure the Accuracy, Precision, Reproducibility, Repeatability, and Biasability of Conclusions Resulting from the ACE-V Process', *Journal of Forensic Identification*, 59 (2), pp. 219–257.

Langenburg, G 2012, 'A critical analysis and study of the ACE-V process', PhD thesis, University of Lausanne

Langenburg, G, Champod, C & Wertheim, P 2009, 'Testing for Potential Contextual Bias Effects During the Verification Stage of the ACE-V Methodology when Conducting Fingerprint Comparisons*', *Journal of Forensic Science*, 54 (3), pp. 571–582.

Montooth, M, S 2019, 'Errors in Latent Print Casework Found in Technical Reviews', *Journal of Forensic Identification*, 69 (2), pp. 125–140.

OIG 2006, 'A Review of the FBI's Handling of the Brandon Mayfield Case', U.S. Department of Justice, Office of the Inspector General, https://oig.justice.gov/special/s0601/final.pdf.

OIG 2011, A Review of the FBI's Progress in Responding to the Recommendations in the Office of the Inspector General Report on the Fingerprint Misidentification in the Brandon Mayfield Case, U.S. Department of Justice, Office of the Inspector General, June.

OSAC 2019, Best Practice Recommendation for Verification in Friction Ridge Examination, Organization of Scientific Area Committees for Forensic Science, Friction Ridge Subcommittee, Version 1.0, September.

Pacheco, I, Cerchiai, B & Stoiloff, S 2014, 'Miami-Dade Research Study for the Reliability of the ACE-V Process: Accuracy & Precision in Latent Fingerprint Examinations', NCJRS, Doc. No. 248534, Award No. 2010-DN-BX-K268, U.S. Department of Justice, https://www.ncjrs.gov/pdffiles1/nij/grants/248534.pdf.

Triplett, M, Cooney, L 2006, 'The Etiology of ACE-V and its Proper Use: An Exploration of the Relationship Between ACE-V and the Scientific Method of Hypothesis Testing', *Journal of Forensic Identification*, 56 (3), pp. 345–355.

Ulery, B, T, Hicklin, R, A, Buscaglia, J & Roberts, M, A 2011, 'Accuracy and Reliability of Forensic Latent Fingerprint Decisions', *PNAS*, 108 (19), pp. 7733–7738, https://www.pnas.org/content/pnas/108/19/7733.full.pdf.

Ulery, B, T, Hicklin, R, A, Buscaglia, J & Roberts, M, A 2012, 'Repeatability and Reproducibility of Decisions by Latent Fingerprint Examiners', *PLoS ONE* 7 (3), e32800. https://doi.org/10.1371/journal.pone.0032800.

Wertheim, K, Langenburg, G & Moenssens, A 2006, 'A Report of Latent Print Examiner Accuracy During Comparison Training Exercises', *Journal of Forensic Identification*, 56 (1), pp. 55–93.

Appendix 3

Bias

Studies have shown that some examiner's conclusions can be influenced by factors other than the details revealed in the impressions. Dror et al. (2006) describe an experiment in which five examiners were told the impressions they were to compare were those FBI examiners wrongly concluded had been made by the same area of skin in connection with the terrorist attacks in Madrid in 2004. However, the impressions were actually ones the examiners participating in the experiment had identified as having been made by the same area of skin approximately five years earlier. The experiment created a context in which the examiners may expect the result of the comparison to be that the impressions had not been made by the same area of skin, but used impressions that each examiner had previously concluded were made by the same area of skin. Under these conditions Dror et al. found that:

- one examiner maintained their previous conclusion;
- one examiner changed their previous conclusion to inconclusive;
- three examiners changed their conclusion to exclusion.

The conclusion of four of the five examiners appears to have been influenced by the erroneous information that the impressions had not been made by the same area of skin. Risinger et al. (2002) suggest this type of behaviour occurs because '. . .the desires and expectations people possess influence their perceptions and interpretations of what they observe'. This effect has been seen in several other studies as well as being cited as a contributory factor in errors made by examiners. Some of the factors that can create desires and expectations include contextual information and the experience and motivation of the examiner.

A3.1 Contextual Information

The Dror et al. (2006) experiment used an extreme context, but other studies have demonstrated that examiners can also be influenced by the knowledge of other examiner's conclusions under more realistic conditions. Several studies have assessed the influence of one examiner's conclusion over another by presenting examiners with a mark and a print that were not made by the same area of skin with the conclusion of another examiner that they were (Langenburg et al., 2009; Langenburg, 2009; Pacheco, Cerchiai & Stoiloff, 2014; Wertheim et al., 2006). Under those conditions, almost all examiners were not biased into repeating the incorrect conclusion of identification – though one study did find that some examiners were influenced to do so if they thought two other examiners had reached that conclusion before them (Pacheco, Cerchiai &

The Forensic Analysis, Comparison and Evaluation of Friction Ridge Skin Impressions, First Edition. Dan Perkins.
© 2022 John Wiley & Sons Ltd. Published 2022 by John Wiley & Sons Ltd.

Stoiloff, 2014). Though examiners appear to have a strong resistance to the influence of another examiner's conclusion of identification, one study found they were more easily influenced if the previous conclusion was exclusion. Langenburg (2009) found that in his study, whilst none of the erroneous identifications was agreed with, all the erroneous exclusions were agreed with. Another of the studies found that if examiners were aware of previous conclusions of inconclusive, this resulted in more inconclusive conclusions than would be seen if they were not aware of that information (Langenburg et al., 2009). Fraser-Mackenzie et al. (2013) reported a greater likelihood an examiner would judge a mark unsuitable for comparison when they thought another examiner had already reached that conclusion, though a previous conclusion of suitable did not make it more likely they would reach that conclusion. Overall, the studies show that examiners acting as verifiers with the knowledge of a previous conclusion tend to be more conservative and produce less definitive decisions. For example, Wertheim et al. (2006) found that examiners did not agree with all the correct identifications the original examiners made.

Some information about the offence the mark relates to has also been shown to be capable of influencing the examiner's conclusion. Dror & Charlton (2006) tasked six examiners with comparing impressions that, unbeknown to them, they had already compared some years before. The examiners had previously identified half the impressions as having been made by the same area of skin and excluded the others as having not been made by the same area of skin. Before comparing impressions they had previously identified, the examiners were given information that would suggest the donor of the print was not responsible for the offence the mark related to (for example that they had been in custody at the time it was committed). Before comparing impressions they had previously excluded, the examiners were given information that would suggest the donor of the print was responsible for the offence (for example that the donor had confessed to it). Two examiners reached all the same conclusions as they had previously, but the remaining four all reached at least one different conclusion (though for one of the examiners, the only change of conclusion occurred with a mark and print that were used as a control and presented without additional contextual information).

Another study found that laboratory practitioners were more likely to conclude the same mark was suitable for identification when it was presented to them as relating to an offence of murder rather than an offence of theft (Earwaker et al., 2015). A separate study tasked two groups of examiners with comparing a mark with a print and provided one group with the information that the mark related to an offence of forgery and the other an offence of murder (Hall & Player, 2008). Though half the examiners who thought the mark related to the more serious offence reported that their analysis had been affected by that knowledge, the results of the study showed that the conclusions reached by the two groups of examiners were consistent. Similarly, an analysis of actual casework from one organisation also found examiner's conclusions of identification were unaffected by their knowledge of the nature of the offence. Langenburg, Bochet, & Ford (2014) found no noticeable difference in the rate of identifications from marks relating to crimes against people (including homicide, robbery and assault) than for those relating to property crimes (including burglary, theft and vandalism). That same study also found that identification rates were equal in cases where the examiner was exposed to a lot of contextual information about the case and had a lot of contact with those investigating it compared to cases where they had minimal contextual information and little or no contact with investigators (Langenburg, Bochet, & Ford, 2014).

The inquiry into the erroneous identifications made by the Scottish Criminal Record Office in 1997 identified contextual information as a factor that may have contributed to one of the errors (Campbell, 2011). The mark was on a tin that had been found with money in it at the address of a person suspected of involvement in the offence. Examiners had access to the information that a clean area surrounded by dust which was the same size as the tin, had been found at the victim's address. The tin was viewed by the police as evidentially significant as, if it had come from the victims address, then it would provide a link between the suspect and the offence. The inquiry concluded that this knowledge may have contributed to the SCRO examiners wrongly identifying the mark on the tin as having been made by the victim.

A3.2 Experience

As well as knowledge of other examiners conclusions and information about the case, the examiner's experiences can also create expectations. One study showed that the way results of searches of marks through a computer database are presented to an examiner can influence their conclusions. After a mark has been searched, the prints the computer considers most similar to the mark will be presented to the examiner in the form of a list. Usually, the prints considered most similar will be at or near the top of the list to maximise the efficiency of the examiners work, and it is the examiner's experience of this that has been shown to be capable of influencing the examiner. Using medium or low-quality marks, Dror et al. (2012) manipulated the order in which prints were presented to examiners in the list. When prints made by the same area of skin as the marks appeared at or near the top of the list, the examiners were more likely to reach the correct conclusion of identification than when those same prints were at or near the bottom. The experiment also found that examiners spent less time comparing prints at or near the bottom of the list. Additionally, Dror et al. reported that examiners were more likely to wrongly conclude that a print was made by the same area of skin as the mark when the print appeared at or near the top of the list. These results led to Dror et al. to question whether examiners acquire and process the information in different ways when comparing prints in different positions or whether 'expectations of finding a match influence their motivation and drive the time, effort, and attention they dedicate to the comparison?'

The examiner may also have expectations of reaching a conclusion of identification as a result of their experiences as a verifier if the organisation that employs them only requires identifications to be verified.

A3.3 Motivation

The desires or motivation of the examiner and the culture that surrounds them can also influence their decision making. Much of the examiner's work may involve comparing impressions at the request of those who are attempting to secure a conviction. This may result in the examiner viewing the conclusion of identification as being the most desirable outcome of their work. Organisations may further reinforce this idea by recognising or rewarding those examiners who identify the most people or by setting targets to increase identifications. Charlton et al. (2010) report examiners articulating a personal interest in solving crime and catching criminals and feeling a direct link between that and a conclusion of identification. Their study also found examiners experienced a positive emotional feeling when they reached such a conclusion and, to a lesser degree, when they found a small area of detail in agreement before any conclusion was reached. Charlton et al. also found that for some examiners, the positive feelings associated with identifications were greater when the marks related to more serious offences. The authors suggest, therefore, that the desire for those examiners to find an identification may be greater in those cases, though Langenburg,

Bochet, & Ford (2014) and Hall & Player (2008) did not see a significant difference in identification rates of the examiners work they considered as a result of the nature of the offence.

As well as a feeling of reward when identification was made, Charlton et al. (2010) reported examiners also described feelings of fear of making mistakes – particularly the fear of concluding that a mark was made by a particular donor when it was not. In comparison to other mistakes, the consequences of this mistake are usually the most severe, both for the person wrongly identified and the examiner. The authors suggest the examiner could therefore be motivated to be more conservative with their conclusions to avoid it. Langenburg et al. (2009) suggest this motivation may be one reason that explains the results in Langenburg et al. (2009) and Langenburg (2009) that showed verifying examiners were least susceptible to being influenced by the knowledge of another examiners conclusion when that conclusion was identification. Similar results can also be seen in Dror & Charlton (2006), where all changes of examiner conclusion in relation to marks where contextual information was present were changes to an exclusion rather than identification.

A3.4 How Bias Can Affect Decision Making

Research in the field of social hypothesis testing indicates that people:

- Have a tendency to test preconceived ideas or beliefs by looking for evidence that would support them. This would not necessarily mean that opposing evidence would not be seen, but it may be that it would be given less weight, discredited or explained away (Nickerson 1998; Lord et al. 1979; Snyder & Gangestad 1981).
- Tend to interpret evidence in ways that would support their beliefs (Darley & Gross, 1983). This may be related to a tendency to remember the strengths of supporting evidence and weaknesses of opposing evidence (Lord et al. 1979).
- Are more likely to question conflicting evidence than supporting evidence (Nickerson 1998; Lord et al. 1979).
- Generally, will require less evidence that supports their belief to conclude that belief is correct than they would require evidence that does not support their belief to conclude that it is incorrect (Pyszczynski & Greenberg 1987).
- Once they have a belief, can be very resistant to changing that belief – even in the face of compelling evidence that it is wrong. This may be related to a tendency to attach more weight to information gathered earlier in a process than that acquired later (Nickerson 1998; Lord et al. 1979).

If these tendencies also apply to the examiner, they may explain some of the findings of the studies. For example, bias may result in the examiner:

- selectively looking for supporting details rather than objectively evaluating all details that are present;
- being more likely to consider supporting details as true representations of the skin of the donor and explaining opposing details as products of distortion;
- being more likely to interpret ambiguous details as supporting ones;
- utilising lower decision thresholds for supporting conclusions (resulting in them being reached sooner and based on less information).

Whilst there are many potential sources of bias that could influence the examiner's perceptions and interpretation of the impressions in front of them, the presence of one or more of these sources does not mean the examiner's conclusion will be altered from what it would have been had the source been absent. According to Nickerson (1998): 'We can be selective with respect to the evidence we seek, and we can tilt the scales when we weigh what we find, but we cannot completely ignore counter indicative evidence of which we are aware'. For the examiner, this assessment appears to be generally accurate when the impressions reveal clear and abundant detail. In that scenario, the potential effects of any bias will likely be constrained by the appearance of the details revealed in the impressions. For example, the changes of conclusion in Dror et al. (2006) and all but one of the changes of conclusion in Dror & Charlton (2006) involved marks that were regarded as difficult or relatively difficult to compare. Whilst effects were seen with marks that were not considered difficult to compare (Dror & Charlton, 2006 & Dror et al., 2012), it appears the examiner is more vulnerable to the effects of bias when the impressions do not reveal clear and abundant details (Dror et al., 2005; Dror et al., 2006; Dror et al., 2012; Langenburg et al., 2009; Fraser-Mackenzie et al., 2013). The comparison of such impressions is often difficult, and Schiffer & Champod (2007) suggest examiners may be more vulnerable to bias with difficult comparisons because '. . .the tendency to rely on additional, though not necessarily relevant information increases when the data present does not offer enough information for a clear decision. It is then that case relevant and not mark relevant information will be used to reach an opinion'. However, even if the impressions are difficult to compare and there is a source of bias present, the conclusion the examiner reaches may not be different from one they would have reached had the source been absent. For instance, the conclusion of one of the five examiners in Dror et al. (2006) and two of the six examiners in Dror & Charlton (2006) were unchanged. This indicates that some examiners are more able to focus objectively on the details in the impressions than others; the bias may have influenced their decision making but not to the extent that it altered their conclusion – as was reported in Hall & Player (2008). As well as their individual susceptibility to bias, whether an examiner's conclusion is affected is dependent on the strength of the bias and its direction; for example, if the bias is pushing the examiner in the 'correct' direction, then the examiner's conclusion should be no different than it would have been had the influence been absent (Dror et al., 2012).

A3.5   Minimising the Effects of Bias

This first step in minimising the effects involves the examiner acknowledging that their observations and conclusions could be influenced by factors other than the details revealed in impressions. The examiner must also be aware of potential sources of bias, understand the effects they can have and the contexts in which the examiner is most susceptible to them. Awareness will not counteract the effects, so the organisation the examiner is employed by will likely utilise specific measures to limit the influence of bias.

One such measure would be to only provide the examiner with access to information that is relevant to the task. For example, there is no reason for an examiner to know that a person whose prints have been nominated for comparison with a mark has previous convictions for the offence the mark relates to. As that information could influence the examiner's conclusion, the organisation may simply not make it available to the examiner. However, taking the same approach with all non-task relevant information is more difficult. For example, the crime type the mark relates to may be considered as non-task relevant, but some organisations require examiners to prioritise cases according to crime type. Similarly, some large organisations may structure their workforce by grouping examiners in teams that only deal with one crime type or only with a serious crime. As a result, though it may not be task-relevant, some organisations will likely need to balance the changes and additional resources needed to prevent examiners from being aware of the crime type with the effects that awareness may have on their conclusions. Currently, the only research available suggests that knowledge of crime type does not have a significant effect on the examiner's conclusions of identification (Hall & Player, 2008; Langenburg, Bochet, & Ford, 2014). As a result, with some information organisations may focus their attention on measures to prevent access to that information in particular scenarios where there is most risk rather than removing it in every case. For example, verification of examiners conclusions is a crucial quality control measure, and the most common way it is performed involves the verifying examiner having knowledge of the previous examiner's conclusion. Studies show that knowledge can influence the verifier, but to 'blind' the verifier to the original examiner's conclusion in all scenarios would likely double the workload of the organisation. As the studies demonstrated that the verifier is more susceptible to the influence of the original examiner's conclusion in some scenarios than others, rather than blind them to that information in every case, the organisation will balance the risks against the additional time required and likely conduct blind verification only in scenarios where the verifier is most susceptible. As well as being used to prevent a verifier being aware of another examiner's conclusion, the organisation may also use blind verification in other scenarios to prevent a verifier being influenced by information another examiner has been exposed to that could influence them – such as crime type.

If information is considered both task-relevant and capable of influencing the examiner, one possible compromise that has been suggested is to employ a 'sequential unmasking' approach (Dror et al., 2015; Krane et al. 2008). Sequential unmasking allows the examiner to have access to all relevant information, but only when it becomes necessary. For instance, where information is withheld, if the examiner makes changes to their analysis, comparison or evaluation after it is revealed, then they would be required to document those changes. Whilst the approach does not prevent the examiner from being influenced by information, it should at least make any influence more transparent.

To minimise some of the influence of bias that may be present as a result of the examiner's motivation, some have recommended that examiners should not be employed by law enforcement agencies (NRC, 2009). Others have pointed out that whilst such a separation may remove some influences; it would also mean the examiner would be in a new environment with its own influences and argue what is most important is that the examiner is as isolated as possible from such influences (Dror, 2013). To that end, organisations may take measures to physically separate examiners from investigators and limit the amount of contact between them.

Some of the same measures used to minimise bias with contextual information can do the same for bias arising from examiner experiences. For example, introducing verification for conclusions other than identification and doing some of those verifications blind can have an effect on a verifier who has an expectation that everything they verify will be an identification because the organisation only verifies identifications. However, addressing some other biases that arise from experience may be more difficult and outside the immediate control of the organisation. For example, Dror (2013) suggests the respondent lists of prints provided to examiners as a result of a search of a computer database be randomised, which would need to be done by the provider of the AFIS.

References

Campbell, A 2011, *The Fingerprint Inquiry Report*, Available at: http://www. thefingerprintinquiryscotland.org.uk/inquiry/3127-2.html.

Charlton, D, Fraser-Mackenzie, P, A, F & Dror, I, E 2010, 'Emotional Experiences and Motivating Factors Associated with Fingerprint Analysis', *Journal of Forensic Sciences*, 55 (2), pp. 383–393.

Darley, J, M & Gross, P, H 1983, 'A Hypothesis-Confirming Bias in Labelling Effects', *Journal of Personality and Social Psychology*, 44 (1), pp. 20–33.

Dror, I, E 2013, 'Practical Solutions to Cognitive and Human Factor Challenges in Forensic Science', *Forensic Science Policy & Management*, 4 (3–4), pp. 1–9.

Dror, I, E & Charlton, D 2006, 'Why Experts Make Errors', *Journal of Forensic Identification*, 56 (4), pp. 600–616.

Dror, I, E, Charlton, D & Peron, A, E 2006, 'Contextual Information Renders Experts Vulnerable to Making Erroneous Identifications', *Forensic Science International*, 156 (1), pp. 74–78.

Dror, E, Peron, A, E, Hind, S & Charlton, D 2005, 'When Emotions Get the Better of Us: The Effect of Contextual Top-down Processing on Matching Fingerprints', *Applied Cognitive Psychology*, 19 (6), pp. 799–809.

Dror, I, E, Thompson, W, C, Meissner, C, A, Kornfield, I, Krane, D, Saks, M & Risinger, M 2015, 'Context Management Toolbox: A Linear Sequential Unmasking (LSU) Approach for Minimizing Cognitive Bias in Forensic Decision Making', *Journal of Forensic Science*, 60 (4), pp. 1111–1112.

Dror, I, E, Wertheim, K, Fraser-Mackenzie, P & Walajtys, J 2012, 'The Impact of Human-Technology Cooperation and Distributed Cognition in Forensic Science: Biasing Effects of AFIS Contextual Information on Human Experts', *Journal of Forensic Sciences*, 57 (2), pp. 343–352.

Earwaker, H, Morgan, R, M, Harris, A, J, L & Hall, L, J 2015, 'Fingermark Submission Decision-Making within a UK Fingerprint Laboratory: Do Experts Get the Marks That They Need?', *Science and Justice*, 167 (4), pp. 239–247.

Fraser-Mackenzie, P, Dror, I & Wertheim, K 2013, Cognitive and Contextual Influences in Determination of Latent Fingerprint Suitability for Identification Judgments. NCJRS, 241289.

Hall, L, J & Player, E 2008, 'Will the Introduction of an Emotional Context Affect Fingerprint Analysis and Decision-Making?', *Forensic Science International*, 181 (1–3), pp. 36–39.

Krane, D, E, Ford, S, Gilder, J, R, Inman, K, Jamieson, A, Koppl, R, Kornfield, I, L, Risinger, D, M, Rudin, N, Taylor, M, S & Thompson, W, C 2008, 'Sequential Unmasking: A Means of Minimizing Observer Effects in Forensic DNA Interpretation', *Journal of Forensic Science*, 53 (4), pp. 1006–1007.

Langenburg, G 2009, 'A Performance Study of the ACE-V Process: A Pilot Study to Measure the Accuracy, Precision, Reproducibility, Repeatability, and Biasability of Conclusions Resulting from the ACE-V Process', *Journal of Forensic Identification*, 59 (2), pp. 219–257.

Langenburg, G, Bochet, F & Ford, S 2014, 'A Report of Statistics from Latent Print Casework', *Forensic Science Policy & Management: An International Journal*, 5 (1–2), pp. 15–37, doi: https://doi.org/10.1080/19409044.2014.929759.

Langenburg, G, Champod, C & Wertheim, P 2009, 'Testing for Potential Contextual Bias Effects During the Verification Stage of the ACE-V Methodology when Conducting Fingerprint Comparisons', *Journal of Forensic Science*, 54 (3), pp. 571–582.

Lord, C, G, Ross, L & Lepper, M, R 1979, 'Biased Assimilation and Attitude Polarization: The Effects of Prior Theories on Subsequently Considered Evidence', *Journal of Personality and Social Psychology*, 37 (11), pp. 2098–2109.

Nickerson, R, S 1998, 'Confirmation Bias: A Ubiquitous Phenomenon in Many Guises', *Review of General Psychology*, 2 (2). pp. 175–220.

NRC 2009, *Strengthening Forensic Science in the United States: A Path Forward*, National Academy Press, Washington, DC.

Pacheco, I, Cerchiai, B & Stoiloff, S 2014, 'Miami-Dade Research Study for the Reliability of the ACE-V Process: Accuracy & Precision in Latent Fingerprint Examinations', NCJRS, Doc. No. 248534, Award No. 2010-DN-BX-K268, U.S. Department of Justice, https://www.ncjrs.gov/pdffiles1/nij/grants/248534.pdf.

Pyszczynski, T & Greenberg, J 1987, 'Toward an Integration of Cognitive and Motivational Perspectives on Social Inference: A Biased Hypothesis-Testing Model', *Advances in Experimental Social Psychology*, 20, pp. 297–340.

Risinger, D, M, Saks, M, J, Thompson, W, C & Rosenthal, R 2002, 'The Daubert/Kumho Implications of Observer Effects in Forensic Science: Hidden Problems of Expectation and Suggestion', *California Law Review*, 90 (1), pp. 3–56.

Schiffer, B & Champod, C 2007, 'The Potential (Negative) Influence of Observational Biases at the Analysis Stage of Fingermark Individualisation', *Forensic Science International*, 167 (2–3), pp. 116–120.

Snyder, M & Gangestad, S 1981, 'Hypothesis-Testing Processes', in: Harvey, J, H, Ickes, W, Kidd, R, F (eds.). *New Directions in Attribution Research*, Volume 3, Lawrence Erlbaum Associates, Hillsdale, New Jersey.

Wertheim, K, Langenburg, G & Moenssens, A 2006, 'A Report of Latent Print Examiner Accuracy During Comparison Training Exercises', *Journal of Forensic Identification*, 56 (1). pp. 55–93.

Appendix 4

Activity Level Propositions

ACE is the framework used by examiners to consider 'source level' propositions – who was or was not the donor of the mark. However, occasionally the examiner may be asked to consider 'activity level' propositions which often relate to how or when the mark was left (Cook et al., 1998). The examiner's opinion on an activity level proposition will typically be sought as a result of the account provided to police by the person who has been identified as the donor of a mark. For example, the police may believe that a mark on a knife was made as a result of the person stabbing the victim, whereas the person identified may claim it was made while they were rendering aid to the victim. The examiner may then be asked whether the evidence offers more support for one proposition than the other.

For the examiner, considering activity level propositions can be very different from considering source level propositions. For example, the examiner will use ACE and specific organisational procedures to minimise errors when considering source level propositions. They will have received training in what the relevant factors are that need to be considered, and the validity of their conclusions will likely have been tested, verified and reviewed on many occasions. In contrast, the examiner may never have had any training in activity level propositions, may not have a process they can follow and may not be aware of the relevant factors for the proposition in question. Depending on the proposition, this may mean that the examiner is not in a better position than anyone else to offer an opinion on which proposition is more likely. However, as activity level propositions often involve interpretation of the orientation of the mark or the area of skin that made it, examiners frequently do have specialist knowledge, which can be crucial in considering those propositions.

Any opinion the examiner does offer on an activity level proposition must be arrived at carefully by considering both propositions in a balanced way. All relevant factors must be identified and considered, and the basis for the opinion should be documented. The opinion should be given in terms of which proposition there is more support for and should be subject to technical review (Champod et al., 2016; Doubleloop Podcast, 2020).

Though the examiner will usually have considered the propositions at the request of the police or prosecutors, they must be aware that seeing themselves as part of a 'side' can introduce a subconscious bias (Dror, 2015; FSR, 2015). Murrie et al. (2013) paid forensic psychologists and psychiatrists to review the same evidence but instructed some they were working for the prosecution and some for the defence. Though it was not the case with every participant, the authors did find evidence that some participants tended to reach conclusions that supported the side they thought they were on. Though there are differences between the relationship many examiners have with a 'side' and the one the participants in the study had (particularly in terms of payment for services and the prospect of future work), the potential for influence exists.

The Forensic Analysis, Comparison and Evaluation of Friction Ridge Skin Impressions, First Edition. Dan Perkins.
© 2022 John Wiley & Sons Ltd. Published 2022 by John Wiley & Sons Ltd.

This appendix considers some issues that are often central to activity level propositions the examiner may be asked to comment on, beginning with the age of the mark, the circumstances at the time it was made, issues involving marks in blood and what can be said about unidentified marks.

A4.1   How Long Has the Mark Been on the Surface?

The identification of the donor of a mark is extremely strong evidence that the person touched the surface the mark was on. However, sometimes that identification will have little evidential value unless it can be established when the person touched the surface. For instance, in 1980, a man was found dead in the basement of his shop in California (Mikes v. Borg, 947 F.2d 353 (9th Cir. 1991)). The man had been attacked with a turnstile post and his shop burgled. Sixteen identifiable marks were found at the crime scene, and six of them were identified as having been made by the same donor. All those marks were found on turnstile posts, one of which was the one used in the offence. The posts were part of a disassembled turnstile unit bought by the victim from a hardware store approximately four months before the offence and kept in the basement of the shop. The prosecution case against the donor was based on the theory that his marks were made on the posts at the time of the offence. The donor was convicted but appealed, claiming that the marks were made prior to the offence. The Court of Appeal considered that the prosecution needed to present evidence that the posts were inaccessible to the donor prior to the offence. The prosecution argued that the posts were inaccessible as they were in a location the general public and the donor did not have permission to enter – the basement of the victim's shop. However, the Court of Appeal found that the evidence submitted by the prosecution did not preclude the reasonable possibility that the donor's marks had been made on the posts prior to the victim buying them. Specifically, the Court found that the marks could have been made on the posts by any person who; touched them while they were in the hardware store, disassembled the turnstile or sold the turnstile to the victim. The Court of Appeal also considered that, though less likely, the marks could have been made on the posts when the turnstile was last in use, either in the hardware store or elsewhere. As a result, the donor's conviction was overturned.

In order to formulate a response to the question of how long the mark could have been on the surface, the examiner must consider a broad range of interconnected factors that are described in this section[1].

Once a mark is left on a surface, the detail it reveals will begin to deteriorate as a result of a range of chemical, biological and physical processes acting on the material the mark is made in. Whilst it seems logical that an older mark will therefore reveal fewer or less clear details than a newer mark, this is not always the case. One reason for this is that the initial quality of the mark is variable. For example, even a small amount of deterioration to a mark that was poor quality to begin with may render it unidentifiable, whereas a good quality mark may remain identifiable for longer. Second, the rate that the deterioration occurs is not the same for every mark but will instead be determined by the interaction of three factors: the constituents of the mark, the surface it is on, and the environment it is in (Bleay, Croxton & de Puit, 2018a).

---

1 The selection of the treatment or development technique used to visualise the mark could also be considered a factor as it can determine whether or not the mark is found at all. As some treatments are more effective in particular scenarios than others, the use of one may result in nothing being developed whereas the use of another may reveal the mark is still on the surface or may result in a better-quality mark being developed than another. However, as they do not affect how long the mark actually remains on the surface they are not considered in this appendix.

A4.1.1 Constituents of the Mark

The constituents of the mark can broadly be divided into two categories: natural secretions and contaminants.

The natural secretions will include material produced by the eccrine glands. The eccrine glands secrete directly onto the friction ridge skin, and so their product is likely to be present to some degree in every mark. Eccrine secretions mainly consist of water but also contain many other constituents, including acids and salts.

The natural secretions may also include material produced by the sebaceous glands. This material will contain very little water and mainly consists of oily, waxy matter. Though the sebaceous glands do not secrete directly onto the friction ridge skin, their product may be present in the mark as a result of the donor touching areas they do secrete onto, such as their hair or face.

The eccrine and sebaceous glands collectively account for most natural secretions that are found in marks[2], and it has been reported that nearly 99% of a mark made in natural secretions will consist of water (NIJ, 2011). However, Kent (2016) reasons that though this may be true for the secretions of the eccrine glands alone, it is not likely to be representative of the water proportion of the total natural secretions in the mark. This is because the sebaceous secretions contain virtually no water, and much of the water in the eccrine secretions will have evaporated or been absorbed back into the skin before contact is made with a surface. Kent suggests that the water content of the natural secretions in a mark is likely to be no more than 20% and could be considerably less. Irrespective of the precise water content, the natural secretions have been found to contain over 300 compounds, and their composition varies from person to person due to factors including age, sex, health, diet, and medication (Bernier, Booth & Yost, 1999; Bleay, Croxton & de Puit, 2018a; Girod, Ramotowski & Weyermann, 2012; Mong, Petersen & Clauss, 1999; Ramasastry et al., 1970). As well as varying between people, the constituents also vary at different times of the day and between different fingers of the same person (Bleay, Croxton & de Puit, 2018a; Bond, 2008; Cadd et al., 2015; Girod, Ramotowski & Weyermann, 2012). Differences in the constituents can significantly affect how quickly the mark deteriorates as, for example, some constituents may be much more or less resistant to the effects of the environmental conditions around the mark.

In addition to the natural secretions, contaminants on the skin of the donor will likely form part of the mark. These may include anything the donor has come into contact with, such as cosmetics, soap, sunscreen, foodstuffs, dirt, grease and blood. Some of these may also affect how quickly or slowly the mark deteriorates. For instance, over time, the water content of marks may be lost through evaporation, but the oils present in cosmetic creams and hand lotions may slow down this process (Bleay, Croxton & de Puit, 2018a).

As a result of all the potential variability involved, no two marks will have exactly the same chemical composition (Bleay, Croxton & de Puit, 2018a).

The amount of the constituents that are transferred from the skin to the surface is highly variable. For example, the quantity of natural secretions on the skin at the time of deposition will be affected by the extent to which the donor is perspiring, which may vary according to their age, gender, weight and fitness level as well as a result of the ambient temperature and the presence of mental or emotional stimuli. Also, exercise may result in an increase in the amount of natural secretions on the skin, whereas handwashing or contact with other surfaces may result in a decrease.

Other factors that may influence how much material is transferred to a surface include the duration of the contact and the amount of pressure used (Cadd et al., 2015; Girod, Ramotowski & Weyermann, 2012; Fieldhouse, 2015; Jones et al., 2001).

---

2 Secretions from the apocrine glands, located principally in the groin and armpits, may also be present.

A4.1.2 Environmental Conditions

Temperature, wind, rain, snow and condensation are some of the environmental factors that may affect how long a mark can remain on a surface. The effects environmental conditions will have on the mark will vary according to its constituents. For example, in a study spanning several years and involving the examination of 20 000 marks on various surfaces exposed to a range of environmental conditions, Baniuk (1990) found that marks with a considerable amount of sebaceous secretions remained identifiable five times longer than marks consisting of principally of eccrine secretions.

Perhaps the most common environmental effect is the drying out of the material the mark is made in as a result of the loss of water due to evaporation. Mong, Petersen & Clauss (1999) presumed in their experiments it was this factor that resulted in a loss of up to 85% of the weight of the material a mark was made within two weeks of its deposition. The rate of evaporation can be affected by temperature, and in general, elevated temperatures will result in more evaporation and so more rapid loss of the water constituents, whereas lower temperatures will tend to preserve them. Moody (1994) found that whilst marks on plastic sealed in evidence bags deteriorated visibly over two and a half years when kept at room temperature, marks on the same surfaces kept at cooler temperatures for the same period did not visibly deteriorate at all.

Whilst the loss of the water content of the mark may be accelerated by increased temperature, other constituents do not evaporate and have been found to be able to withstand extreme temperatures. Vandiver (1976) reports a mark being found on gasoline can left at the scene of an arson that had been exposed to temperatures sufficient to melt the can's soldered seam. Bradshaw et al. (2008) found that though increased temperatures (and exposure time to those temperatures) reduced the number of marks developed, under laboratory conditions, identifiable marks were developed after exposure to temperatures of up to 600 °C.

In some cases, extreme temperatures at a crime scene can actually assist in the preservation of marks. Harper (1938) found that if the surface became covered in soot before the mark could evaporate, the soot, which can be removed without damaging the mark, protected the mark and allowed it to be developed even after exposure to temperatures that caused permanent damage to the surface. Also, studies have reported that exposure to heat can result in the residue of the mark being 'baked' onto metal surfaces (Bradshaw et al., 2008; Moore et al., 2008). Where this occurs, the resulting mark may be much more robust than it was prior to exposure to heat and, therefore, may be able to remain on the surface for an extended period without significant deterioration. Spawn (1993) describes several excellent quality marks found on a metal light fixture that had been above the point of origin of a fire. The marks were fixed to the surface, and Spawn reports that aggressive rubbing of the marks had no effect on their appearance.

Another common environmental factor that can affect how long a mark remains on a surface is exposure to water. As the secretions of the eccrine glands mainly consist of water or water-soluble constituents, contact with water may result in the loss of those constituents. However, as the secretions of the sebaceous glands contain very little water and are generally not water-soluble, marks rich in those secretions (or in non-water-soluble contaminants) may remain identifiable even after prolonged immersion in water. For instance, under the conditions of their experiment, Sutton et al. (2014) were able to develop identifiable marks on surfaces that had been submerged in natural aquatic environments for 13 weeks. Marks surviving prolonged emersion in water have also been encountered in casework. Batey et al. (1998) describe a mark developed and identified on a plastic bag found wrapped around part of a homicide victim that was estimated to have been in a river for 18 hours before being recovered. In another casework example, a television stolen in a burglary was dumped in a river where it remained in rapidly running water until its discovery over three weeks later. Several palm marks were found on the wooden underside of the television and identified as belonging to a man who admitted committing the offence and dumping the television in the river on the night of the burglary (Devlin, 2011; Vandiver, 1976). In another case, the aquatic environment may have actually preserved a mark in what may be the longest known period of time a mark has survived underwater. In 1993 divers recovered gold coins from the wreck of the S.S. Brother Jonathan that had been sunk off the coast of California in 1865 (Becker, 2009). The coins were covered in corrosion, and after it had been removed, an 'intact and discernible' mark was found on one of them, 128 years after the vessel was sunk.

Marks may also be exposed to or immersed in other liquids that may have a different effect on them than water. Maslanka (2016) immersed sebaceous-rich marks in milk, red wine, soft drink, beer, orange juice and soapy tap water for up to 24 hours. After that time, all the marks were of reasonable to excellent quality except those in the soapy tap water that had begun to degrade after only one hour. Wood & James (2009) report that immersion in petrol in their study resulted in no marks being developed after one hour of submergence. Masterson & Bleay (2021) were able to develop one-day old marks on glass slides after they had been exposed to corrosive substances such as sulfuric acid or potassium hydroxide (chemicals that may be used in 'acid attacks') and Wilkinson et al. (2004) were able to develop marks after exposure to various chemical warfare agents (including sulfur mustard, sarin, lewisite I and dimethyl sulfate).

Other environmental factors which may affect how long a mark remains identifiable on a surface include ambient humidity, air currents, sunlight and exposure to airborne particles (Cadd et al., 2015; De Alcaraz-Fossoul et al., 2015; De Alcaraz-Fossoul et al., 2016; De Alcaraz-Fossoul et al., 2017; De Alcaraz-Fossoul et al., 2019; Girod, Ramotowski & Weyermann, 2012; González Amorós & de Puit 2014).

Under the right conditions, a mark protected from elevated temperatures, rain and deterioration from other environmental factors may remain on a surface indefinitely. For example, Moses (in Moses Daluz, 2015) describes a case where a mark in a safe was identified six years after it was deposited. Batey et al. (1998) reported the development of a 6-year-old mark on a plastic bag that had been used to conceal a homicide victim, and Tweedy (2010) conducted an experiment in which identifiable marks on plastic bags were developed after seven years. The FBI (in Olsen, 1978) reported a case where police were able to develop and lift good quality marks from a metal box that had been in storage for eight years, and Hunter (1997) describes a 25-year-old mark that was developed and identified on a rubber glove used in a homicide. In a more extreme case, Moynihan (1973) reports the discovery of a mark made in mud on a deer antler unearthed from a chalk shaft by the British Museum in 1973. The antler was radio-carbon dated as being up to 5000 years old, and it is believed the mark was made by Neolithic man using the antler as a tool to mine chalk and flint in the shaft.

As a result of the adverse effects that environmental factors can have on a mark, in general, it may be expected that a mark deposited outside may not remain identifiable as long as a mark deposited inside (or in a similarly sheltered environment). Bunter (2014) experimented with eccrine and sebaceous rich marks from the same donor on interior, and exterior surfaces in the United Kingdom. Bunter found that eccrine marks on interior surfaces were identifiable after six months, and some sebaceous rich marks were identifiable after a year. With the exterior tests, the eccrine marks outside were gone within 20 days, and only one of the sebaceous marks was identifiable after 20 days (and none were after 40 days). O'Brien (1984) conducted experiments to establish how long marks would remain on the outside of a back door at the scene of a homicide in New Zealand. Marks were deposited by a range of donors on the inside and outside of the door and were examined over the course of six weeks. After 16 days, none of the marks on the outside of the door remained; however, some marks on the inside of the door were still there after 40 days.

Azoury et al. (2004) deposited marks on different surfaces both inside and outside in two different geographical areas in Israel and found that no identifiable marks were developed on surfaces outside after three months, whilst identifiable marks were developed on all surfaces kept inside after eight months.

However, there have been instances where marks have remained identifiable for significant periods of time despite exposure to the elements. For example, Medlin (in Midkiff, 1993) describes identifiable marks being recovered from glass jars that were found outside in a remote section of a military reservation. A skeleton was found next to the jars, and the marks were identified as having been made by the person despite having been outside for eight months in all types of weather. Greenlees (1994) describes a case where marks on the outside of a window in Australia were identified after having been deposited two years earlier and been exposed to temperatures of up to 40 °C. Davidson (in Girod et al., 2016) reports a good quality mark that survived more than nine years on the external surface of a door exposed to western Australia weather before being developed with powder. Bunter (2017) describes a case where a mark remained on an exterior security gate for 10 years in the United Kingdom before being identified.

One reason some marks may be more able to withstand exposure to the elements than others is the presence, type and proportion of contaminants contained in their constituents. For example, Bunter (2014) experimented with marks made in different contaminants, including linseed oil and material present on the fingers after the donor had eaten sausage and chips. The experiment took place in the United Kingdom and the marks were made outside, on a sheet of glass that was exposed to various weather conditions. Despite this, the marks made in both linseed oil and 'sausage and chips' remained identifiable over 2.5 years after they had been made (Figure A4.1).

A4.1.3   Surface Factors

What the surface is made of, whether it is rough or smooth, porous or non-porous, soft or hard, wet or dry or hot or cold can all affect how long a mark can remain identifiable on it. De Alcaraz-Fossoul et al. (2013, 2015, 2016, 2017 & 2019) observed marks on plastic deteriorating much more quickly than those on glass and Baniuk (1990) found that marks survived for longer on smooth surfaces such as glass, metals and china as opposed to rough surfaces (Figure A4.2).

Sometimes some of the constituents of the mark will chemically react with the surface to produce a deposit that is far more durable than the original mark. For example, the salts present in marks can corrode metal surfaces to produce what are known as 'etched' marks (Lind, 1972). The corrosion may result in either pits being etched into the surface or the product of the corrosion building up on the surface (Bleay, Croxton & de Puit, 2018a). Sometimes the corrosion will result in a visible mark, but sometimes it will only be possible to detect the mark with the use of specialised techniques (Bond, 2009).

Cohen et al. (2012a) describe a case in Israel in which marks deposited years previously had also reacted with a surface. A burglary occurred at a college, and marks on an interior painted aluminium window frame were found, developed with powder and lifted. The marks were identified and when questioned, the donor denied any involvement in the offence. He was, however, the owner of a window factory and was able to supply evidence that his factory had supplied the college windows more than two and a half years before. Cohen et al. visited the scene and observed that the material the marks were made in appeared to have been 'fixed' to the frame. Furthermore, when they experimented with making fresh marks, those also adhered strongly to the frame. Cohen et al. found that these marks could not be erased by abrasion with a brush or finger in a way that would normally remove marks. They also observed that the same marks were robust enough to be developed again and again. Based on the suspect's testimony and the nature of the marks, the decision was taken not

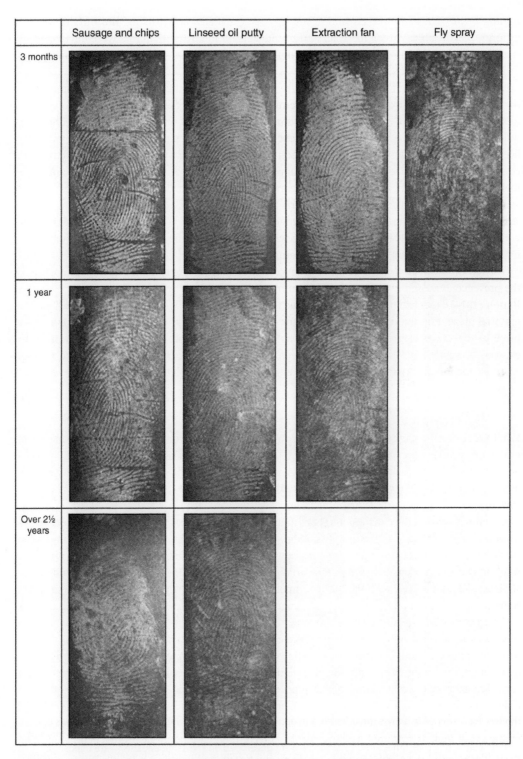

Figure A4.1   The marks that remained identifiable in the different contaminants after periods of up to 2.5 years on exterior surfaces. *Source:* Reprinted with permission from Bunter (2014).

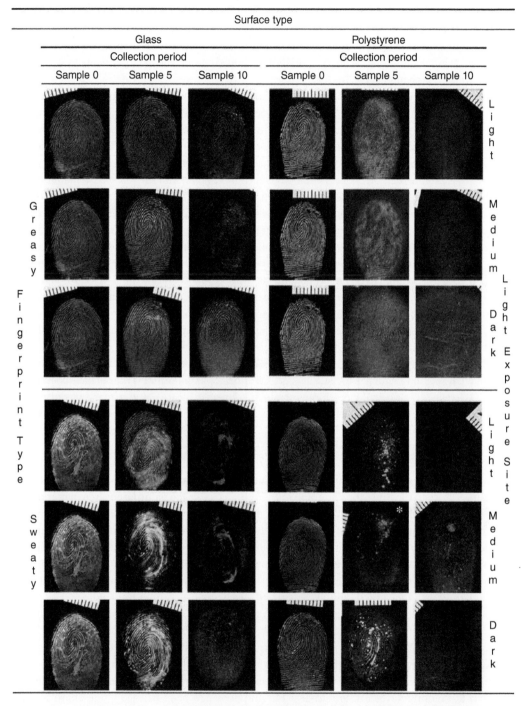

Figure A4.2   This table shows the difference in appearance between 'sweaty' (eccrine-rich) or 'greasy' (sebaceous-rich) marks made by the same donor on either glass or polystyrene when exposed to different lighting conditions over time. 'SAMPLE 0' marks were developed on the same day they were deposited, 'SAMPLE 5' marks were developed after seven weeks, and 'SAMPLE 10' marks were developed after 24 weeks. *Source:* Reprinted with permission from De Alcaraz-Fossoul et al. (2013).

to prosecute the man. To investigate the 'fixing' effect, Cohen et al. conducted a series of laboratory experiments and concluded that the effect was the result of interaction between a sebaceous element of the residue and the surface. Other reactions between a surface and the constituents of the mark can be caused by contaminants such as insect repellants that are capable of dissolving varnish on some surfaces (Ashbaugh, 1999).

Because some constituents of marks will be absorbed into porous surfaces, marks made on those surfaces may be less vulnerable to being rubbed off by contact or to deterioration by environmental factors than marks made on non-porous surfaces. Rimmer (in Bradshaw et al., 2008) reported results of a study on Molotov cocktails that found whilst long-term immersion in fuel tended to destroy the marks on the glass, the marks on the porous label were still able to be developed. The protection offered by porous surfaces is also demonstrated by numerous accounts of marks on such surfaces remaining identifiable for decades. Beaudoin (2011) reports the development of twenty-one-year-old marks on paper recovered during the investigation of a homicide. Involdstad (1978) describes a case where a 30-year-old mark on the page of a book was used to identify the body of a Norwegian sailor who died in police custody in Australia. Hazen (1984) reports the FBI developing a 40-year-old mark on a postcard that was used to identify a World War II collaborator, and the UK Home Office (2003) described developing potentially identifiable marks on documents over 50 years old. Bouzin et al. (2020) and Bleay et al. (2018b) were able to develop marks on 80-year-old and 90-year-old documents, respectively, and Wertheim (2003) reports marks being developed on documents known to be over 100 years old.

If the surface is soft, rather than being made as a result of a transfer of material from skin to surface, marks can also be made as a result of the skin being impressed into the surface. As they become part of the surface, these indented (or 'plastic') marks may then be as resistant to environmental deterioration factors as the surface itself. For example, identifiable marks have been found on 10 000-year-old clay artifacts ('Fingerprints at Boncuklu', 2014), and a mark has also been observed on one of the world's oldest known ceramics ('The Venus of Dolní Věstonice I' clay figurine) which has been dated as having been fired between 27 and 24 thousand years ago (Králík, Novotny & Oliva, 2002).

Depending on the conditions they are exposed to, indented marks can also survive for significant periods of time as part of less robust surfaces than ceramics. Lambourne (1984) examined Nicolas Poussin's 'Dance to the Music of Time' and found the surface of the painting covered in marks. The marks, which may have belonged to Poussin or one of his assistants, were made in the primer at the time it was applied almost 400 years ago. In 2003, a 2000-year-old mark in cream was found in a sealed container unearthed during the excavation of a Roman temple in London ('Roman fingerprints', 2003). Marks have also been found in plaster on the walls of houses excavated in Iran that are over 3200 years old (CAIS, 2005), and Koller, Baumer & Mania (2001) report the discovery of a mark indented in a type of glue that has been dated as being over 80 000 years old.

Often it may be obvious that a mark has been indented into a surface, but with some marks, it may be more difficult to tell. Bunter (2007) describes a case involving the identification of two marks on an internal door at the scene of a burglary at a sports club. When questioned the man identified denied involvement in the offence but stated that he used to frequent the club some four years before the offence took place. The police established that, as well as being cleaned several times in the months prior to the burglary, and the door had been painted white approximately 18 months before the offence – which seemed to suggest that the marks were unlikely to have been made four years earlier. Bunter was employed to review the fingerprint evidence, and when he examined the door, he observed that the marks were still visible and could not be removed by rubbing them with his finger. Bunter concluded that this was an indication that rather than leaving the marks at the time of the burglary, the man may have made them in the paint when it was still wet – although this also did not fit with the man's account that he had not been inside the club for four years. Bunter attempted to lift the marks himself, and as the tape was removed from the surface, some of the white paint came away with it, uncovering a creamy yellow layer of paint that also revealed the same marks but in much greater detail – suggesting it was, in fact, this layer of paint that the man had touched before it had dried. When presented with these findings, the prosecution decided not to proceed with the case.

The nature of the surface or where the mark is on the surface, can also affect how long it remains in an identifiable condition. For example, some marks on non-porous surfaces may be quite fragile and easily rubbed off by contact, so if they are left on surfaces, such as door handles that are routinely contacted, they may not remain in place very long. Alternatively, marks may remain in place longer on other surfaces that are less frequently contacted or those that are sheltered from exposure to environmental factors.

The nature of the surface may also be significant as, for example, some surfaces are more likely to be cleaned frequently than others. Sampson & Moffett (1994) describe a case in which a woman was found murdered on the tiled floor of her bedroom. A palm mark located on the tile adjacent to the victim was identified and, based on the mark and the fact that property belonging to the victim was found in his premises, the donor was arrested and charged. The man's defence was that he had attended the apartment in the capacity of a handyman some six months before the murder and left the mark at that time. However, several witnesses gave evidence that the victim was an extremely fastidious housekeeper and mopped the floors of the apartment at least twice a week. As a result, the police wanted to know whether the mark on the tile could have remained on the floor if it had been moped. To answer that question Sampson & Moffett conducted experiments over the course of a year using marks they deposited on the same type of tile under similar temperature and humidity conditions to those in the apartment to see whether a mark could be developed after being 'damp wiped' by various methods. They found that marks they left and did not wipe remained intact up to a year after deposition on the tile, but that all marks that were wiped were destroyed.

O'Brien (1984) reports a case in New Zealand where cleaning was also relevant. Marks were found in 71 different locations at the scene of a homicide, and those in eight of the locations were identified as having been made by a man who had previously lived in the premises. The man claimed that he had not been back to the premises since he left six months before the offence took place. The tenant who occupied the premises after the man left was particularly house-proud and was able to provide specific evidence on the areas she cleaned when she moved in and moved out. Examiners compared all of the marks with every person who had lived in the house or was known to have been there between the time the man lived there and the time of the offence. There were no marks identified for any of the people who lived or visited the house before the house-proud tenant lived there – all the marks were identified for people who had been in the house after the woman had moved out. So, the chances of only the man's marks surviving, in eight different locations when none of his friends or families had, was considered almost nil.

However, the examiner should not consider that marks could not remain identifiable on a surface that has been cleaned. Etched marks on metal surfaces may be unlikely to be removed by cleaning, and Cohen et al. (2012b) experimented with different cleaning products on the surfaces they had already established marks could be 'fixed' to and reported that under the conditions they tested, most of the products did not remove the marks. Cohen et al. report that the concentration of detergent in a cleaning product is a major factor when considering whether it can remove a mark that is 'fixed' to the surface. As well as the cleaning agent and the concentration of detergent in it, other factors that may affect whether a mark can remain on a surface after that surface has been cleaned include the constituents of the mark and how thorough the cleaning was. For example, McRoberts

(1994) describes a case in which the same premises was burgled twice within three months. During the second examination of the premises, the same mark that had been powdered and lifted in connection with the first offence was powdered and lifted a second time even though the victim had washed the exterior window it was on by spraying it with a water hose. Hill (2008) also reported a similar case in which the same mark was lifted a second time after cleaning had taken place. Hoye (1977) reports marks being left on a milk bottle during a burglary which were developed and identified as having been made by the burglar despite the bottle being washed-up after the offence. Bleay et al. (2019) studied marks on metal surfaces that had been exposed to either washing and scrubbing in a mixture of water and detergent using a nylon-bristled brush or exposed to a solvent (acetone) which had been rubbed vigorously across the surface with a paper tissue and reported that despite the aggressive nature of the cleaning method marks may still be developed.

The temperature of the surface can also be a factor in how long a mark remains identifiable on it. A surface that becomes hot may cause rapid evaporation and drying out of the mark, though on some surfaces, even if the mark itself is destroyed by extreme heat, it may be possible to identify its donor. Moore et al. (2008) exposed marks in blood on ceramic tiles to 900 °C for 15 min, and though the material the marks were made in was burned off, identifiable marks were still developed. It is thought that this was possible as a result of the protection provided by the mark to the surface underneath it before it was burned off. This is believed to have caused chemical differences between the surface underneath the mark and its surroundings that can be detected by a process known as Vacuum Metal Deposition, which allows the mark to be visualised.

The temperature of the surface can also be a factor as, whilst some constituents of the mark will adhere better to a surface that is cooler than the human body, such conditions may also result in the formation of condensation, which may affect the water-soluble constituents (Champod et al., 2016; Johnson, 1973). Some surfaces, such as chocolate, may also be melted by contact with the skin at a higher temperature and result in an indented mark.

The presence of contaminants or coatings on a surface can also affect how long a mark will remain on it. For example, though they are designed to prevent marks being left on the surfaces they cover, Forchelet & Bécue (2018) observed in their experiments that anti-fingerprint coatings may benefit the longevity of marks left on them by preserving the sharpness of the ridges. Also, rather than a mark being created as a result of material being transferred from the skin to the surface, if a surface is covered in dirt or dust, contact with skin may result in the ridges removing that material which can create a mark that is particularly fragile.

A wet surface may make it more difficult for constituents of the mark to adhere to it than a dry one (Johnson, 1973), though Castelló, Francés & Verdú (2013) were able to develop good quality marks two weeks after they were deposited on surfaces that were underwater at the time.

A4.1.4 Conclusion

Broadly, three types of approaches have been proposed or used to provide an estimate of how long a mark has been on a surface.

Examiner's use of the first approach, involving the consideration of the appearance of the mark and/or how it reacts with the development media, has been widely criticised (Bunter, 2015; Girod et al., 2016; Wertheim, 2003). In general, as its constituents deteriorate over time, the quality of a mark lessens, and studies have observed changes in appearance, including the ridges becoming thinner, fainter and discontinuous (Baniuk, 1990; Barros, Faria & Kuckelhaus, 2013; De Alcaraz-Fossoul et al., 2015; De Alcaraz-Fossoul et al., 2016; De Alcaraz-Fossoul et al., 2017; De Alcaraz-Fossoul et al., 2019; Popa, Potorac & Preda, 2010). However, whilst it makes sense for recently deposited marks to exhibit better quality than marks that have been on a surface for an extended period – this is not always the case. For example, faint marks with thin, discontinuous ridges may be found that have only recently been deposited as a result of there being very little material on the donor's skin at the time the surface was contacted. Also, the changes in appearance that have been observed are not only dependent on the amount of time the mark has been on the surface but also on its constituents and the environmental and surface factors. Therefore, even if the quality of the mark was good at the time of deposition, it may have deteriorated much more rapidly than expected as a result of those factors to produce a mark that appears to have been on the surface much longer than it has been. Conversely, the same factors can result in a mark that has been on a surface for a long time having the appearance of a freshly deposited mark (Figures A4.3 and A4.4).

(a)   (b)   (c)

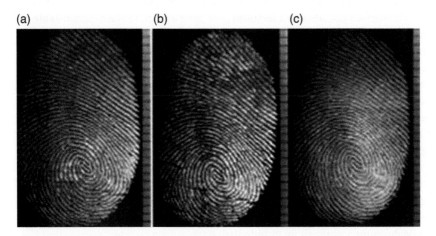

Figure A4.3 These marks were made in natural secretions on glass slides and stored in a box for varying periods of time before being photographed. There is no significant difference in the quality of the marks despite; (a) being photographed 10 minutes after deposition, (b) being photographed one day after deposition and (c) being photographed five weeks after deposition. *Source:* Reprinted with permission from Girod et al. (2016).

(a)   (b)   (c)

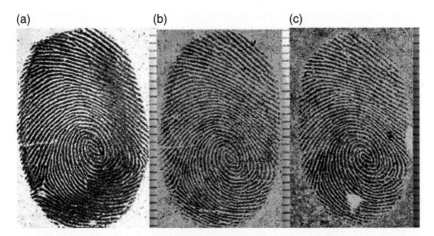

Figure A4.4 These marks were made in natural secretions on glass slides and stored in a box for varying periods of time before powdered and lifted. Whilst the amount of background interference increased over time which decreased the contrast of the mark with the surface, no significant difference can be observed in the quality of the marks despite; (a) being powdered 10 minutes after deposition, (b) being powdered one hour after deposition and (c) being powdered six weeks after deposition. *Source:* Reprinted with permission from Girod et al. (2016).

As a result, there are many examples from studies and casework of marks known to have been on surfaces for different amounts of time that appear similarly or marks that have the opposite appearance of what may be expected. For example, Baniuk (1990) describes fresh marks on a glass surface that appear to be 'old' as a result of exposure to temperatures of 200 °C. Balloch (1977) lifted marks from the interior surface of a window each month for three months and, when the marks were compared, reported it was impossible to determine visually which had been on the surface the longest. Belcher (1982) experimented with marks on bottles and found that no difference could be detected in appearance in powdered marks that had been recently deposited or had been on the surface for three months, similarly Barnett & Berger (1977) found that it was not possible to determine that a mark was fresh or several weeks old by examining it. In a casework example, McRoberts (1994) reports that a second lift of the same mark from an exterior window three months after the first resulted in a mark of better quality than the original. De Alcaraz-Fossoul et al. (2015) described marks on glass that appeared 'fresh' and exhibited no detectable degradation after being on the surface for six months. Azoury et al. (2004) found that in experiments they conducted, no correlation could be found between the amount of time a mark had been on a surface and its appearance. The authors report that some of the marks that had been on surfaces kept inside for nine months were of high clarity and contrast. In a further experiment, the authors tasked twenty experienced fingerprint technicians with estimating the age of the developed marks based on their appearance and found that they only got it right for around 20% of them. Moody (1994) observed that marks made on plastic that was then sealed in plastic evidence bags for 2.5 years were indistinguishable from those made on the same surface and sealed for four months. In another casework example, Greenlees (1994) references a burglary in Australia where two 'good quality' marks were developed on the outside glass surface of a window that was under an open-ended carport. The marks were identified as having been made by a police officer who initially became a suspect in the offence as he denied ever having been to the address. Records later revealed that the officer had been to the address two years earlier to assist the owner, who had locked himself out of the property. The officer then did recall attending the address and remembered pushing against all the windows to gain entry. The window was re-examined, and further marks were found and identified as belonging to a police cadet who, records showed, was with the officer the day he attempted to gain entry to the premises. The examiners who looked at the marks initially believed that they were unlikely to have been there for two years. One of the reasons for this view was their observation that the marks were of good quality and showed little or no fragmentation of ridge detail. However, the marks had actually been on the window through two summers with temperatures up to 40 °C and winters with temperatures dropping to 1 or 2 °C.

Because the appearance of the mark can be affected by factors other than how long it has been on a surface, its appearance alone is not a reliable means of estimating how long it has been on a surface. However, there have been several attempts to develop a more scientific age estimation method based on the appearance of the mark. For example, in Poland, a database of 20 000 marks deposited both inside and outside on smooth, non-porous surfaces and subjected to various environmental conditions over time has been used by examiners to gain experience of the changes in the appearance of marks (Baniuk, 1990; Girod et al., 2016; Howorka, 1989). When an estimation of how long a mark has been on a surface needs to be made, new marks from the person identified in a particular case are made on a corresponding surface and are subjected to similar environmental conditions over time based on information recorded at the time the mark in question was found (Girod et al., 2016). Images of these marks taken at different time periods are then compared with the mark in question, and a probabilistic estimation of how long the mark has been on the surface is provided by the examiner based on their knowledge and statistics from the database (Girod et al., 2016). However, the approach has limitations in that it does not account for variations within the constituents of the donors marks, can only be used for marks found on smooth, non-porous surfaces and has only been tested by one organisation (Cadd et al., 2015). Other studies have also observed changes in the appearance of marks over time and attempted to take into account some of the other factors that can affect their appearance, though all require further work and testing (Barros, Faria & Kuckelhaus, 2013; De Alcaraz-Fossoul et al., 2015; De Alcaraz-Fossoul et al., 2016; De Alcaraz-Fossoul et al., 2017; De Alcaraz-Fossoul et al., 2019; Faria & Kuckelhaus, 2013; Popa, Potorac & Preda, 2010).

The other aspect of this approach involves considering the way the mark reacts to the development media and is based on the idea that older marks are more difficult to develop or enhance than more recently deposited marks. Many marks are developed by the application of powders that principally adhere to the liquid content of the mark and as that content will generally be lost over time, it makes sense that older marks would not accept powder as readily as fresh marks. However, if at the time it is deposited, the mark is not rich in liquid content or is exposed to an environmental factor that accelerates the loss of liquid, then a recently deposited mark may also not readily accept powder. Conversely, some marks that have been on a surface for an extended period may accept powder readily. For example, Azoury et al. (2004) found some nine-month-old marks accepted powder like fresh marks did in their experiments. Similarly, Barnett & Berger (1977) found that it was not possible to differentiate between fresh marks and those that are several weeks old based on how they develop when the powder is applied. Further evidence of the ability of some marks to accept powder despite having lost some of their liquid content comes from several reported instances of the same mark being powdered, lifted and then powdered and lifted again as a result of a second burglary occurring at the same premises at a later date (Clements, 1986; Hill, 2008; McRoberts, 1994). In the case, Clements (1986) describes, even after being powdered and lifted once, the mark remained on the surface for another year before it was powdered, lifted and identified again for the same person (who was still in prison from the first burglary at the time of the second). Illsley (1984) was able to powder and lift the same marks developed with cyanoacrylate (super glue) 559 times before the marks became unidentifiable. Also, though the mark may be expected to dry out over time, Perkins (2016) observed marks made in a contaminant that may be commonly found on the skin (a liquid bandage) remaining wet and able to be smeared by a brush 17 days after deposition on an outside glass surface and 42 days after deposition on an interior glass surface. Similarly, Bunter (2014) reported marks made after the donor had eaten crisps were still wet after being outside for 20 days.

The deterioration of a mark over time may also make it more difficult to lift from a surface. However, some marks may lift as easily as fresh marks even though they have been on the surface for a long time. For example, Moses (in Moses Daluz, 2015) describes a mark found inside a safe during the examination of a burglary. The mark looked 'fresh' and lifted off the surface easily even though it had been made by a person who had died six years before the mark was discovered.

In summary, whilst the quality of many marks may deteriorate significantly over time, the quality of others may not, and so the conclusion cannot be drawn that because a mark is poor quality, it is 'old' or that because a mark is good quality, it is 'fresh'. Similarly, older marks may be more difficult to develop or enhance, but the fact that a mark was difficult to develop or enhance does not mean it is 'old' or that marks that are easy to enhance are 'fresh'.

The second type of approach used to estimate how long a mark may have been on a surface involves the use of well-designed experiments to put limits on how long the mark may have been there. Typically, the experiments make use of a variety of donors to leave marks on the same or similar surfaces as the mark in question that are then exposed to the same or similar environmental factors to see how long the marks can remain identifiable. To produce a reliable estimate, care must be taken not only to mirror the circumstances of the case in question but also, as far as it is possible, to consider the unknown variables such as the constituents of the mark at the time of deposition. For example,

Schwabenland (1992) conducted an experiment to establish how long a mark found on an aluminium can at the scene of a double homicide in California could have been there. The can was found in a field near the two victims who it was established had died within the last 24 hours. The person identified admitted being at the scene but claimed he was there one week earlier. The experiment considered the time the mark was suggested to have been deposited on the can by the donor and took account of the environmental conditions over that period. Schwabenland found that marks would only have lasted on the surface for between 24 and 48 hours rather than the week the donor claimed. However, the experiment was criticised because it was conducted under the premise that the constituents of the material the mark was made in were natural secretions and a particular contaminant (the donor claimed to have been eating fired chicken at the time the mark was made). McRoberts & Kuhn (1992) point out that only one donor was used to leave the test marks, and the constituents of their natural secretions may have been very different from those produced by the donor of the mark in question. Similarly, McRoberts & Kuhn also criticised the experiment for not testing marks with different balances of eccrine and sebaceous secretions or marks made in contaminants other than fried chicken. McRoberts & Kuhn (1992) caution that if this approach is used to provide an estimate of the amount of time a mark may have been on a surface, the experiments used must 'consider all relevant factors, follow scientific protocols, and limit conclusions to the extent supported by the study'.

The third type of approach involves the analytical study of specific constituents of the mark. Over several decades various methods have been proposed, most involving the measurement of chemical changes in the constituents of the mark over time (Cadd et al., 2015; Dikshitulu et al., 1986; Duff & Menzel, 1978). This approach suffers from the fact that chemical changes are highly dependent on environmental factors, and to date, no constituent has been discovered that is present in most marks, deteriorates over time in a predictable way and is not significantly affected by environmental conditions.

Muramoto & Sisco (2015) devised an alternative approach involving the measurement of the movement of particular constituents in a mark over time. The authors found that over time material from the ridges diffuses into the furrows and that this movement occurs at a predictable rate. By measuring the extent of the diffusion, the authors were able to reliably estimate the age of marks made on a clean, non-porous surface. However, at the moment, the approach has not been tested on different surfaces under different environmental conditions and is only effective on marks less than four days old.

A further approach involving the measurement of the reduction in the electrical charge of the mark has been developed but also requires further testing with marks on different surfaces and under different environmental conditions (Watson et al., 2011).

In conclusion, how long a mark can remain identifiable on a surface is dependent on the interaction of a wide variety of factors and whilst some marks may only stay on the surface for a short period of time, some may remain indefinitely. Though the surface the mark is on is known and in some cases, for example, with marks found indoors, the precise environmental conditions may also be known, the quality, composition and amount of material that makes up the mark at the time it was left will not be. Therefore, the precise effects that the surface and environmental factors will have had on the mark are also unknown. Research is ongoing to test existing, and develop new approaches that may lead to a viable method that can be implemented in the future but currently, there is no validated and accepted method to determine how long a mark has been on a surface.

A4.1.5 Circumstantial 'Ageing' of Marks

Whilst it is not possible to scientifically determine how long a mark has been on a surface, in some cases, the examiner may be able to establish that the mark was most likely to have been made before or after a certain event. For example, if a mark is developed on a newspaper, then it is unlikely that mark was made on the paper prior to the date on the paper. Similarly, if an object the mark is found on has a date of manufacture on it then that date likely provides an upper limit for the age of the mark. Cowger (1993) provides an example involving a piece of paper belonging to the complainant of a crime that they purchased in a sealed package and opened two days before the offence. If the mark in question is found on the paper, then, unless the donor could have touched the paper prior to it being packaged, the mark is likely to be no more than two days old. If a mark is made on, or extends onto, the broken edges of a piece of glass, then the mark must have been made after the glass was broken. Olsen (1978) describes an instance where the time an injury occurred to the donor was known, and so the appearance of a scar caused by that injury in a mark demonstrated that the mark was made after the injury took place. There are also several studies that have established it may be possible to determine whether a mark found overlapping ink on a document was made before or after that document was printed on with ink (Attard-Montalto et al., 2014; Bright et al., 2012; Montalto, Ojeda & Jones, 2013). This may be pertinent to an investigation in that it could show whether a piece of paper was blank when the donor handled it or whether it was endorsed with incriminating material.

The following case studies detail examples where examiners were able to reach an opinion about whether a mark was made before, after or at the same time as a certain event by considering the specific circumstances of the case.

Ellis (2005) reports an instance of a palm mark being found on the inside of a window that was broken and used as the entry and exit point to facilitate a burglary. The mark was in a position that was consistent with where a person's hand may be placed when climbing out of the premises. However, the premises was a service station and the marks position was also consistent with it being left legitimately by the donor when he was a customer. Therefore, the evidential value of the mark was dependent on whether it was left before or after the glass was broken. Ellis observed several features within the mark that allowed him to argue that it was left after the glass was broken and not before.

First, where the mark meets the broken edge of the glass, the powder that has developed the ridges has also adhered to the edge of the glass resulting in a dark line. Ellis also observed that the ridges in the mark became wider adjacent to the broken edge (indicating heavier downward pressure in that area) and that some of those ridges curved towards each other forming the appearance of a recurve close to the edge. To ascertain whether his observations could be used to indicate whether the mark was made before or after the glass was broken Ellis conducted experiments involving the deposition of marks on broken glass, and on glass that was subsequently broken.

Ellis was not able to replicate the effects he had seen in the mark in question in any of the tests he did where the glass was broken after the mark was deposited, but he did see them in the experiments where the mark was deposited after the glass was broken. Ellis ultimately attributed his observations about the ridge flow adjacent to the broken edge to differences in downward pressure caused by the hand of the donor being partially off the edge of the glass. In regard to the black line, Ellis suggests that as it does not extend beyond the mark in either direction, it was likely created by powder adhering to material that was left by the donor's palm as it overlapped the broken edge. Ellis argues that had the mark been made while the glass was intact, there would be no reason for this line to be present as each ridge would simply run off the broken edge (as was seen in his experiments with marks made on glass before it was broken).

Saunders (1993) describes a case involving superimposed marks in which it was possible to offer an opinion as to which mark had been made first. Two palm marks were found on a safe at the scene of commercial burglary and had each been identified as having been made by a different suspect in the offence. One suspect was an employee of the company that had been burgled who denied involvement in the offence but stated he had been required to move the safe as part of his duties with the company. If that were true, and assuming the other mark was made at the time of the offence, then the employee's mark should be 'under' the other mark. After examining the marks and conducting numerous tests, the FBI were able to determine that the employee's mark had been deposited first. The principal support for this opinion came from the examination of superimposed marks deposited on several nonporous surfaces and developed with powder or cyanoacrylate (superglue). The FBI found that under those conditions, whilst the ridges of a mark that was deposited last would be continuous over the ridges of the earlier mark, under high magnification, it could be seen that the ridges of the earlier mark would end when they met a mark deposited over them. As a result of the FBI finding in this case, the charges were dropped against the employee and the other suspect pled guilty.

The London Metropolitan Police (MPS) were able to show that a palm mark supported the prosecution case that it was made during, rather than prior to an offence as the donor of the mark claimed. In 2007 two men robbed a bank in south London with one pointing a firearm at the staff. During the robbery, one of the staff activated the counter security screen, which struck the arm of the man as he was reaching over the counter. The men abandoned the robbery at that point and exited the bank. Approximately two months later, an engineer attended the bank as a result of a defect with the security screen. The engineer opened the secure cabinet that housed the mechanism and discovered most of the missing money and the firearm, both of which had fallen inside unnoticed when the robber was struck by the screen.

A palm mark and DNA were found on the firearm, and both were identified as having been made by the same person.

When questioned about the evidence, the man identified denied being involved in the robbery. He did not dispute that he touched the firearm but claimed he was offered it before the offence took place, and he touched it when he 'pushed it away' from him to indicate he did not want it.

The Crown Prosecution Service (CPS) reviewed the evidence (including CCTV footage that was considered weak and of limited value) and proposed that proceedings against the individual be discontinued as there was not a realistic prospect of obtaining a conviction. The CPS view was that the palm mark and DNA evidence showed that the individual had touched the firearm at some time (which he admitted) but did not place him within the bank. The CPS asked if there was any additional evidence that could be considered.

The palm mark had been identified around the grip of the firearm, and the examiner was asked if they could comment on whether the mark had been disturbed by subsequent touching of the firearm after it had been deposited. After studying the image of the mark, the examiner observed that the mark was very clear and that there were no other marks superimposed over the top of it, nor was any part of it distorted or smudged. After studying the firearm, the examiner observed that the mark was not etched into the metal of the grip and was in a position and orientation that was consistent with holding the firearm as if to point it at someone. The examiner also observed two additional, previously undiscovered marks on the barrel of the firearm. Both of which were subsequently identified as having been made by the donor of the palm mark.

As a result of their examination of the firearm, the examiner was able to offer the opinion that the last person to have handled the grip area of the firearm was the donor of the mark.

Another MPS case involved the murder of two students in their flat in London in 2008. The examination of the scene took several days, involved numerous Crime Scene Examiners (CSEs) and resulted in the discovery of many marks. One of which was a palm mark on the frame of a bed that was identified for a man who, when interviewed, claimed that he had burgled the flat but had nothing to do with the murders. The mark had been found as a result of a light source examination, and the CSE who found it documented its location on their examination report. Finger marks in blood were also found on the same surface by a different CSE during a separate examination and documented on their examination report. The finger marks did not contain enough detail to allow them to be identified or even compared with the prints of the donor of the palm mark, but the examiner was able to use the descriptions provided by the CSEs to establish that the marks were in close proximity to, and directly above the palm mark.

The donor of the palm mark was charged with the offence, and the examiner was asked to comment on whether the finger marks in blood could have been made by the same person and at the same time as the palm mark. The examiner attended the scene to look at the bed frame and took with them an outline of the shape and size of the hand of the individual whose palm mark had been identified and found that this overlay fitted over the finger and palm marks. The examiner ultimately produced a statement to say that in their opinion, the marks in blood were consistent with being finger marks, and because of their shape and orientation and their spatial relationships to each other and the palm mark, they were likely to have been made at the same time and by the same person as the palm mark. When confronted with this evidence, the man changed his story to say that he had burgled the flat that day and taken the victims bank cards and pin numbers then gone to the bank to withdraw cash. When the pin numbers did not work, he returned to the flat and discovered that the victims were dead. In his new account, he claimed he must have got blood on his hands as a result of his attempts to administer first aid to the victims. The man was found guilty and sentenced to 40 years in prison.

A4.2 What Were the Circumstances at the Time the Mark Was Made?

Marks may provide information about the position and location of the hands and fingers at the time the marks were made as well as the approximate amount and direction of force used to deposit them.

The orientation of a mark can, for example, indicate it was likely made by someone who was behind the counter of a shop rather than in front (as may be expected if they were a customer), and the location and orientation of marks high up on a wall may be consistent with being made by a person climbing through a window and reaching down, but not with that person touching the wall whilst at ground level. The illustration in Figure A4.5 was produced to show where a mark was found to a jury in rape trial. The man identified as being the donor of the mark denied the offence and claimed that he had not been on the bed on which the victim was raped. However, providing it was made after the bed was assembled, the area of skin that made the mark along with its location and orientation would indicate its donor was likely to have been on the bed when it was made.

Likewise, the position and orientation of marks on the neck of a bottle, such as those in Figure A4.6, may offer more support to a proposition that the donor touched the bottle to use it as a weapon rather than to drink out of it.

Ashbaugh (1999) and de Ronde et al. (2019b) describe how indicators of downward pressure or movement in a mark may offer more support to one hypothesis about the circumstances in which it was made than another. For example, a mark made as a result of the donor hanging from a window ledge to gain entry to a property should reveal indicators of heavy downward pressure, which may not be present if the mark was made as a result of the donor touching the ledge in other circumstances. Ashbaugh also describes a case where the orientation of marks, as well as indicators of movement in them, were consistent with the donor sliding open a window from the outside rather than being left legitimately from the inside.

The area of skin revealed in the mark can also sometimes indicate the circumstances at the time the mark was made. The marks in Figure A4.7 are on the outside surface of a door. If an offence took place inside the room and the person identified as the donor of the marks had legitimate access to

Figure A4.5   A mark made by this area of skin in this location with this orientation is very likely to have been made by someone who was on the bed at the time it was made. *Source:* Courtesy of Mayor's Office for Policing and Crime, London.

Figure A4.6   This illustration was produced to show a jury the position of a thumb mark found at the scene of a rape during which the victim was struck with a bottle. The man identified denied the offence and claimed that he had only touched the bottle to drink out of it; however, the location and orientation of the mark is more consistent with the bottle being gripped to use as a weapon rather than to drink out of. *Source:* Courtesy of Mayor's Office for Policing and Crime, London.

Exhibit LSO/22

(a)　　　　　　　　　　(b)　　　　　　　　　　(c)

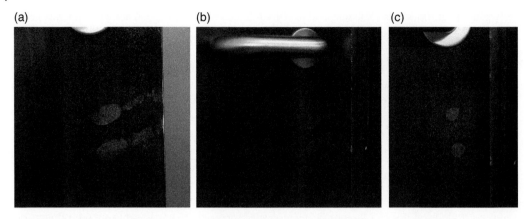

Figure A4.7　The position of the marks in (a) and the areas of skin that made them are consistent with them having been made while the door was open rather than closed, as shown in (b). An attempt to leave marks with the same areas of skin while the door was closed would result in marks with a very different appearance, as shown in (c).

the area outside the room but denied being inside it, then the presence of their marks on the outside of the door would not be significant. However, the marks were made by the person's left index and middle fingers and include areas made by the medial and proximal phalanges, which extend to the edge of the door. When the door is closed, it can be seen that most of the parts made by the medial phalange, and all of the parts made by the proximal phalange are covered by the doorjamb. Therefore, assuming the marks were made after the door was fitted, they could only have been made when the door was open and would be difficult to make without the donor being in the room. Similar marks may also be found around the opening edge of the boot or trunk of a motor vehicle and may be consistent with having been made while the boot was open due to the presence of the protruding bumper.

The position of marks was found to be an indicator of the circumstances at the time the marks were made in an experiment conducted by de Ronde et al. (2019b). The authors carried out a study to see whether the position of marks on a pillowcase could be used to suggest whether they were made by a person changing the case or using the case to smother someone. The authors found meaningful differences with regard to the location of the marks between the different scenarios. The marks made as a result of changing the case showed a much more random distribution over the case on both sides and were more highly distributed around the opening. Conversely, the marks made by a smothering action showed a high density in the middle of the front side, whereas on the back, almost no marks were found and those that were found were mostly around the opening.

The authors developed a model for this study using a Bayesian network which, under the conditions of the study, was accurate 98.8% of the time in evaluating which of the two activities was taking place when the marks were made. A Bayesian network is a way of estimating the probability of something which is dependent on several variables that can influence each other. The model can be used to evaluate the location of marks on any two-dimensional item and its use was described in de Ronde (2021) to assess whether the activity of writing a letter can be distinguished from reading a letter. In this experiment de Ronde (2021) tasked 110 participants with reading and writing A4 letters. The study found that for read letters, the marks were mostly distributed on the left and right edges on both sides of the paper, whereas for written letters, almost all the marks were on the front side and mostly distributed in the middle of the paper. Using the Bayesian network, the authors were able to ascertain which letters related to which activity with 98% accuracy under the conditions of the study.

De Ronde (2021) also describes a similar experiment to examine whether the location of marks could be used to determine whether a knife was used to stab someone or to cut food. Participants were filmed carrying out both activities (a Styrofoam plate was used for the stabbing scenario), and the authors observed differences in the way the knives were held. Specifically, 54% of those carrying out the stabbing scenario held the knife in an overhand position in which the knife was at shoulder height, and the wrist was above the elbow, which resulted in the thumb being near the backside of the knife handle. This way of holding the knife was not observed in the cutting food scenario, and the authors found that under the conditions of their experiment, thumb marks found on the backside of the knife provided support for the stabbing scenario. De Ronde (2021) also observed two participants rotated the knife, so the blade was facing upwards in the stabbing scenario, which was not seen in the cutting food scenario. Also, 13 participants put their fingers on the blade of the knife while cutting food, whereas only one participant did so during the stabbing scenario. Though it was found that the most likely areas of skin to leave marks in both scenarios were the same, de Ronde reported that when marks were found in all three of the areas of the knife that were considered in this study (the blade, the handle and the backside), this provided support for the food cutting rather than stabbing scenario.

Bunter (2017) describes a case where the location of marks was initially seen to support the prosecution case that they were made at the time of the offence, but upon further examination, was agreed to be consistent with the donor's account that they were made innocently some 10 years earlier. In 2014 the victim encountered a masked man striking the locked security gate that protected his front door. The man did not manage to open the gate and fled when he was seen by the victim. The police examined the gate and found a palm mark and three-finger marks in the upper left corner that were identified as having been made by a local man. The police and fingerprint examiner's view was that the marks were in positions consistent with having been made in an attempt to force open the gate. When questioned the man stated he had no knowledge of the offence. The man answered 'no comment' when asked how his fingerprints came to be found on the gate and was subsequently charged with the offence. However, the man later gave an account that he may have touched the gate some 10 years previously during its construction or fitting at the premises as part of the work experience he carried out. Bunter was asked by the defendant's solicitor to review the fingerprint evidence – which was the only evidence against the man. Bunter examined the marks and visited the scene to look at the gate. Bunter found that whilst the palm mark could have been made at the time of the offence, it would have been virtually impossible to leave the fingermarks while the gate was fixed to the wall (which it had been for around 10 years). Bunter's view was that the marks were consistent with having been made by someone gripping the bar and curling their fingers around it, which would not be possible with the gate in place as it was at the time of the offence. Upon further examination of the gate, Bunter identified several other marks on the other end of the same bar that had been made by the defendant's left hand that were also consistent with gripping the gate. Bunter considered that both left- and right-hand marks were consistent with the defendant gripping the gate at a time when it was not fixed to the wall. Bunter observed that the marks were extremely robust and difficult to remove and appeared to have been made in paint or a surface coating that was applied during the manufacturing process. After their examination, Bunter and the police examiner prepared a joint statement that declared that the marks were consistent with being left by the defendant at a time prior to the offence when he handled the gate before it was fixed to the premises. As a result, the prosecution dropped the case against the man.

A4.3 Blood Marks

When considering activity level propositions with blood marks, it must first be ascertained through testing that the material the mark was made in is blood. For example, Derris Lewis spent 18 months in prison after being accused of killing his twin brother in the bedroom they had shared for four years. Because Mr Lewis had access to the room, the presence of his marks there would be expected, but as one of them was believed to be in blood, it was considered to be significant. However, when tested the mark was found not to have been made in human blood but was believed to have contained animal proteins that reacted with the chemical treatment used to make marks in blood visible (Triplett, 2019; Welsh-Huggins, 2009; 'Prosecutor to dismiss', 2009). Even if the mark is made in blood, it does not necessarily mean that it was made at the time of the offence, and as with marks made in other materials, though approaches have been proposed there is no validated and accepted method to determine how long a blood mark has been on a surface (Bremmer, Bruin & van Gemert, 2012; Bleay, Croxton & de Puit, 2018a; Cadd et al., 2018).

The evidential value of a mark made in blood may hinge on how the mark was made. For example, was the mark made as a result of blood on the skin of the donor being transferred to the surface or was it made by the donor's 'clean' skin touching a surface that was covered in blood? Alternatively, was it made by the donor leaving a mark, not in blood on a surface that was later exposed to blood that made the mark appear as if it had been made in blood?

If blood is covering the entirety of a distal phalange when a mark is made, then it may be obvious from the size and shape of the mark and the fact that the blood does not extend beyond its perimeter that the blood was likely transferred from finger to surface rather than being already on the surface. However, Huss et al. (2000) considered whether there might be indicators that could allow an examiner to differentiate between marks made by a finger placed into a drop of blood (smaller than the area of the digit) on the surface from ones made after the same size drop was deposited on the finger before the surface was contacted. The authors found that it was not always possible to tell which had occurred by examining the resulting mark. They reported that the appearance of the mark was the same if the blood was wet, but where the drop was applied to the surface, if it had dried somewhat, then a 'ghost' image of the blood drop may be visible in the mark.

Huss et al. (2000) also experimented by making marks with a clean finger on a surface covered with a wet blood smear. The authors found that these marks all had a distinctive 'halo' or clear outline around them resulting from downward pressure forcing the blood to the perimeter of the mark.

Geller, Volinits & Wax (2017) considered a slightly different issue – whether a mark in blood could be left as a result of the donor's wet fingers contacting a dry bloodstain on a surface and then transferring that blood to a secondary surface. The question arose from an investigation into a homicide in which the victim was found dead on a construction site. The police suspected that the deceased was shot by his business partner in an office and then transported to the site in a company van. Two marks were found on the van; both were made in the victim's blood and were identified as having been made by the victim's business partner. The victim's business partner stated that he had inspected the van after it had been brought back to the office from the site by an employee. He claimed that he had washed his hands prior to the inspection and that he must have touched a dry bloodstain that transferred blood to his wet fingers and allowed him to leave marks in the victim's blood. The Israel Police Latent Fingerprint Laboratory conducted experiments by depositing human blood on the same type of car from six donors. Samples of blood were dropped onto car parts, smeared and then left to dry before fingers with different levels of moisture touched the stain and then a clean part of the car. The laboratory found that under these conditions dry, or slightly dampened fingers were unable to transfer blood from the stain to the surface. However, a finger that was wetted with cold tap water and immediately rubbed against the stain was able to deposit a blood-contaminated mark on the clean surface. The laboratory observed that all the experiments resulted in the creation of a reverse colour mark in the stain as a result of the lifting of the blood by the ridges. As in this case there were no such marks in any of the stains on the car, the laboratories findings contributed to the conviction of the defendant.

Much of the published blood mark research focuses on the question of whether a latent mark not made in the blood can appear to have been made in blood as a result of being exposed to it after deposition (Creighton, 1997; Geller et al., 2018; Huss et al., 2000; Lin et al., 2019; Praska & Langenburg, 2013; Reitnauer, 2011).

Creighton (1997) describes a case that began with a man reporting his car had been stolen. When the car was found, a woman was discovered dead inside it from wounds resulting in considerable blood loss. A visible ('patent') mark in the victim's blood was found on the outside of the car and was identified as having been made by the owner of the car. To answer the question as to whether blood applied to a latent mark could 'develop' it to make it appear like a visible blood mark, Creighton carried out a series of experiments using a similar surface and marks rich in eccrine material, sebaceous material or made in grease. The marks were left on the surface for different periods of time before human blood was applied to them. Creighton observed that with the water-soluble eccrine rich marks, the blood would adhere to the entire surface uniformly making the existing mark invisible. However, with the non-water-soluble sebaceous or grease rich marks, the blood would be repelled by the mark and flow around it, resulting in a void where the mark was. Therefore, whilst the application of blood could highlight the existence of a latent mark on the surface, it did not give the mark the appearance that it had been made in blood as none of the details within the void was made visible. Whilst some of the other studies also reported similar voids in certain circumstances, unlike Creighton, they all found that latent marks could be developed by the application of blood.

Huss et al. (2000) reported marks on non-porous surfaces were developed when a swab soaked in beef blood was dragged over them. The authors observed that the marks were made visible as a result of the blood running into the furrows, which resulted in the appearance of a visible mark in blood that is in reverse colour. The fact that the marks in Creighton's experiment were not made visible by exposure to blood, whereas those in Huss et al. were, may be attributable to the orientation of the surfaces the marks were on. Creighton tested marks on the hood of a motor vehicle that was in an upright (vertical) position, whereas Huss et al. experimented with surfaces at different orientations. Huss et al. found that as a surface was moved from horizontal to vertical, less blood stayed in the furrows until a point was reached where all the blood was repelled – as in the experiment by Creighton.

Praska & Langenburg (2013) and Geller et al. (2018) also found that marks could be made visible by exposure to blood; however, unlike those in Huss et al., not all of them were in reverse colour. In their experiments, the authors poured diluted blood over marks on a horizontal surface and found that sometimes the blood deposited on the ridges of sebaceous rich latent marks to produce visible marks with red ridges (Figure A4.8). Praska & Langenburg (2013) also found that pouring whole blood over latent marks on a horizontal surface or dipping vertical surfaces into whole blood for longer than a few seconds resulted in the marks becoming visible beneath a layer of blood. Lin et al. (2019) described a similar type of mark appearing as if it was behind a red filter in some of their experiments.

Praska & Langenburg compared the 'faux' blood marks developed with diluted blood with real marks made as a result of a transfer of blood from the skin to the surface and observed some differences in appearance that could be used to distinguish between them. Specifically, they noted

(a)   (b)

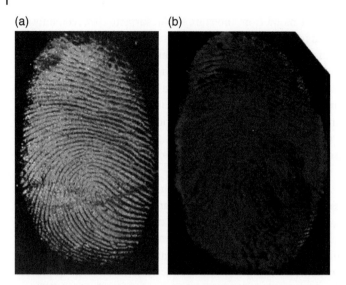

Figure A4.8 The mark in (a) was made in sebaceous secretions on a horizontal glass surface and made visible as a result of diluted blood being poured over it. The mark in (b) is a genuine blood mark that was made as a result of a transfer of blood from the skin of the donor to the surface. *Source:* Reprinted with permission from Praska and Langenburg (2013).

that whilst genuine blood marks had an uneven distribution of blood with places where it pooled in the furrows, marks that were developed with blood were much more uniform and consistent in appearance (Figure A4.8). The authors also observed that genuine blood marks were darker, and the contrast between the ridges and furrows was less distinct. A further difference was that whilst the edges of real blood marks were 'sealed' as a result of downward pressure repelling blood to the edges of the mark and creating a solid perimeter, the perimeters of the 'faux' marks were not solid and were just marked by the ridges extending to the edges and stopping whilst the furrows blended in with the surface.

Not all marks in blood at a crime scene will be visible. For example, some made with only a very small amount of blood on the ridges will be no more visible than marks in natural secretions and will therefore need to be developed by the application of chemical treatment. The chemicals used do not react with natural secretions, and so where they are used and a mark becomes visible, there is a tendency to assume the mark was made in the blood (as in the Derris Lewis case). Two studies tested whether these chemical treatments would also develop marks that were not made in blood but had been exposed to it (Praska & Langenburg, 2013; Reitnauer, 2011). Both studies found that it is possible for marks made in sebaceous secretions, exposed to blood and then treated with chemicals, to be developed and mistaken for real blood marks. Praska & Langenburg (2013) also observed that where blood had been applied to a surface and flowed around a mark to create a void with no details in it (as in Creighton, 1997), subsequent chemical treatment could make some of the details visible and make the mark difficult to distinguish from a real blood mark. The authors also observed that if chemical treatments were applied to latent marks developed by blood, it was much more difficult to tell the difference between them and marks actually made in blood after treatment than before (Praska & Langenburg, 2013). The authors, therefore, recommended that photographs be taken of visible marks at a crime scene before any treatments are applied.

The fact that most of the latent marks that were developed by blood in the studies were reverse colour does not mean that the discovery of a reverse colour mark in the blood means it is likely that the mark was actually a latent mark that was developed by blood. For example, Langenburg (2008) was able to produce reverse colour marks as a result of a transfer of blood from the skin to the surface by touching the surface multiple times with the same area of skin. Whilst the first contact(s) transferred the blood from the summits of the ridges; once there was no more wet blood on the ridges, the following contacts (providing significant downward pressure was used) resulted in blood in the furrows contacting the surface to produce a reverse colour mark. Langenburg also observed a significant difference in the time it takes for blood on the ridges to dry in comparison to blood in the furrows. This effect meant that reverse colour marks could be produced by waiting for the blood to dry on the ridges and then depositing a mark with enough downward pressure to allow the wet blood in the furrows to transfer to the surface. Geller et al. (2018) actually tested whether the colour (reverse or true) could be used to determine how an apparent blood mark was made. The authors tested three different scenarios and found that the results were not always as may be expected. First, they looked at marks made by a finger with blood on it, contacting a clean surface. The authors found that as expected, the marks were in true colour as the ridges were red and darker than the furrows, but they also recognised that under different conditions, such as those reported by Langenburg (2008), the results could be different. Second, Geller et al. tested marks made in natural secretions that were developed by exposure to blood on an inclined surface. The author's expectation was that the marks would be in reverse colour as a result of the blood collecting in the furrows as occurred in Huss et al. (2000). However, what they found was more in keeping with the results of Praska & Langenburg (2013) in that the blood actually deposited on the ridges and resulted in their colour being darker than the furrows. Lastly, Geller et al. considered marks made by a clean finger on a surface contaminated with blood. Their expectation was that the mark would be reverse colour as a result of blood flowing into the furrows during deposition and making them darker than the ridges. However, the authors found that there were many variables that could affect the appearance of a mark made in these circumstances, including downward pressure, the amount of time the finger was in contact with the surface, the velocity of the finger as it was lifted from the surface and the thickness of the blood. As a result, in only about 25% of the tests were the ridges lighter than the furrows, and this occurred mostly when the downward pressure was low and the blood layer was thin. Geller et al. also observed some ridges of marks made in this way changing from reverse to true colour along their length. Overall, Geller et al. concluded that though it could offer important information, it was not possible to determine the mechanism used to create the mark solely by using its colour.

References

Ashbaugh, D, R 1999, *Quantitative-Qualitative Friction Ridge Analysis*, CRC Press LLC, Boca Raton, Florida.

Attard-Montalto, N, Ojeda, J, J, Reynolds, A, Ismail, M, Bailey, M, Doodkorte, L, de Puit, M & Jones, B, J 2014, 'Determining the Chronology of Deposition of Natural Fingermarks and Inks on Paper Using Secondary Ion Mass Spectrometry', *Analyst*, doi: https://doi.org/10.1039/C4AN00811A.

Azoury, M, Rozen, E, Uziel, Y & Peleg-Shironi, Y 2004, 'Old Latent Prints Developed with Powder: A Rare Phenomenon?', *Journal of Forensic Identification*, 54 (5), pp. 534–541.

Balloch, S, R 1977, 'The Life of a Latent', *Identification News*, July, p. 10.

Baniuk, K 1990, 'Determination of Age of Fingerprints', *Forensic Science International*, 46, pp. 133–137.

Barnett, P, D & Berger, R, A 1977, 'The Effects of Temperature and Humidity on the Permanency of Latent Fingerprints', *Journal of the Forensic Science Society*, 16, pp. 249–254.

Barros, M, R, Faria, B, E, F & Kuckelhaus, S, A, S 2013, 'Morphometry of Latent Palmprints as a Function of Time', *Science and Justice*, 53 (2013) pp. 402–408.

Batey, G, W, Copeland, J, Donnelly, D, L, Hill, C, L, Laturnus, P, L, McDiarmid, C, H, Miller, K, J, Misner, A, H, Tario, A & Yamashita, A, B 1998, 'Metal Deposition for Latent Print Development', *Journal of Forensic Identification*, 48 (2), pp. 165–175.

Beaudoin, A 2011, 'Oil Red O: Fingerprint Development on a 21-Year-Old Col d Case', *Journal of Forensic Identification*, 61 (1), pp. 50–59.

Becker, R, F 2009, *Criminal Investigation*, Third Edition, Jones and Bartlett Publishers, Sudbury, Massachusetts.

Belcher, G, L 1982, 'Relative Dating of Fingerprints', *Fingerprint Whorld*, 7 (27), pp. 72–73.

Bernier, U, R, Booth, M, M & Yost, R, A 1999, 'Analysis of Human Skin Emanations by Gas Chromatography/Mass Spectrometry. 1. Thermal Desorption of Attractants for the Yellow Fever Mosquito (Aedes aegypti) from Handled Glass Beads', Publications from USDA-ARS/UNL Faculty. Paper 945. http://digitalcommons.unl.edu/usdaarsfacpub/945.

Bleay, S, M, Croxton, R, S & de Puit, M 2018a, *Fingerprint Development Techniques, Theory and Application*, John Wiley & Sons Ltd., Chichester, UK.

Bleay, S, Fitzgerald, L, Sears, V & Kent, T 2018b, 'Visualising the Past – An Evaluation of Processes and Sequences for Fingermark Recovery from Old Documents', *Science & Justice*, 59 (2), pp. 125–137.

Bleay, S, M, Kelly, P, F, King, R, S, P & Thorngate, S, G 2019, 'A Comparative Evaluation of the Disulfur Dinitride Process for the Visualisation of Fingermarks on Metal Surfaces', *Science & Justice*, 59, pp 606–621.

Bond, J, W 2008, 'Visualization of Latent Fingerprint Corrosion of Metallic Surfaces', *Journal of Forensic Science*, 53 (4), pp. 1556–4029.

Bond, J, W 2009, 'Visualization of Latent Fingerprint Corrosion of Brass', *Journal of Forensic Science*, 54 (5), pp. 1034–1041.

Bouzin, J, T, Merendino, J, Bleay, S, M, Sauzier, G & Lewis S, W 2020, 'New Light on Old Fingermarks: The Detection of Historic Latent Fingermarks on Old Paper Documents Using 1, 2-indandione/ zinc', *Forensic Science International, Reports*, 2, 100145.

Bradshaw, G, Bleay, S, Deans, J & NicDaeid, N 2008, 'Recovery of Fingerprints from Arson Scene: Part 1 - Latent Fingerprints', *Journal of Forensic Identification*, 58 (1), pp. 54–82.

Bremmer, R, H, de Bruin, K, G & van Gemert, M, J, C 2012, 'Forensic Quest for Age Determination of Bloodstains', *Forensic Science International*, 216, pp. 1–11.

Bright, N, J, Webb, R, P, Bleay, S, Hinder, S, Ward, N, I, Watts, J, F, Kirkby, K, J & Bailey, M, J 2012, 'Determination of the Deposition Order of Overlapping Latent Fingerprints and Inks Using Secondary Ion Mass Spectrometry (SIMS)', *Analytical Chemistry*, 84 (9), 4083–4087.

Bunter, S 2014, '*How Long Can an Identifiable Fingerprint Persist on an Exterior Surface*', *CSEye*.

Bunter, S 2007, 'Fingerprints in Paint', *Fingerprint Whorld*, 34 (130), pp. 21–23.

Bunter, S 2015, 'How Long can a Fingerprint Persist on an External Surface? – A Case Study', *Fingerprint Whorld*, 40 (157), pp. 5–14. http://www.keithborer.co.uk/uploads/assets/files/ Drainpipe%20Case%20Study%20for%20FP%20Whorld%20KBC%20Website%20Version%202015.pdf.

Bunter, S 2017, 'Location, Location, Location: Misinterpretation of Fingerprints on a Security Gate – A Case Study', *Fingerprint Whorld*, 42 (163), pp. 8–25.

CAIS 2005, The Circle of Ancient Iranian Studies, 15th June, http://www.cais-soas.com/News/2005/ June2005/15-06.htm.

Cadd, S, Islam, M, Manson, P & Bleay, S 2015, 'Fingerprint Composition and Aging: A Literature Review', *Science and Justice*, 55, pp. 219–238.

Cadd, S, Li, B, Beveridge, P, O'Hare, W, T & Islam, M 2018, 'Age Determination of Blood-Stained Fingerprints Using Visible Wavelength Reflectance Hyperspectral Imaging', *Journal of Imaging*, 4 (141), pp. 1–11.

Castelló, A, Francés, F & Verdú, F 2013, 'Solving Underwater Crimes: Development of Latent Prints Made on Submerged Objects', *Science and Justice*, 53, pp. 328–331.

Champod, C, Lennard, C, Margot, P & Stoilovic, M 2016, *Fingerprints and Other Ridge Skin Impressions*, Second Edition, CRC Press, Taylor & Francis Group, Boca Raton.

Clements, W, W 1986, 'Latent Fingerprints – One Year Later', *Fingerprint Whorld*, 12(46), p. 54.

Cohen, Y, Rozen, E, Azoury, M, Attias, D, Gavrielli, B & Elad, M, L 2012a, 'Survivability of Latent Fingerprints Part 1: Adhesion of Latent Fingerprints to Smooth Surfaces', *Journal of Forensic Identification*, 62 (1), pp. 47–53.

Cohen, Y, Azoury, M & Elad, M, L 2012b, 'Survivability of Latent Fingerprints Part 2: The Effect of Cleaning Agents on the Survivability of Latent Fingerprints', *Journal of Forensic Identification*, 62 (1), pp. 54–61.

Cook, R, Evett, I, W, Jackson, G, Jones, P, J & Lambert, J, A 1998, 'A Hierarchy of Propositions: Deciding Which Level to Address in Casework', *Science & Justice*, 38 (4), pp. 231–239.

Cowger, J, F 1993, *Friction Ridge Skin, Comparison and Identification of Fingerprints*, CRC Press, Boca Raton, US.

Creighton, J, T 1997, 'Visualization of Latent Impressions After Incidental or Direct Contact With Human Blood', *Journal of Forensic Identification*, 47 (5), pp. 534–541.

De Alcaraz-Fossoul, J, Patris, C, M, Muntaner, A, B, Feixat, C, B & Badia, M, G 2013, 'Determination of Latent Fingerprint Degradation Patterns—A Real Fieldwork Study', *International Journal of Legal Medicine*, 127, pp. 857–870.

De Alcaraz-Fossoul, J, Patris, C, M, Feixat, C, B, McGarr, L, Brandelli, D, Stow, K & Badia, M, G 2015, 'Latent Fingermark Aging Patterns (Part I): Minutiae Count as One Indicator of Degradation', *Journal of Forensic Sciences*, doi: https://doi.org/10.1111/1556-4029.13007.

De Alcaraz-Fossoul, J, Feixat, C, B, Tasker, J, McGarr, L, Stow, K, Carreras-Marin, C, Oset, J, T & Badia, M, G 2016, 'Latent Fingermark Aging Patterns (Part II): Color Contrast Between Ridges and Furrows as One Indicator of Degradation', *Journal of Forensic Sciences*, doi: https://doi.org/ 10.1111/1556-4029.13099.

De Alcaraz-Fossoul, J, Feixat, C, B, Carreras-Marin, C, Tasker, J, Zapico, S, C & Badia, M, G 2017, 'Latent Fingermark Aging Patterns (Part III): Discontinuity Index as One Indicator of Degradation', *Journal of Forensic Sciences*, doi: https://doi.org/10.1111/1556-4029.13438.

De Alcaraz-Fossoul, J, Feixat, C, B, Zapico, S, C, McGarr, L, Carreras-Marin, C, Tasker, J & Badia, M, G 2019, 'Latent Fingermark Aging Patterns (Part IV): Ridge Width as One Indicator of Degradation', *Journal of Forensic Sciences*, 64 (4), doi: https://doi.org/10.1111/1556-4029.14018.

de Ronde, A 2021, 'What fingermarks reveal about activities', Thesis, VU University Amersterdam.

de Ronde, A, van Aken, M, de Puit, M & de Poot, C 2019b, 'A Study into Fingermarks at Activity Level on Pillowcases', *Forensic Science International*, 295, pp. 113–120.

Devlin, B, E 2011, Recovery of Latent Fingerprints after Immersion in various Aquatic Conditions, A Research Project Submitted to the Graduate Faculty of George Mason University in Partial Fulfilment of the Requirements for the Degree of Master of Science, Forensic Science, http://ebot.gmu.edu/bitstream/handle/1920/7484/Devlin_thesis_2011. pdf?sequence=1&isAllowed=y.

Dikshitulu, Y, S, Prasad, L, Pal, J, N & Rao, C, V, N 1986, 'Aging Studies on Fingerprint Residues Using Thin-Layer and High Performance Liquid Chromatography', *Forensic Science International*, 31, pp. 261–266.

DoubleLoop Podcast 2020, Activity Level, Episode 212, January 18.

Dror, I, E 2015, 'Cognitive Neuroscience in Forensic Science: Understanding and Utilizing the Human Element', *Philosophical Transactions of The Royal Society B*, 370 (1674), pp. 1–8, https://royalsocietypublishing.org/doi/full/10.1098/rstb.2014.0255.

Duff, J, M & Menzel, E, R 1978, 'Laser-Assisted Thin-Layer Chromatography and Luminescence of Fingerprints: An Approach to Fingerprint Age Determination', *Journal of Forensic Sciences*, 23 (1), pp. 129–134.

Ellis, E, L 2005, 'Latent Print on Glass Surface: Deposited Before or After Breakage?', *Journal of Forensic Identification*, 55 (1), pp. 36–46.

Fieldhouse, S 2015, 'An Investigation into the Effects of Force Applied During Deposition on Latent Fingermarks and Inked Fingerprints Using a Variable Force Fingerprint Sampler', *Journal of Forensic Science*, 60 (2), 422–427.

'Fingerprints at Boncuklu', 2014, Boncuklu Project, August 27th, 2:05pm, http://boncuklu.org/fingerprints-at-boncuklu/.

Forchelet, S & Bécue, A 2018, 'Impact of Anti-Fingerprint Coatings on the Detection of Fingermarks', *Journal of Forensic Identification*, 68 (3), pp. 348–368.

Forensic Science Regulator 2015, *Cognitive Bias Effects Relevant to Forensic Science Examinations*, FSR-G-217, Issue 1, https://www.gov.uk/government/publications/cognitive-bias-effects-relevant-to-forensic-science-examinations.

Geller, B, Volinits Y & Wax, H 2017, 'Can Dry Bloodstains Provide a Source for a Blood-Contaminated Fingermark?', *Journal of Forensic Identification*, 67 (3), pp. 355–360.

Geller, B, Leifer, A, Attias, D & Mark, Y 2018, 'Fingermarks in Blood: Mechanical Models and the Color of Ridges', *Forensic Science International*, 286, pp. 141–147.

Girod, A, Ramotowski, R & Weyermann, C 2012, 'Composition of Fingermark Residue: A Qualitative and Quantitative Review', *Forensic Science International*, 223, pp. 10–24.

Girod, A, Ramotowski, R, Lambrechts, S, Misrielal, P, Aalders, M & Weyermann, C 2016, 'Fingermark Age Determinations: Legal Considerations, Review of the Literature and Practical Propositions', *Forensic Science International*, 262, pp. 212–226.

González Amorós, B & de Puit, M 2014, 'A Model Study into the Effects of Light and Temperature on the Degradation of Fingerprint Constituents', *Science and Justice*, 54, pp. 346–350.

Greenlees, D 1994, 'Age Determination-Case Report', *Fingerprint Whorld*, 20 (76), pp. 50–52, http://scafo.org/library/100702.html.

Harper, W, W 1938, 'Latent Fingerprints at High Temperatures', *Journal of Criminal Law and Criminology*, 29 (4), pp. 580–583.

Hazen, R, J 1984, Significant Advances in the Science of Fingerprints, presented to International Conference on Fingerprints, London, November 27 to 30.

Hill, S 2008, 'Two Cases with the Same Latent Print Evidence', *Journal of Forensic Identification*, 58 (1), pp. 42–45

Howorka, H 1989, 'Questions relating to the Determination of the Age of objects assuming Relevance in Criminal Investigations', *Fingerprint Whorld*, 15 (57), pp. 23–28.

Hoye, C 1977, 'Ridge Persistency', *Fingerprint Whorld*, 3 (10), p. 42.

Home Office 2003, Police Scientific Development Branch Crime Investigation Sector, Fingerprint Development and Imaging Update, November, Publication No. 26.

Hunter, J, L 1997, 'Fingerprint Evidence with Coomassie Blue – After 25 Years', *Fingerprint Whorld*, 23 (89), pp. 91–93.

Huss, K, Clark, J & Chisum, W, J 2000, 'Which Was First - Fingerprint or Blood?', *Journal of Forensic Identification*, 50 (4), pp. 344–350

Illsley, C, P 1984, 'Superglue Fuming and Multiple Lifts', *Identification News*, 34 (1), pp. 6–15.

Involdstad, H 1978, 'How Long Will a Fingerprint Last?', *International Criminal Police Review*, 316, pp. 87–88.

Johnson, P, L 1973, 'Life of Latents', *Identification News*, April, pp. 10–13.

Jones, N, E, Davies, L, M, Russell, C, A, L, Brennan, J, S & Bramble, S, K 2001, 'A Systematic Approach to Latent Fingerprint Sample Preparation for Comparative Chemical Studies', *Journal of Forensic Identification*, 51 (5), pp. 504–515.

Kent, T 2016, Water Content of Latent Fingerprints –The Myth, *Forensic Science International*, 266, pp. 134–138.

Koller, J, Baumer, U & Mania, D 2001, 'High-Tech in the Middle Palaeolithic: Neandertal-Manufactured Pitch Identified', *European Journal of Archaeology*, 4 (3), pp. 385–397.

Králík, M, Novotny V, & Oliva, M 2002, Fingerprint on the Venus of Dolní Vestonice I, *Anthropologie*, XL/2, pp. 107–113.

Lambourne, G 1984, *The Fingerprint Story*, Harrap, London.

Langenburg, G 2008, 'Deposition of Bloody Friction Ridge Impressions', *Journal of Forensic Identification*, 58 (3), pp. 355–389.

Lin, S, Luo, Y, Xie, L, Yu, Y, & Mi, Z 2019, 'Faux Blood Fingermark on Pistol: Latent Fingerprint Developed by Whole Blood', *Journal of Forensic Sciences*, 64 (6), pp. 1913–1915.

Lind, S, E 1972, 'Corrosion of Metals by Human Sweat and its Prevention', *Corrosion Science*, 12, pp. 749–755.

Masterson, A & Bleay, S 2021, 'The Effect of Corrosive Susbtances on Fingermark Recovery: A Pilot Study', *Science & Justice*, 61, pp. 617–626.

Maslanka, D, S 2016, 'Latent Fingerprints on a Nonporous Surface Exposed to Everyday Liquids', *Journal of Forensic Identification*, 66 (2), pp. 137–154.

McRoberts, A, L, 1994, Editorial Comment: 'Latent Fingerprints One Year Later', *The Print*, 10 (7), p. 6, http://www.scafo.org/images/theprint/Vol10Issue7_Aug1994.pdf.

McRoberts, A, L & Kuhn, K, E 1992, 'A Review of the Case Report– "Determining the Evaporation Rate of Latent Impressions on the Exterior Surfaces of Aluminum Beverage Cans"', *Journal of Forensic Identification*, 42 (3), pp. 213–218.

Midkiff, C, R 1993, 'Lifetime of a Latent Print How Long? Can You Tell?', *Journal of Forensic Identifictaion*, 43 (4), pp. 386–392.

Mong, G, M, Petersen, C, E & Clauss, T, R, W 1999, *Advanced Fingerprint Analysis Project: Fingerprint Constituents*, United States: N. p., 1999. Web, doi: https://doi.org/10.2172/14172, https://www.osti.gov/biblio/14172-advanced-fingerprint-analysis-project-fingerprint-constituents.

Montalto, N, A, Ojeda, J, J & Jones, B, J 2013, 'Determining the Order of Deposition of Natural Latent Fingerprints and Laser Printed in Using Chemical Mapping with Secondary Ion Mass Spectrometry', *Science and Justice*, 53 (1), pp. 2–7.

Moody, E, W 1994, 'The Development of Fingerprint Impressions on Plastic Bags Over Time and Under Different Storage Temperatures', *Journal of Forensic Identification*, 44 (3), pp. 266–269.

Moore, J, Bleay, S, Deans, J & NicDaeid, N 2008, 'Recovery of Fingerprints from Arson Scenes: Part 2 - Fingerprints in Blood', *Journal of Forensic Identification* 58 (1), pp. 83–108.

Moses Daluz, H 2015, *Fundamentals of Fingerprint Analysis*, CRC Press, Boca Raton.

Moynihan, M 1973, 'The antler riddle: has Lambourne of the Yard been called in 5000 years too late?', *The Sunday Times*, December 9, 1973.

Muramoto, S & Sisco, E 2015, 'Strategies for Potential Age Dating of Fingerprints Through the Diffusion of Sebum Molecules on a Nonporous Surface Analyzed Using Time-of-flight Secondary Ion Mass Spectrometry', *Analytical Chemistry*, 87 (16), 8035–8038.

Murrie, D, C, Boccaccini, M, T, Guarnera, L, A & Rufino, K, A 2013, 'Are Forensic Experts Biased by the Side That Retained Them?', *Psychological Science*, 24, pp. 1889–1897.

NIJ 2011, *The Fingerprint Sourcebook*, www.nij.gov.

O'Brien, W, J 1984, 'My Fingerprints? Of course, I lived there!', *Australian Police Journal*, 38 (4), pp. 142–147.

Olsen, R, D, S, R 1978, *Scott's Fingerprint Mechanics*, Charles C Thomas, Illinois.

Perkins, D 2016, 'The Use of a Liquid Bandage to Prevent the Deposition of Friction Ridge Detail Impressions', *Journal of Forensic Identification*, 66 (4), pp. 309–315.

Popa, G, Potorac, R & Preda, N 2010, 'Method for Fingerprints Age Determination', *Romanian Journal of Legal Medicine*, 2, pp. 149–154, https://pdfs.semanticscholar.org/488b/8e2319cb768603cb06713 5b49f85c8b5a9fc.pdf.

Praska, N & Langenburg, G 2013, 'Reactions of Latent Prints Exposed to Blood', *Forensic Science International*, 224, pp. 51–58.

'Prosecutor To Dismiss All Charges Against Derris Lewis' 2009, Franklin County Prosecuting Attorney, 8th June, https://prosecutor.franklincountyohio.gov/press-releases/ prosecutor-to-dismiss-all-charges-against-derris-l.

Ramasastry, P, Downing, D, T, Pochi, P, E & Strauss, J, S 1970. 'Chemical Composition of Human Skin Surface Lipids from Birth to Puberty', *The Journal of Investigative Dermatology*, 54 (2), pp. 139–144.

Reitnauer, A, R 2011, 'Is it a Latent or a Patent?: Development of a Latent Print on Drywall', *Fingerprint Whorld*, 37 (145), pp. 208–214.

'Roman fingerprints found in 2000-year-old cream' 2003, *The Guardian*, 28th July, 18:15 BST, https:// www.theguardian.com/uk/2003/jul/28/artsnews.london.

Sampson, W, C & Moffett, G, C 1994, 'Lifetime of a Latent Print on Glazed Ceramic Tile', *Journal of Forensic Identification*, 44 (4), pp. 379–386.

Saunders, J, C 1993, Macroscopic Examination of Overlapping Latent Prints on Non-Porous Items, *Journal of Forensic Identification*, 42 (2), pp. 138–143.

Schwabenland, J, F 1992, 'Determining the Evaporation Rate of Latent Impressions on the Exterior Surfaces of Aluminum Beverage Cans', *Journal of Forensic Identification*, *42* (2), pp. 84–90.

Spawn, M, A 1993, *Effects of Fire on Fingerprint Evidence*, The Spawn Group LLC, http://www.scafo. org/library/110201.html.

Sutton, R, Grenci, C & Hrubesova L 2014, 'A Comparison on the Longevity of Submerged Marks in Field and Laboratory Conditions', *Journal of Forensic Identification*, 64 (2), pp. 143–156.

Triplett, M 2019, Fingerprint Dictionary, http://fprints.nwlean.net.

Tweedy, C 2010, 'How long can Fingerprints Survive on – Plastic bags?', *FDIAI News*, April-June, pp. 18–21.

Vandiver, J, V 1976, 'Fingerprints', *Identification News*, May, pp. 3–4.

Watson, P, Prance, R, J, Beardsmore-Rust, S, T & Prance H 2011, 'Imaging Electrostatic Fingerprints with Implications for a Forensic Timeline', *Forensic Science International*, 209 (1-3), pp. 41–45.

Welsh-Huggins, A 2009, 'No longer accused, man ready to mourn slain twin', *The Seattle Times*, August 7, https://www.seattletimes.com/nation-world/no-longer-accused-man-ready-to-mourn-slain-twin/.

Wertheim, K 2003, 'Fingerprint Age Determination: Is There Any Hope?', *Journal of Forensic Identification*, 53 (1), pp. 42–48.

Wilkinson, D, Hancock, J, Lecavalier, P & McDiarmid, C 2004, 'The Recovery of Fingerprint Evidence from Crime Scenes Contaminated with Chemical Warfare Agents', *Journal of Forensic Identification*, 55 (3), pp. 326–361.

Wood, M, A & James, T 2009, 'ORO. The Physical Developer replacement?', *Science and Justice*, 49, pp. 272–276.

Appendix 5

Errors

Examiner errors can usually be categorised as being clerical or technical in nature.

A5.1 Clerical Errors

Clerical errors are those in which the examiner's conclusion is correct, but the way they have recorded that conclusion is not. The most common clerical error occurs when the examiner identifies a mark with one print but accidentally records it as being identified with a different print from the same person, e.g. they identify the mark with a print made by the left middle finger but record it as the right middle finger. Alternatively, an examiner working with two sets of fingerprints may identify a mark with a print from one set but accidentally record it as an identification with the other set or they may identify mark 'A' but accidentally record the identification as having occurred with mark 'B' in the same case. Though occasionally it may be difficult to differentiate such errors from technical errors in which the examiner has recorded their conclusion accurately, but that conclusion is erroneous, in most cases, it will be obvious when a clerical error has occurred as the details in the print identified will bear little or no similarity to those in the mark.

The organisation the examiner is employed by will have procedures in place to detect such errors but in the event of those procedures failing, clerical errors can have extremely serious consequences, i.e. if undetected, some could result in the arrest and conviction of an innocent individual in the same way that a technical error could.

A5.2 Technical Errors

In the course of their work, the examiner will make various technical decisions. They will decide whether a mark is suitable for identification or whether it is suitable for searching through a computer database. They will also decide when the use of the inconclusive conclusion is appropriate. With marks that reveal an abundance or paucity of clear details, these decisions will usually be consistent among examiners, but with marks, in between those two ends of the scale, different examiners may make different decisions with the same mark. If, for example, the examiner decides a mark is not suitable when it is or decides the result of a comparison is inconclusive when the details revealed are sufficient to support a conclusion of identification or exclusion, the opportunity to identify or exclude its donor may be lost. Though suitability and inconclusive decisions can be

The Forensic Analysis, Comparison and Evaluation of Friction Ridge Skin Impressions, First Edition. Dan Perkins.
© 2022 John Wiley & Sons Ltd. Published 2022 by John Wiley & Sons Ltd.

incorrect, where examiners reach different decisions with the same marks, the disagreement is usually treated as a difference of opinion rather than an error by one examiner. Usually, the only incorrect decisions that are treated as technical errors are erroneous exclusions and erroneous identifications (SWGFAST, 2012).

With marks found at crime scenes, the 'ground truth' of who actually made them will usually never be known, so if one examiner concludes identification and another exclusion, at least one other examiner will be involved, and the correct conclusion will be determined by consensus.

A5.2.1   Erroneous Exclusion

An erroneous exclusion or false negative occurs when an examiner excludes impressions that should have been identified. The examiner's opinion is that it is extremely unlikely the impressions were made by the same area of skin, but the details revealed indicate that it is extremely likely they were made by the same area of skin.

Erroneous exclusions have the potential to result in the perpetrator of an offence remaining unidentified and free to commit further offences.

There are two ways an examiner can make an erroneous exclusion. First, they can fail to compare the mark with a print made by the corresponding area. This may occur because the examiner mistakenly thinks the mark could only have been made by an area of skin with a particular pattern on it, or because they think the mark was made by a particular area (e.g. a digit when it was made by an area of palm) or because the examiner does not realise the mark was made by an area of skin that is not revealed in the set of fingerprints they have.

An examiner can also make an erroneous exclusion if they compare the mark with a print made by the same area but fail to recognise that. This could be because the mark is being compared with the correct area at the incorrect orientation, but more commonly is due to the examiner failing to see the corresponding details or seeing some but also seeing others that appear not to be in agreement and placing too much weight on those details.

There are many factors that can contribute to the examiner making an erroneous exclusion. The report 'Latent Print Examination and Human Factors' (NIST, 2012) lists over 150 such factors under the headings of Examiner actions, Conditions that affect performance, Supervisory issues and Organisational influences. For example, an analysis of the erroneous exclusions made at one organisation found that a third of them were made with marks that were made by little fingers (Ray & Dechant, 2013). Ray & Dechant suggest that this may be due to examiners having an expectation that marks are rarely made by these digits, which results in them spending less time and effort in comparing marks to prints made by those areas.

Ulery et al. (2011) found that, in general, most marks that were erroneously excluded in their study were distorted and or revealed details that gave the appearance of a different ridge flow than that present in the print made by the same area of skin. Marks with which it is difficult to determine the orientation or area of skin that was most likely to have made them may also be more likely to be erroneously excluded, as may marks that do not reveal fixed points such as cores and deltas. A lack of clarity in details in the mark or print may also make an erroneous exclusion more likely, particularly if it affects the area containing the examiners target group of characteristics. It has also been observed that erroneous exclusions are more common with cases that feature a lot of marks (Montooth, 2019; Ray & Dechant, 2013). Ray & Dechant (2013) observed this was even more pronounced when most of the conclusions that led up to the error were correct exclusions and suggest this may have led to the examiner developing an expectation that the remaining marks would also be excluded and meant they were more likely to overlook similarities.

The organisation that employs the examiner may utilise various methods to detect erroneous exclusions, including verification, technical review and using a computer to re-compare the impressions (Langenburg, Hall & Rosemarie, 2015).

A5.2.2   Erroneous Identification

An erroneous identification or false positive occurs when an examiner identifies impressions that should have been excluded. The examiner's opinion is that it is extremely likely the impressions were made by the same area of skin, but the details revealed indicate that it is extremely unlikely they were made by the same area of skin.

Because of the potentially grave consequences for the person wrongly identified, examiners generally regard erroneous identifications as the most serious error they can make. Erroneous identifications are rare, and those that do occur will usually be discovered by the organisation the examiner is employed by as a result of a verification process. However, very occasionally, they may be erroneously verified and result in the arrest and/or conviction of the person wrongly identified. Arguably the most famous cases of erroneous identifications involved the fingerprints of Shirley McKee, Marion Ross, and Brandon Mayfield.

A5.2.2.1   Shirley McKie & Marion Ross

In 1997 Marion Ross was found dead in her home in Scotland as a result of multiple stab wounds. The 428 marks found at the scene were submitted to the Scottish Criminal Records Office (SCRO) in Glasgow (Campbell, 2011).

The subsequent investigation determined that David Asbury was someone the Police wished to speak to in relation to the homicide. Police Officer Shirley McKie attended the address of Mr Asbury and, while there, handled a tin that was to become a key exhibit (Campbell, 2011). As a result, Ms McKie's fingerprints were added to a list that were then compared with all the marks found at both the scene of the homicide and Mr Asbury's address.

One mark found at the scene of the homicide and designated 'Y7' was identified for Ms McKie (Figure A5.1).

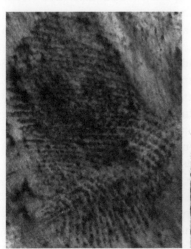

Figure A5.1   The mark 'Y7' which was found on a doorframe at the scene of the homicide and the left thumbprint of Shirley McKie. Reprinted with permission from 'The Price of Innocence', Shirley McKie, Birlinn, 2007.

Figure A5.2 The mark 'Q12 Ross' which was found on a tin at the address of David Asbury and the print made by the right index finger of Marion Ross. *Source:* Ed German.

A second mark, designated 'QI2 Ross' and found on the tin at Mr Asbury's address, was identified for the victim Miss Ross (Figure A5.2).

Mr Asbury was tried and convicted of the murder of Miss Ross in 1997. Ms McKie gave evidence at Mr Asbury's trial but did not accept that she made the mark Y7 and denied ever being inside Miss Ross's house. Ms McKie was subsequently prosecuted for perjury on the basis that she had lied under oath about having been inside the victim's house during Mr Asbury's trial. At Ms McKie's trial in 1999, two examiners engaged by the defence disputed the identification of Y7, and the jury found Ms McKie not guilty.

After that decision, Mr Asbury also engaged an examiner and appealed his conviction, which was subsequently quashed.

Ms McKie had begun civil proceedings against the Police and the SCRO that were settled in 2006 on the basis of a payment of £750 000 to Ms McKie with no admission of liability (Campbell, 2011). Most of the experts involved in the identifications left the employment of the SCRO in 2007 on agreed terms – but continued to maintain that the identifications were correct.

Sir Anthony Campbell was appointed in 2008 to hold an independent public inquiry into the identification and verification of both marks. 'The Fingerprint Inquiry Report' was published in 2011 and found that both Y7 and QI2 Ross had been erroneously identified.

A5.2.2.2   The Inquiry's Findings

Those examiners who think Y7 was made by Ms McKie believe the mark is made up of two parts. Their view is there is movement or superimposition between the top and bottom parts, and so the details in apparent disagreement in the top half are unreliable. However, they contend there is sufficient detail in the bottom part for identification (the four officers involved found at least 16 characteristics in agreement in this area).

The examiners who believe that Y7 was not made by Ms McKie contend that it is all one mark and that the details in apparent disagreement in the top part mean that Ms McKie was not the donor of the mark.

The Inquiry found that the lower part of the mark cannot be relied upon as (at most) six characteristics bear a degree of similarity to Ms McKie's print (Campbell, 2011). This was an inadequate number to have supported a finding of identification under the requirement in place at the time for 16 characteristics to be in agreement for an identification to enter the criminal justice system.

Those who think QI2 Ross was made by Miss Ross could not agree on an interpretation of the details that they saw, and the Inquiry found that as only three characteristics appear to be in agreement, the SCRO were wrong to identify the mark as having been made by Miss Ross (Campbell, 2011).

The Inquiry found that the following factors may have contributed to the erroneous identifications:

- **Examiners '100% certainty' in their conclusions**
  Examiners believed that they could be 100% certain in the conclusions they reached. The inquiry found that this belief may have compromised the independence of the verifications of the marks by influencing the verifiers to support the conclusions of the previous examiners. Furthermore, the belief within the SCRO that senior examiners were infallible may also have made it more difficult for verifying examiners to challenge either identification, as the first examiners to identify both Y7 and Q12 Ross were senior examiners. These beliefs may also provide an explanation as to why, even though some of the examiners involved re-examined Y7 on several occasions, their re-examination led to them confirming their original conclusion.

- **The ethos in the SCRO fingerprint bureau**
  SCRO examiners considered their bureau as amongst the best in the world and believed they could identify marks that others may regard as unsuitable for identification. One practice that may have contributed to this involved examiners 'pushing' the number of characteristics in agreement they could find in order to reach the sixteen that were the legal minimum requirement at the time. Examiners were used to reaching a conclusion that impressions were made by the same area of skin before they had found sixteen characteristics in agreement. Therefore, they were also used to having to find more characteristics to support the conclusion they had already arrived at. The inquiry found that this approach can lead to the examiner interpreting ambiguous details in the mark in a way that is consistent with those in the print.

- **An inappropriate hierarchical philosophy**
  The second examiner to compare Y7 could only find 10 characteristics in agreement – less than the legally required minimum of sixteen. The practice at the time allowed the SCRO to simply find another examiner to verify the identification. The inquiry considered that the fact that one examiner could not find sufficient agreement to meet the legal requirement should have provided an opportunity for the other examiners who did to reflect on their conclusions. The inquiry found that the reason this did not happen may have been due to the fact that the examiner who did not find sixteen characteristics in agreement had less experience than all of the others involved, and the view in the SCRO at the time was that more experience equalled a higher skill level.

- **Inappropriate tolerances**
  Examiners used too wide a tolerance for the details they could see in the mark to appear differently in the print and still be considered in agreement.

- **Reverse reasoning**
  The inquiry found that the examiners used details in the print to inform their view of details in less clear areas of the mark. For example, the inquiry considered it difficult to believe that some of the characteristics relied upon would have been picked out in an unguided analysis of the mark. Due

to the large number of marks in the case, reverse reasoning may also have occurred as a result of the examiners becoming familiar with the details in the prints prior to analysing all the marks.

- **A failure to analyse the entirety of the mark**

  Instead of analysing the whole mark, some examiners focused their attention only on the target group of characteristics they were intending to use. The inquiry found that this introduced the risk that the comparison may be confined to that area and not take into account the remainder of the detail in the mark and print. The inquiry considers that this can lead to the examiner using details in the print to inform their view of details in less clear areas of the mark that they had not already analysed.

- **Circular reasoning and generalised thinking**

  Some examiners subscribed to the view that as long as sixteen characteristics were found in agreement in the impressions, then any additional details that differed in appearance must be explainable as products of distortion. The inquiry found that this circular reasoning meant the examiners did not properly evaluate whether there were factors that could have accounted for the apparent differences in appearance or whether they were indicators that the mark and print were not made by the same donor. However, the inquiry also found that even those examiners who did not subscribe to this form of circular reasoning did not properly evaluate the differences in the appearance in the impressions. This may have been due to the fact that examiners were trained to talk about distortion factors in generalities and not to consider whether, for example, a type of movement could have reasonably explained the specific differences in the appearance in the upper part of Y7 (Figure A5.3).

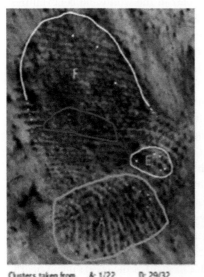

Figure A5.3 Independent examiner Arie Zeelenberg produced this image to illustrate some of the differences in appearance that were explained by examiners as having been caused by distortion. Each of the five coloured areas shows characteristics that were considered to be in agreement by one of the SCRO examiners. Zeelenberg argued that if the top of the mark had moved relative to the bottom (as was claimed) then the green clusters and red clusters should be in different positions in relation to each other in the print – but they are not. Zeelenberg also points out that the clusters that have moved, have not done so in a consistent direction. Beginning with their positions in the print, the pink cluster has leapfrogged the blue cluster and moved from the right of the red cluster to the left. However, whilst the pink cluster has moved further away from the yellow cluster, the blue cluster has moved closer to it. *Source:* Ed German.

- **Slack working practices**

  Some examiners carried out verifications of comparisons entirely using a comparator (a mechanical device that enlarges photographs of the mark and print and projects images of them onto screens upon which the examiner can mark details they consider to be in agreement). The inquiry found that as the image on the screen may not include all the mark or print, this practice may introduce a temptation to the examiner to only compare the part displayed and not look at the entirety of the impressions. Also, once they had reached a conclusion using a comparator, some examiners would leave markings indicating the characteristics they considered to be in agreement on the screens for the verifier to see. The inquiry found that the markings could be a form of 'peer pressure' and could compromise the independence of verification.

- **A failure to address the distinction between elimination and identification**

  A mark could be 'eliminated' as having been made by a victim or a police officer involved in the investigation (as in the case of Y7) without the examiner needing to find sixteen characteristics in agreement. Except for the second examiner who was replaced by another, the four examiners who initially compared Y7 identified it as having been made by Ms McKie to the sixteen-characteristic standard. After Ms McKie challenged the identification, several other examiners were tasked with comparing the impressions, but only one of these was able to find sixteen characteristics in agreement. The others concluded the mark was identified to Ms McKie but only to the lower 'elimination' standard. This meant that by the time of Ms McKie's trial, SCRO examiners were equally divided; there were five who had found sixteen characteristics in agreement and five who had not. Though those examiners may have been no less 'certain' about that conclusion than one in which they did find sixteen characteristics in agreement, the distinction was relevant to the criminal justice system. The inquiry found the fact that several examiners had not been able to find sixteen characteristics in agreement should have presented an opportunity for the SCRO to reflect on the robustness of their evidence. Additionally, it should also have been disclosed to the criminal justice system so that the reliability of the evidence against Mr Asbury and Ms McKie could be properly evaluated.

- **Contextual information from the police**

  Some information communicated to the examiners by the police contained contextual detail that may have influenced or placed pressure on the examiners to 'get results'. For example, in reference to Q12 Ross, the information that a 'shape in dust' had been found on the bedside table of the victim that was believed to be the same size and shape as the tin of money found in Mr Asbury's house. The inquiry recognised that there is some research to show that contextual information can sub-consciously influence the conclusions of examiners and found that the influence of this information on some of the examiners may have contributed to the erroneous identification of Q12 Ross.

A5.2.2.3 Brandon Mayfield

In 2004 a series of explosive devices planted on trains in Spain killed 191 people and injured nearly 2000 others. The Spanish Police (SNP) recovered a plastic bag containing several detonators from a stolen van found near a train station (OIG, 2006). Two identifiable marks were found on the bag, and images of them were sent to the FBI via Interpol. The FBI searched the marks through their computer system and identified one of the marks as having been made by Brandon Mayfield (Figure A5.4).

When notified of the identification, SNP examiners expressed concerns about it which led to them issuing a document that became known as the 'Negativo Report' in which they stated their position that the mark had not been made by Mr Mayfield.

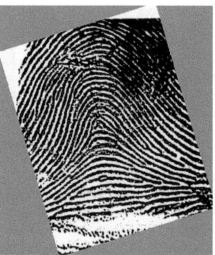

Figure A5.4 One of the marks found on the plastic bag and the left index print of Brandon Mayfield. *Source:* Ed German.

In response, the FBI reviewed the identification, confirmed their view it was correct and dispatched an examiner to Spain to meet with the SNP examiners – at this point, three FBI examiners had identified the mark for Mr Mayfield. At the meeting, the SNP stated that though they had found similarities in the impressions, they had found details in the upper left of the mark that did not match the print made by Mr Mayfield (OIG, 2006). The FBI explained their view that these differences were either due to wrinkles in the plastic bag or the mark actually being comprised of two separate marks. The SNP examiners' view was the mark had been made by one continuous area of detail, and so the details in the upper left meant that it could not have been made by Mr Mayfield. A further difference between the FBI and the SNP was that the FBI examiners had relied on third level details to support their identification, whereas the SNP had not. As a result of the meeting, the SNP examiners agreed to re-examine the impressions, though by this time, Mr Mayfield had already been arrested.

The SNP examiners could not reconcile the differences in the impressions and sought the view of another Spanish examiner who also disagreed with the identification.

Press reports circulated indicating that the SNP had doubts about the FBI identification, and Mr Mayfield's attorney argued that the primary basis for his detention (the FBI identification of the mark) was in doubt. As a result, another American examiner was engaged to review the identification, and that examiner also identified the mark for Mr Mayfield. On the same day that examiner identified the mark for Mr Mayfield, the SNP notified the FBI they had identified the mark for a different person – Ouhnane Daoud (Figure A5.5) (OIG, 2006).

Daoud had emerged as a suspect in the bombings as a result of an SNP raid on an apartment building in Madrid in which several suspects blew themselves up, killing a policeman in the process. The SNP found documents inside the address with Daoud's name on them and were subsequently able to locate a set of his fingerprints that had been taken as a result of an immigration violation (OIG, 2006).

The SNP identification led the FBI to re-examine their identification of Mr Mayfield again, and this time they identified the mark for Daoud rather than Mr Mayfield. The FBI withdrew their identification of Mr Mayfied, who was then released and subsequently filed a civil action naming the FBI examiners as individual defendants that resulted in a formal apology from the US Government and a settlement of $2 million (OIG, 2006).

Figure A5.5 The mark on the plastic bag and the right middle print of Ouhnane Daoud. *Source:* Ed German.

The US Department of Justice, Office of the Inspector General (OIG) published 'A Review of the FBI's Handling of the Brandon Mayfield Case' in 2006. The OIG found six major contributing causes to the erroneous identification of Mr Mayfield.

1) **The unusual similarity of the impressions**

   This similarity can be illustrated by the fact that of the 15 characteristics one of the FBI examiners found in agreement in the mark and Mr Mayfield's print, ten of those same characteristics were also found to be in agreement with Ouhnane Daoud's print by another examiner. The additional American examiner who agreed with the FBI identification stated he had never seen or heard of two impressions having eight or more characteristics in agreement that had not been made by the same area of skin. Though there were differences in the impressions that should have prevented the erroneous identification, the OIG concluded that the correspondence of ten characteristics in impressions from different sources is an 'extremely unusual event'.

   The OIG consider that the prospect of encountering such unusually similar impressions is a particular hazard of using computerised databases of prints. As a result, they recommended that special care be taken by examiners when comparing impressions that are respondents from searches because of the increased danger of encountering a close non-match.

2) **Bias from the prints**

   The OIG concluded that once the FBI examiners had noticed the unusual similarity of some of the characteristics in the mark with those of Mr Mayfield, a premature expectation that the impressions had been made by the same person led to a process of circular reasoning beginning. This process resulted in the examiners using details in the print to inform or adjust their interpretation of details in the mark.

   The OIG found that five of the characteristics in the mark FBI examiners found to be in agreement with the print of Mr Mayfield were not present in the print of Ouhnane Daoud (the actual donor of the mark) and therefore were not in the mark. The examiners interpreted two of these characteristics as ridge endings, but as they occurred at the edge of the mark, there was no evidence that the ridges on either side did actually converge in front of them. In the prints of Daoud, it can be seen that neither are ridge endings – both are continuous ridges. A third of the five characteristics was also interpreted as an ending despite there being no convergence of ridges on either side of it. The remaining two characteristics were very unclear, and the OIG concluded the FBI examiners would not have had confidence that they were actually characteristics without the erroneous reference of Mr Mayfield's print.

Further evidence of the influence of Mr Mayfield's print on the examiner's interpretation of the characteristics can be found in the way the initial FBI examiner encoded the mark to allow it to be searched. Before a mark can be searched, and so before any responding prints are seen, the examiner is required to designate the location and type of some of the characteristics for the computer. Of the seven characteristics that were designated by the FBI examiner, the type or location of five were interpreted differently after the examiner saw the print of Mr Mayfield. All five were altered in a way that was consistent with the appearance of characteristics in the print of Mr Mayfied. Three were changed from ridge endings to bifurcations, one from a bifurcation to an ending and one was moved so that it occurred on the ridge below the one it was originally seen on. When compared with the print of Ouhnane Daoud, the original interpretation of four out of five of these characteristics was correct. The examiner said that during comparison, his practice was to ignore how he had initially seen the details in the mark and reconsider them from a new perspective. The OIG considered that this approach meant that the examiner could discard information gathered during the original analysis without even asking or answering why that analysis was wrong. The OIG found that the lack of a procedure requiring examiners to document their interpretation of details in certain marks prior to seeing the print meant that the examiner may lose track of whether they saw a detail first in print or mark or that their interpretation of it had changed.

3) **Faulty reliance on third level details**

The OIG found that none of the level three details relied upon by the examiners had any correspondence to the print of Ouhnane Daoud and, therefore, could not have been in the mark. The OIG also reported there were several indicators that those details were not reliable; for example, examiners used two subsidiary ridge 'dots' that appeared to be present in one of the rolled prints taken from Mr Mayfield. However, these details were not present in the plain impression of the same area on the same set of prints, nor could they be seen in another set of Mr Mayfied's prints. So, details that some examiners considered to be very persuasive did not appear in most sets of Mr Mayfield's prints and even in the one print they did appear to be in, they were significantly further away from the nearest characteristic in the mark than in print.

Also, independent examiners interviewed by the OIG disputed the FBI examiners contention that there was a similarity in the size and shape of a pore in the mark and in Mr Mayfield's print. The OIG also questioned the practice of relying on such details due to their lack of reproducibility and discriminating power (given that there is a pore on every ridge unit that makes up a ridge). Some independent examiners interviewed by the OIG were unsure whether the mark even had enough clarity to support any reliance on any level three details.

Overall, the OIG found that the FBI examiners did not appear to use 'fair reasoning' (IEEGFI, 2004) when comparing and evaluating the third level details in the mark with those in the print of Mr Mayfield. Instead, they appeared to 'cherry pick' the details that could be seen as similar in the impressions and ignore the many others that could not be. The FBI examiners appeared to use the details they considered to be in agreement to support the identification while attributing those that did not appear to be in agreement to the lack of reproducibility of third level details. The OIG also considered that as the FBI examiners did not try to observe third level details in the mark until after they began the comparison, it appears that the details in the print were used to determine which in the mark were reliable.

4) **Inadequate explanations for differences in appearance**

The OIG found it difficult to review the examiner's explanations for differences in appearance between the mark and Mr Mayfield's print as the FBI procedure at the time did not require examiners to record such information. However, after considering the recollections of the examiners, the OIG did find that there was limited support for their explanations. The OIG found that if true, the examiners explanation that the characteristics in apparent disagreement in the upper left part of the mark were caused by that area having been made by a different person or a different area of Mr Mayfield's skin required the examiners to believe an extraordinary set of coincidences. The only support the examiners could point to for this area not being part of the same area of skin that made the rest of the mark was the gap between the two areas. However, the ridge flow across the gap was consistent – there were no misaligned or crossing ridges and the downward pressure used appeared similar on both sides of the gap. Therefore acceptance of the FBI explanation required the examiners to believe that either; Mr Mayfield touched the surface a second time with an area of skin that was not recorded on any of the sets of prints (because the FBI had been unable to identify it as a separate area), or another person coincidentally did so in a way that the resulting mark lined up consistently with the ridge flow and downward pressure used with the rest of the mark with no misaligned or crossed ridges.

The OIG also found other differences that were inadequately explained, such as differing distances between characteristics in the mark and print. The FBI examiners attributed these differences to slippage or twisting of the area of skin during deposition of the mark. However, the independent examiners interviewed by the OIG found that there were no indicators of significant slippage or twisting in the mark and that the consistent width of the furrows throughout indicated that this type of movement had not occurred. Also, the independent examiners pointed out that skin could not flex or stretch over the small distances to a degree sufficient to explain these differences and that the distances were not consistent across the mark. The OIG concluded that the cumulative impact of all these differences should have been sufficient to preclude identification.

5) **Failure to assess the poor quality of similarities**

One of the FBI examiners produced an enlargement of the mark and print of Mr Mayfield that recorded fifteen characteristics in agreement. Of those, when compared to the print made by the actual donor of the mark, it can be seen that five did not exist at all. When considering the type of the remaining ten, the examiner only correctly determined the type of three. Furthermore, the examiner's interpretation of the type of some of the characteristics had, in fact, changed between the time he encoded the mark for searching on the computer and seeing the print of Mr Mayfield. The examiner felt able to change his interpretation without explanation as the poor quality of the characteristics permitted either interpretation of their type (bifurcation or ridge ending).

The OIG found that even assuming ten characteristics were in agreement according to their orientation, location and relative position, the fact that the examiner was unable to accurately establish agreement in the type of most of them meant that they should have been given less weight in the evaluation stage than those that were unequivocally the same type.

6) **Failure to re-examine the mark following the 'Negativo Report'**

The OIG found that the notification that the SNP had doubts about the identification of Mr Mayfield should have prompted the FBI to establish the basis of the SNP's concerns and appoint a new examiner to review the identification. The OIG concluded that the FBI laboratory's overconfidence in the skill and superiority of its examiners prevented it from taking the report as seriously as it should and carrying out these actions. Though the identification had already occurred at the time of the report, in not re-examining it as a result of the report, the OIG consider the FBI missed an opportunity to detect its error.

A5.2.2.4 Summary

It is clear that some of the same factors are cited in both of these investigations. For example, as there were details in disagreement in all the marks, arguably the principal contributory factor in all the errors were the examiners inappropriate and unsubstantiated reasoning that those differences were caused by distortion. Similarly, the examiners use of the print to inform or guide their interpretation of details in the mark is cited in both. The influence the knowledge of one examiner's conclusion can have on another was also cited.

Both investigations to establish the factors that contributed to these errors were conducted in a different country, by different people at a different period in time. Had they been investigated at the same time by the same people, it may be that many more of the same factors would have been cited in each investigation. For instance, the FBI investigation cites the unusual similarity of the details, but clearly, many examiners found the details in the SCRO errors also to be unusually similar. Similarly, had the practice of designating marks as complex been commonplace at the times of these errors, it is likely that some or all of the marks would have been designated as such as its clear that at least some of the examiners in every case found the comparisons to be challenging. Looking at the appearance of the three marks, it can be seen that none of them reveals an abundance of clear details.

In addition to the factors that contributed to the causes of the errors, the investigations also reveal factors that they all have in common that may assist the examiner in recognising instances where there is a higher risk of a mistake occurring. In each of the cases:

- **Only one mark was 'identified' for each person[1]**
  It makes sense that if the examiner has identified multiple marks made by different areas of skin for a person, there is less chance of them all being erroneous identifications than if there is only one mark.
- **Each mark had details that appeared to be in agreement as well as those that appeared to be in disagreement**
  It also makes sense that generally, the risk of error with such marks is greater than with those revelling similar amounts of detail that is all in agreement or all in disagreement.

Also, though it will not assist the examiner in identifying when they may be more at risk in making an erroneous identification, it is worth noting that another thing the errors all had in common was that they were all verified multiple times. All of these verifications were 'open' (the verifier was aware of the conclusion of the previous examiner). It may be that had blind verification been common practice at the times of these errors, they may have been detected. But it is also worth noting that all the errors were only discovered as a result of comparison by an examiner external to the organisation involved, and these were also likely open (and in fact, some external examiners did also agree with the errors in both cases).

A5.3. Error Rates

A key factor for the criminal justice system in evaluating the significance of examiners conclusions is how often those conclusions are wrong. In response to questions about error rates, examiners have asserted that there are two relevant rates: the rate for examiners and the rate for the 'methodology'.

---

1 Other marks were identified by the SCRO for Miss Ross at her address but no other marks were identified for her on items found at the address of the accused.

Examiners have explained that whilst the former is unknown, as there is no record of the total number of comparisons examiners have conducted versus mistakes they have made, the error rate of the methodology is 'zero' (Cole, 2005; Wertheim, 2001). This viewpoint is based on the idea that if the methodology is properly applied, the examiner will not make errors (Cole, 2005; Grieve, 2001; Wertheim, 2001). However, many would argue that ACE is not specific enough to qualify as a methodology (Champod et al., 2016; Haber & Haber, 2009; Langenburg, 2011; NIST, 2012; NRC, 2009). Rather than following a series of explicit steps and making decisions according to quantifiable criteria, throughout ACE, the examiner makes subjective decisions based on what they see and how they interpret that according to their training and experience (Langenburg, 2011; NIST, 2012). As a result, the examiner cannot be separated from the methodology. As Cole (2005) explains, to have an error rate for the methodology without the examiner is 'akin to calculating the crash rate of an automobile, provided it is not driven'. Therefore, what is relevant for the criminal justice system is the overall error rate, and that is not zero.

Cole (2005) compiled a list of twenty-two erroneous identifications that have been discovered either in the United Kingdom or the United States since 1920. Triplett (2021) maintains a record of errors that includes those that occurred or came to light after Cole's list was published, as well as ones from other countries. Triplett documents approximately 50 erroneous identifications that may have occurred globally in the last 100 years. There is no official record of every global erroneous identification, but even assuming this figure is roughly accurate, without knowing how many correct identifications occurred over that period it is impossible to say how frequently they occur. However, there are some sources that can provide information about error rates, one of which are published records of examiners casework conclusions.

Rairden et al. (2018) reviewed two years of casework at the Houston Forensic Science Centre (United States) involving 12 examiners, 2535 cases and 12,363 marks. All conclusions of identification (3232) were verified, 53% of exclusion conclusions were verified (825), and between 70 and 79% of inconclusive conclusions were verified. There were 132 marks where the verifier reached a different decision to the original examiner. No erroneous identifications were detected, but 22 erroneous exclusions were discovered at the verification stage. Though there were no instances where one examiner reached a conclusion of identification, and the other reached a conclusion of exclusion, there were 41 marks with which one of the two examiners reached a conclusion of identification and the other a conclusion of inconclusive. After consultation, 25 of these were reported as identifications and 16 as inconclusive. Similarly, there were 26 marks with which one examiner reached a conclusion of exclusion and the other a conclusion of inconclusive (15 of which were reported as inconclusive and 11 as exclusion). The other 43 marks on which different decisions were recorded involved one examiner deciding a mark was no value and the other deciding it was value (26 were considered value and 17 no value).

Montooth (2019) analysed three years of casework from 10 examiners working in one laboratory system involving a total of 3399 cases. As in Rairden et al. (2018), no erroneous identifications were discovered, though in 2682 identifications, there were 14 differences of opinion between the original examiner and verifier (all of which were found during technical review).

In every case, the original examiner concluded identification and the verifier concluded inconclusive and the original examiner re-valuated their conclusion and reached a conclusion of inconclusive.

Unfortunately, other than these two studies, there is very little casework data in the public domain, but even if there were, the major limitation of using real casework to consider error rates is the actual identity of the person who made the mark is unknown and without this 'ground truth' there is no way to know whether an examiners conclusion is right or wrong. The fact that multiple examiners reach the same conclusion with casework marks is evidence that conclusion is reproducible, but not necessarily that it is correct. For example, all the erroneous identifications already described were each verified by multiple examiners, and the use of studies where the ground truth is known has also provided examples of consensus conclusions that are incorrect (Langenburg, 2009; Langenburg, Champod & Wertheim, 2009; Pacheco, Cerchiai & Stoiloff, 2014; Wertheim, Langenburg & Moenssens, 2006).

Another source that can provide information on the frequency of errors is proficiency tests. The first external proficiency tests for examiners were administered by Collaborative Testing Services (CTS) in the 1980s who remain the largest provider of such tests (Cole, 2005). The ground truth for the impressions is known, and tests typically include several marks and sets of prints, some of which will be made by the same donor and some of which will not. Cole (2005) reviewed the CTS data for the period 1983–2004 and calculated an overall false-positive rate (FPR) of around 0.8%. Cole found that the FPR varied over each test in the period from 0.0 to 4.4%, with the highest rate occurring twice – the last time in 1995.

The 1995 CTS test consisted of 7 marks that were to be compared with 4 sets of fingerprints (Grieve, 1996). The marks had been made in blood and five of seven exhibited some details in reverse colour. Five of the marks had been made by donors of the sets of prints and the other two had not. Grieve (1996) reports that 156 examiners took the test, and there were 48 erroneous identifications, 29 were on one of the marks that had not been made by a donor of any of the prints (Figure A5.6).

This mark had been made by the twin brother of the donor of one of the sets who had a similar configuration of characteristics in one of his prints to that of his brother. Nineteen percent of examiners erroneously identified this close non-match, but there were also erroneous identifications on all the other marks, including 13 on the five of which the donor's prints were present – some of which may have been clerical errors (Grieve, 1996).

Garrett & Mitchell (2018) reviewed the CTS data from 1995 to 2016 and reported that the FPR ranged from 1 to 23%, with an average rate of 7% for the period. However, this FPR cannot be compared directly with Cole's for the earlier period as they were not calculated in the same way. Whilst Cole (2005) arrived at his FPR by dividing the number of false positives by the number produced by multiplying the number of examiners who took each test by the number of marks in the test, Garrett & Mitchell (2018) calculated their FPR by dividing the number of participants who made

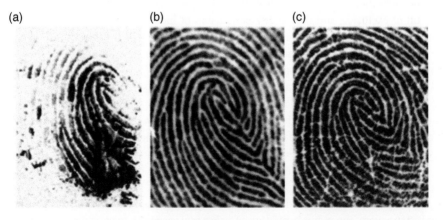

(a)   (b)   (c)

Figure A5.6   29 examiners erroneously identified the mark in (a) as having been made by the same area of skin that made the print in (b). The mark was actually made by the area of skin that made the print in (c), which was not included in the test. *Source:* Reprinted with permission from Collaborative Testing Services, Inc.

at least one error by the total number of participants. So, for the 1995 test, which Cole (2005) calculated as 4.4%, Garrett & Mitchell (2018) calculate as 22%. If the data published by Garrett & Mitchell (2018) is used to calculate an FPR for the period of 1995–2016 in the same way that Cole (2005) did it, then the average is approximately 0.6%.

CTS publishes summaries of the results from the last five years on its website, which allow FPRs to be calculated (Table A5.1).

What can be seen from the data is that some participants make a lot more errors than others. For example, in 2014 – 517/518, 85% of participants correctly identified all the marks that could be identified, but 71% of the false positives were made by 4% of the participants who made between 3 and 11 false positives each on the 12 marks. Similarly, in five of the other tests, between 31 and 63% of all false positives were made by between 0.4 and 1% of participants (2015 – 517/518, 2018 – 5171/2/5, 2017 – 5171/2/5, 2016 – 517/518, 2018 – 5161/2/5). It is also clear that false positives are much more common with some marks than others. In 2014 – 515/516, 86% of false positives occurred on one mark, and four other tests all featured one mark that accounted for between 31 and 63% of all false positives (2015 – 515/516, 2015 – 517/518, 2015 – 515/516, 2020 – 5171/5).

Using Cole's (2005) calculation, for the 14 tests in the period 2014–2020, there is an overall FPR of approximately 0.6%, which is consistent with Cole's (2005) 0.8% for 1983–2004 and Garrett & Mitchell's (2018) 0.6% for 1995–2016.

Whilst the ground truth for the marks in proficiency tests is known and there is much more data in the public domain than there is for casework, there are still significant limitations with using proficiency tests to estimate error rates. The marks used are widely viewed as being less challenging to compare than those found in casework (Garrett & Mitchell, 2018; Koehler, 2008; Saks & Koehler, 2005). A study by Koertner & Swofford (2018) found that the marks in CTS tests were

Table A5.1 A summary of the data from the last five years of CTS tests extracted from http://collaborative-testing.com/.

	Marks	Marks made by a donor whose prints are in the test	Participants	Not Identified	False positives	FPR (%)
2014–515/516	12	12	424	26	21	0.4
2014–517/518	12	10	587	118	178	2.5
2015–515/516	11	9	536	67	48	0.8
2015–517/518	11	11	509	97	36	0.6
2016–515/516	12	9	431	20	51	1
2016–517/518	12	10	487	55	42	0.7
2017–5161/2/5	12	10	360	17	5	0.1
2017–5171/2/5	11	9	438	8	8	0.2
2018–5161/2/5	12	11	353	49	19	0.4
2018–5171/2/5	12	9	479	37	24	0.4
2019–5161/2/5	12	9	331	17	8	0.2
2019–5171/2/5	11	9	369	59	5	0.1
2020–5161/5	12	9	344	15	5	0.1
2020–5171/5	12	11	420	41	19	0.4

generally of higher quality, less complex and not representative of the quality levels observed in routine casework.

Another difference from casework is that the participants are aware they are taking a test which may affect how they carry out their work. Some argue that knowledge results in greater performance accuracy, and so the error rates may be higher in casework (Haber & Haber, 2014). The format of the test also requires the examiner to work in a way that may be different from their normal practice for casework; for example, the only conclusions allowed are 'Identified' or 'Not Identified' (which may mean inconclusive or excluded). The use of these conclusions also means that exclusion rates cannot be calculated. Another limitation of proficiency tests is that though the examiner who takes the test and their organisation will know the result, the results are published anonymously, so do not include who the participants were, what their job is, what that job actually involves and the extent of training they have had. This means it is impossible to ascertain whether those who take the tests are broadly representative of the general fingerprint community. For example, an experienced examiner who is trained to competency and involved in comparing impressions every day may have a very different error rate to a new examiner who is not yet signed off to do unsupervised work or an experienced examiner whose role means they do not routinely conduct comparisons.

Because of these limitations, some feel that proficiency tests should not be used to estimate error rates; in fact, CTS themselves advise in all of their reports that the results of tests cannot be interpreted as an overview of the quality of the work within the fingerprint profession (Haber & Haber, 2014; Wertheim, Langenburg & Moenssens, 2006).

The source of information that is considered to provide the best estimates for error rates are studies that measure the accuracy of examiners' conclusions. Studies can suffer from many of the same limitations as proficiency tests; however, most studies take steps to mitigate those limitations. For example, many studies make efforts to ensure that the marks used are generally agreed to be broadly representative of those found in casework (Pacheco, Cerchiai & Stoiloff, 2014; Ulery et al., 2011). However, some studies also intentionally use marks that are considered to be more challenging which likely increases the error rates in those studies (Koehler & Liu, 2020; Liu et al., 2015; Neumann et al., 2013). Similarly, as with proficiency tests, the participants are aware they are taking part in a study, and so it has been argued that the actual FPR in casework may be higher (PCAST, 2016). However, unlike proficiency tests, most studies are conducted in anonymous conditions, and so it could also be argued that participants may be less motivated and careful as they know there will be no consequences of an error (OSAC, 2016). CTS proficiency tests only allow the examiner to conclude marks are 'Identified' or 'Not Identified', and although some studies do limit conclusions, many allow participants to use the conclusions they are familiar with to more accurately replicate casework conditions. Lastly, unlike proficiency tests, many studies publish information about the examiners who took part, for example, whether they were trainees or qualified, how much experience they have and how frequently they compare impressions. This data allows an assessment of how representative of the general population the participants are.

What follows is a brief description in chronological order of the studies from which an error rate can be established. PCAST (2016) is of the opinion that of those published before 2016, only the studies by Pacheco, Cerchiai & Stoiloff (2014) and Ulery et al. (2011) are appropriate to assess the reliability of fingerprint comparison. However, though some of those described here may not provide significant information in that regard on their own and may not even report an error rate, they all do provide information showing that rates may be higher or lower in certain circumstances than others. For example, though they lack the numbers of marks and/or examiners needed to calculate meaningful rates, the studies by Dror & Charlton (2006), Dror, Charlton & Péron (2006) and Hall & Player (2008) provide information about how contextual information can affect error rates. Because research has shown there is a significant difference between the performance of lay people and those with some training and experience, studies that focus on lay people exclusively are not included in those described (Langenburg, 2006; Langenburg, Champod & Wertheim, 2009; Mnookin et al., 2016; Schiffer & Champod, 2007; Tangen, Thompson & McCarthy, 2011).

1) Evett & Williams (1995)

The study was carried out as part of a review of the 16-point standard in place in the United Kingdom at the time to assess how variable different examiners' conclusions were when considering the same impressions. 130 examiners from England and Wales with at least 10 years of experience were asked to compare ten pairs of impressions. Nine of the pairs were casework examples from the London Metropolitan Police Service of varying quality. Because they were from casework, the ground truth was unknown, but examiners involved in the design of the study considered that all nine comparisons should result in conclusions of 'full' or 'partial' identification.[2] The tenth pair of impressions featured a mark whose donor was known and a print made by a different person. This print had been found by searching a print of the actual donor of the tenth mark through a computer database to find a respondent with similar details – a 'close non match'. The examiners were asked to decide whether the result of each comparison was: a full identification, a partial identification, an exclusion or whether the result was inconclusive.

The study found that there was significant variation between examiners in whether a comparison was a full or partial identification and sometimes whether it was an exclusion. For example, with one comparison, 54% thought the result was a partial identification, 38% thought it was inconclusive and 8% thought it was excluded. The authors reported no false positives but 10 false negatives (nine of which were on one comparison).

2) Dror, Charlton & Péron (2006)

The purpose of this study was to investigate whether examiners can objectively focus on the details in impressions without being misled by extraneous information. Five examiners, selected from an international pool with an average of 17 years' experience, were each asked to compare one pair of impressions. The examiners knew they would be tested at some point within a year without their knowledge but did not know when. Each pair of impressions came from casework, so the ground truth was unknown, but each had been previously compared and identified by one of the participants in the study five years earlier (making it unlikely they would remember the impressions). The authors of the study also engaged two examiners to verify each identification independently of the study. Each examiner was re-presented with the pair they had previously identified by a colleague but were told that the print belonged to Brandon Mayfield and the mark was the one the FBI had erroneously identified for him (participants were pre-screened to establish they were not familiar with the impressions in that case). The examiners were allowed to compare the impressions in the way they normally would, taking as much time as they needed. The authors found that only one examiner reached a conclusion of identification, three reached a conclusion of exclusion, and one decided the result was inconclusive.

---

2 A full identification was one with at least 16 characteristics in agreement, a partial identification was an identification that could not be used in the criminal justice system as it had fewer characteristics in agreement.

3) Dror & Charlton (2006)

The purpose of this study was to examine the possible role that psychological and cognitive factors may have in causing errors. The participants were six examiners, each with a minimum of five years' experience, who were selected from a pool of international examiners. Each examiner was provided with a different set of 8 pairs of impressions. Each set included four pairs that the examiner had previously reached a conclusion of identification with and four pairs they had reached a conclusion of exclusion with. Within each set, two pairs were considered relatively difficult to compare, and two were considered relatively not difficult. The impressions were from casework, so the ground truth was not known, but the participants themselves had reached definitive conclusions with them, and the authors engaged the services of two other examiners who confirmed they agreed with the previous conclusions of the examiners. The examiners were not aware they had already compared the impressions, and they were presented with them along with contextually biasing information designed to influence them to reach the opposite conclusion to the one they previously reached for four of the pairs. For example, if the examiner had previously reached a conclusion of identification with a pair, they were told that the donor of the print was in custody at the time of the offence. Alternatively, for the pairs, they had reached conclusions of exclusion with they were told the donor of the print had confessed to the crime. The comparisons were provided to each examiner by their director or head of the bureau and were told they would be used for an assessment project. The examiners were permitted to compare the impressions in the way they normally would and had an unlimited amount of time to do so.

There was one false positive and four false negatives. There was also one instance where the examiner changed their previous conclusion to inconclusive. Two of the errors (including the only false positive) came on comparisons that were controls in which no additional contextual information was presented to the examiner. Four of the six examiners reached at least one different decision than they had previously though three of the six different decisions were made by the same examiner. Five of the six different decisions occurred with impressions that were considered difficult to compare.

4) Wertheim, Langenburg & Moenssens (2006)

This study used comparison training exercises to assess the accuracy of examiners. 108 examiners who took part in training courses were given packets of impressions of varying difficulty that each contained ten marks and eight sets of prints. Prints of the donors who made all the marks were contained in each packet, so there were no comparisons that should result in exclusion conclusions (the examiners were aware of this). The aim of the courses was to improve the comparison ability of the examiners, so examiners were encouraged to 'push' their comparison abilities by attempting more difficult comparisons. As they identified each mark, the participants were instructed to indicate how confident they were in the identification using a scale of low, moderate or high. The high level of confidence was used to signify the examiner would report the identification in casework; the others indicated the identification was beyond their comfort/ability level. The courses were open to examiners of any level of experience, so the authors separated the results for those with more than one year (92 examiners) and those with one year or less. The authors also classified errors according to whether they considered them to be technical or clerical in nature and at which confidence level they were made.

The authors consider that the data of greatest interest is the 5861 identifications made at the highest level of confidence by examiners with more than a years' experience, in which there were 61 errors.

Two of these were deemed false positives which, if divided by the total number of identifications, produces an FPR of 0.034%. The clerical error rate was 1.01%, and of these, two involved the wrong set of prints but the correct area of skin, with the others being the right set of prints but the wrong area. The study found that over half of the examiners with more than a years' experience did not make any errors, whereas 8 committed three each. Overall, the examiners with a year or less experience committed more errors.

The authors also carried out a follow-up study in which 25 examiners (16 of whom had more than one year's experience) were asked to act as verifiers. There were two packets of marks and prints for verification, each containing one of the false positives made by examiners with more than a years' experience in the initial study, along with another clerical error and eight correct identifications. The verifying examiners were provided with a worksheet indicating the previous examiners' conclusions and invited to agree or disagree with them. Of the 50 possible errors (25 participants each receiving two incorrect conclusions), 49 were detected. The only one that was not was one of the false positives, and the examiner acting as a verifier was a trainee with less than a years' experience who was not performing unsupervised casework.

5) Hall & Player (2008)

The study was conducted to establish whether the introduction of an emotional context would affect the judgement of examiners with a poor-quality mark. Participation in the study was only open to fully trained examiners who had passed the assessments required to be recognised as an expert. Seventy such experts from the London Metropolitan Police Service Fingerprint Bureau with an average experience as an expert of 11 years took part. Most of the examiners were conducting casework on a daily basis, with the other 12 in managerial roles. The examiners were assigned to either the 'low-emotional context' group or the 'high-emotional context' group. All the examiners were tasked with comparing the same mark with the same print, but the low context group were given information that the offence the mark related to was forgery, and the high context group were given information that it related to an offence of murder. The mark was made by the same donor as the print, but the comparison was judged by a sample group of examiners to reveal details that were at the borderline of an identification conclusion. The examiners worked according to their usual conditions, there was no time limit imposed, and the research was conducted anonymously. The examiners were asked to decide whether their conclusion with the comparison was; identification, not an identification, insufficient for comparison or insufficient detail to establish identity (some detail in agreement but not enough to reach a conclusion of identification).

The design of the experiment meant that it was not possible to record a false positive, but there was one false negative. Nineteen percent of the examiners reached an identification conclusion, 41% decided there was some detail in agreement but not enough to identify, 39% decided the mark was not suitable for comparison and 1% that it was not identified.

The authors reported that the crime type contexts had no significant effects on the examiner's final conclusions. However, whilst the numbers of identifications did not vary significantly between the groups, 50% of examiners given the murder context recorded in a feedback sheet that they thought the severity of the offence had affected their analysis, whereas only 6% of examiners in the forgery group recorded the nature of that offence had influenced their analysis.

6) Langenburg, Champod & Wertheim (2009)

The purpose of this study was to assess whether examiners could be influenced by extraneous contextual information during a verification process. It was conducted at a conference, and the participants were deceived into believing they were participating in an experiment to measure variation between examiners during comparisons. Forty-three examiners with an average experience level of 11 years took part.

Examiners were separated into a control group, a low bias group and a high bias group, each of which received a set of six pairs of impressions. Three of the pairs had been made by the same person and three had not. The marks were a combination of easy, medium and difficult examples as agreed by a consensus of examiners. One of the prints used in one of the pairs that had not been made by the same source was a close non-match found as a result of a search of a computer database. The control group received the impressions with no additional contextual information, whereas the low bias group received them with a worksheet of conclusions they were told had been reached by another examiner and were required to state whether they agreed or disagreed. The high bias group got a similar worksheet but were also told by a prominent, internationally known examiner that these were his conclusions from an actual case. The prominent examiner also provided a commentary of his thoughts on the impressions. Once the experiment was complete, the examiners were debriefed, and at that point, many in the bias groups stated they had 'caught on' and realised the intentions of the study – an awareness that the authors consider critical in interpreting the results.

The authors report that there were one false positive and three false negatives. All four errors were committed by different examiners who were in the control group, and all three false negatives were made on the same pair of impressions, which were classed as difficult. The false positive occurred with a pair that was classed as easy, and none of the examiners recorded a false positive with the close non-match.

At the beginning of the study, examiners categorised themselves as being 'trained to competency and performing latent print casework', 'Certified Latent Print Examiner' or 'other' (e.g. in training, no longer performing casework due to management duties, AFIS operator only) and the authors note that two of four errors (including the false positive) were committed by examiners in the latter category.

Though the extraneous contextual information provided did not result in any false positives, examiners exposed to it produced fewer definitive conclusions and more inconclusive responses, whereas those in the control group made more definitive conclusions and, as a result, more errors.

7) Langenburg (2009)

The purpose of this study was to measure the reliability of the examiner's conclusions. The participants were six examiners from the same organisation in Minnesota who were all routinely performing casework and had between 6 and 35 years' experience. The marks used in the study were graded by a consensus of examiners to be of varying levels of difficulty and included close non-matches.

The experiment was split into 3 phases; in phase one, the examiners were given the same 60 marks to compare with eight sets of prints. In phase 2, each examiner was given another 30 marks with eight sets of prints. Once they had recorded their conclusions, the examiner was to give their 30 to another examiner (who had done a different 30) to verify. However, before they passed their work over, some of their correct conclusions were changed to false positives to see if they would be verified. At least one of each participants conclusions was changed to indicate they had reached a conclusion of identification with a close non-match. In phase 3, all errors committed by participants in phases 1 and 2 were re-presented to them amongst some of their correct conclusions with no indication that errors were present. In this phase, each mark was presented side-by-side with a print to see if the examiner would maintain their incorrect conclusions.

The study found that there was one false positive in phase 1 and two in phase 2 though all three errors were considered by the authors to be clerical (they were all for the wrong finger or hand for the correct set of prints). None of these false positives or the seven deliberate control false positives was verified by other examiners in phase 2, and none were repeated by the examiners who made them in phase 3. There were also three false negatives in phase one, two of which occurred with the same mark which had accidentally been included in the study in the reverse direction. There were also six false negatives in phase 2, with four of the six examiners committing one and one of them committing two. All six were incorrectly verified by other examiners. Of the false negatives that were re-presented to the examiner who made them in phase 3, three of seven were repeated.

8) Tangen, Thompson & McCarthy (2011)

The purpose of this study was to measure the accuracy of examiners. Thirty-seven qualified Australian examiners were asked to compare 36 pairs of impressions. Each pair was presented on screen, and the examiners were required to record whether it was a 'match' or 'no match' (inconclusive conclusions were not permitted). A third of the pairs were made by the same donor, a third were made by different donors but were comprised of prints that were close non-matches, and the remaining third were also not made by the same donor but were randomly paired. The authors report there were no false positives with the randomly paired different source donors and 3 with the close non-matches resulting in an FPR for that category of 0.68% (calculated by dividing the number of false positives by the total number of comparisons carried out with close non-matches (444).

9) Ulery et al. (2011)

This study was the first large scale experiment designed to measure the accuracy and reliability of the examiner's conclusions. The 169 examiners with an average of 10 years' experience and almost all from the United States each compared approximately 100 pairs of impressions. The marks were selected to reflect the range of difficulty encountered in casework and included prints that were close non-matches found by searching some of the marks through a large computer database. The participants were sent a DVD containing the impressions and were given several weeks to complete the test.

The authors report that six false positives occurred with the 4083 pairs of non-matching impressions examiners considered to be of value for identification which equates to an FPR of 0.15%. The six errors were made by five examiners (3%), and two of them occurred with the same mark but with different paired prints.

The study found there were 450 false negatives with the 5969 pairs of matching impressions examiners considered to be of value for identification which equates to an FNR of 7.5%. 85% of examiners made at least one false negative. The authors report that a process analogous to blind verification would have detected all false positives (as there were no cases of different examiners making the same error) and most false negatives. The authors found that the likelihood of either type of error varies according to the difficulty of the impressions and the examiner involved. The study also found that examiners frequently differed on whether marks were suitable for reaching a conclusion or not.

The authors conducted a follow-up study (Ulery et al., 2012) under similar conditions in which some of the original examiners were re-presented with 25 pairs of impressions they had already compared in the original experiment. The examiners were not informed they had already compared the impressions, and the follow-up study took place seven months after the original study. The authors report that the FPR and FNR of this re-test were consistent with the original study.

10) Langenburg, Champod & Genessay (2012)

The purpose of this study was to evaluate how examiners would incorporate information from new tools into their decision-making process. For example, during the analysis stage, some of the examiners were provided with an interpretation of the characteristics the mark revealed produced by a consensus of examiners. 176 examiners with a wide range of experience participated in the study, including some trainees. The examiners worked at their own pace to compare 12 pairs of impressions online using the PiAnoS platform. Seven of the pairs were made by the same donor and were determined in pilot testing to be challenging comparisons that revealed significant distortion and/or apparent differences. The other five pairs were all made by different donors and were all close non-matches in which the prints had been found by searching a computer database to find the most similar impression to maximise difficulty. The authors report that during pilot testing, most of these pairs produced errors and disagreements amongst examiners.

Overall, there were 23 false positives which, when divided by the 880 different source pairs produces an FPR of 2.6%. There were 70 false negatives resulting in an FNR of 5.7%. The study found that experienced examiners had significantly lower false positive rates than trainee examiners. The authors also report that the examiners who were exposed to the consensus information from other examiners demonstrated more accuracy in the characteristic selection and in the conclusions reported. Another factor that appeared to impact error rates was whether the examiner chose to document and annotate their work or not – those who did made significantly fewer errors.

11) Neumann et al. (2013)

This study was conducted to provide a better understanding of the concept of sufficiency. 146 US examiners and trainee examiners were provided with 15 pairs of marks and prints. The marks were chosen to represent challenging comparisons, and 12 of the pairs were made by the same source. The prints for the three different source pairs were close non-matches specifically selected to be as similar as possible to the marks they were paired with using a computer database. The examiners carried out their comparisons using the PiAnoS platform and were able to work at their own pace, starting and stopping as required. The authors report there were 13 false positives, all 13 of which were made on two marks (with 11 of them being made on the same mark in Figure A5.7).

Figure A5.7 11 examiners erroneously identified this mark as having been made by the same area of skin as this print. *Source:* Reprinted with permission from Neumann et al. (2013).

Figure A5.8   19 examiners erroneously excluded this mark as not having been made by the same area of skin as this print. *Source:* Reprinted with permission from Neumann et al. (2013).

There were 96 false negatives with at least one occurring on 11 of 12 same source pairs with 19 of them occurring on the mark in Figure A5.8.

12) Pacheco, Cerchiai & Stoiloff (2014)

The largest study of its type to date, its purpose was to evaluate the reliability of examiners and to determine error rates. 109 US examiners from 76 different law enforcement agencies with at least one years' experience were given 80 marks (including palm marks) of varying difficulty that were judged to be representative of casework.

Examiners in the first two phases of the study (ACE phases) were presented with a selection of the marks and asked to compare each with three sets of prints. The prints of the person who made the mark were present for 56 of the marks. In the third phase (ACE-V phase), examiners were asked to act as either the first or second verifier and presented with conclusions from the first two phases (including errors).

There were 3138 examinations conducted where the prints of the donor who made the mark were present and 1398 conducted where they were not.

The authors report there were 42 false positives which they calculated to equate to an FPR of 3% when divided by the total number of different source trials (1398).[3] However, that calculation was widely criticised, and the authors subsequently acknowledged that it was incorrect (Ausdemore, Hendricks, & Neumann, 2019; OSAC, 2016; Pacheco, Cerchiai & Stoiloff, 2019; Wilkinson, Richard & Hockey, 2018). The error originated from the fact that the number of false positives was divided by the number of trials where the prints of the donor of the mark were not included. This is accurate if each mark is only being compared with one print (as in Ulery et al., 2011, for example) because then it is not possible to make a false positive. However, as this experiment required each mark to be compared with three sets of prints, even in the trials where one of those sets of prints was made by the same donor as the mark, it is still possible to make a false positive. Therefore, the correct FPR denominator should be 4536 – the total number of trials where a false positive was possible (all the same source and all the different source examinations). When calculated this way, the FPR is 0.93% including

---

3 The 3.0% FPR included inconclusive conclusions that occurred with the 1398 trials but the authors also calculated another FPR of 4.2% if the 403 inconclusive conclusions were not included.

inconclusive conclusions (or 1.14% excluding the 849 inconclusive conclusions). Twenty-eight of 109 examiners committed at least one false positive, 19 committed 1, 6 committed 2, 2 committed 3 and one committed 5.

The authors report that 35 of the 42 false positives appear to be clerical in nature (all 35 were identifications for the wrong area of skin of the right person whereas the other seven were for the wrong person), but they were unable to determine that with certainty. Of the seven that appear to be true false positives, one examiner committed three, another committed two and the remaining two were by two separate examiners – so they were committed by 4 of the 109 examiners. The false positives occurred on four separate marks that were considered on average as moderately difficult to compare.

The study reported 235 false negatives from 3138 same source trials (including inconclusive conclusions) resulting in a false negative rate of 7.5% (8.7% without inconclusives).

The error rates for the ACE-V trials were lower, with an FPR of 0% (no false positives were verified) and an FNR of 2.9%. However, the study also tested the examiners when they thought they were the second verifier by presenting them with the original examiner's conclusion and a verifying examiner's agreeing conclusion. Under those conditions, there were 329 second verifications – 244 when the source was present and 85 when the source was not present. The authors report there were three false positives resulting in an FPR of 3.37% (including four inconclusive conclusions) and 15 false negatives resulting in an FNR of 5.08% (including 51 inconclusive conclusions).

13) Liu et al. (2015)

This study was designed to assess the accuracy of judgments made by examiners. Approximately 40 Chinese examiners were asked to compare five pairs of impressions online using the PiAnoS platform. The marks were selected to represent difficult cases and included close non-matches. Two of the pairs were made by the same source and three were not. The authors reported three examiners made a false positive with one of the marks that were considered a close non-match (this mark was also used in Neumann et al. (2013) and is shown in Figure A5.7).

14) Mnookin et al. (2016)

The main objective of this study was to identify and quantify fingerprint details that are predictive of identification difficulty and accuracy.

The study comprised of two separate experiments with examiners. In experiment one 56 examiners with a range of experience from 1 to 25 years were asked to compare sets of 20 pairs of impressions at a conference in the United States. The marks were considered to be similar to those encountered in casework, and half of the pairs were from the same source, and half were close non-matches found as a result of searches of a computer database. The examiners completed the comparisons on an online program and had three minutes to record whether each pair was a match or non-match (inconclusive conclusions were not permitted, but the examiners were required to record the confidence level they had in their chosen conclusion). Most participants completed two sets of 20 different pairs. There were 37 false positives from 1144 pairs of impressions not made by the same source resulting in an FPR of 3.2%. Nineteen of the examiners made at least one false positive. Nineteen of the non-matching pairs were judged to be matching by one examiner, six by two examiners and two by three examiners. Twelve examiners made one false positive, three made two, three made three and one made ten. There were 163 false negatives from 1148 pairs of impressions made by the same source resulting in an FNR of 14.2%.

The second experiment with examiners involved 34 participants with between 1 and 36 years' experience recruited via personal contact with the authors. The experiment was an extension of experiment one with different examiners and in a more realistic setting. The examiners completed the comparisons online, had unlimited time to do so and were able to conclude pairs were inconclusive. The authors report performance was slightly better than the first experiment with an FPR of 0.25% and an FNR of 10%.

Overall, the authors observed there were more errors of both types with more difficult comparisons, and there was very little agreement on whether comparisons were judged inconclusive or not.

15) Koehler & Liu (2020)

The purpose of this study was to assess the error rate of examiners when tasked with comparing close non-matching impressions.

A test was taken by examiners from 125 unspecified organisations, sometimes by one examiner from the organisation or sometimes by a collaborative group. The authors had permission from national authorities to conduct the study as a mandatory proficiency test.

The comparisons were conducted using the PiAnoS platform and involved five pairs of impressions, three of which were made by the same source, with the remaining two being close non-matches. The examiners had five days to complete the comparisons, and unlike most studies, they knew that any errors could have serious consequences for them and their organisation because it was a mandatory test. Of the two close non-matches, with one, 17 of 107 organisations that provided a definitive conclusion (not inconclusive) erroneously identified the mark, resulting in an FPR of 15.9% (this mark was the same one used in Neumann et al. (2013) and Liu et al. (2015) and is shown in Figure A5.7). With the other, 27 of 96 organisations did the same resulting in an FPR of 28.1%.

16) Eldridge, De Donno & Champod (2021)

This study was conducted to assess error rates with palm impressions. 226 trained examiners took part who were considered to be representative of examiners working in operational organisations. There were 526 pairs of impressions that were assessed as being of similar quality and complexity as those used in casework. Each pair was assessed by about 23 examiners, and each examiner assessed an average of 37 pairs – all of which were done using the same protocols as their normal casework. There were 1785 no match pairs, and 5855 match pairs. The authors report that 12 false positives were recorded, resulting in an FPR of 0.67% calculated by dividing the number of false positives by the total number of trials where the impressions were not made by the same person. Eight examiners committed false positives, with four making one each and the other four making two each. There were also 552 false negatives resulting in an FNR of 9.46%.

Though studies are considered to provide the best estimates of error rates, like casework and proficiency tests, they still have limitations. For instance, as well as those that have few participants or marks, some only use participants from one organisation. Haber & Haber (2014) argue that for the participants to truly be a representative sample, they would need to be chosen randomly from all examiners in the profession. Also, the design of some studies may be more or less challenging for the examiner than casework. For example, some ask the examiner to compare each mark with one print, whereas others require them to compare each mark with multiple sets of prints. The ratio of matching to non-matching impressions used in the study also can affect the error rates as there may be limited opportunities to make a particular type of error, and the ratio may be very different from that found in casework (Haber & Haber, 2014). It has also been argued that the conditions under which most studies are conducted may be more conducive to accurate performance than those encountered in casework (Haber & Haber, 2014). For instance, in a study, the examiner typically will not be dealing with deadlines, interruptions, bias, and distractions which may decrease accuracy. The conditions of some studies also differ from those of casework in that they do not allow examiners to use the full range of conclusions. Another limitation is that the error rates produced by most studies differ from casework in that they do not account for the fact that almost all conclusions of identification are verified by another examiner before they are reported.

It has also been argued that the way the studies deal with inconclusive conclusions may mean that the error rates they produce underrepresent those in casework (Dror & Scurich, 2020; Max, 2019). As in casework, inconclusive conclusions are not treated as errors; however, whilst such conclusions can be appropriate, they can also be inappropriate. Ulery et al. (2011) found that over one-third of all conclusions in their study were inconclusive and reported it was not unusual for one examiner to reach an inconclusive conclusion while another reached a conclusion of identification with the same impressions. As well as different examiners reaching inconclusive conclusions with impressions others have identified or excluded, in Ulery et al. (2012), the authors found that about 10% of the time, the same examiners reached different conclusions when asked to re-compare the same marks – with most of the changes producing inconclusive conclusions. Dror & Scurich argue that the knowledge that such conclusions will not be treated as errors may encourage examiners to use the inconclusive conclusion as a way out of making a decision that in casework they may have to make. A similar argument can be made for 'no value' decisions (Max, 2019).

As well as their limitations, another issue with using the studies to estimate error rates is it is hard to compare them to each other as not only is their design different, but so is the way they calculate or present the rates. For example, some studies include inconclusive responses in their calculations of FP and FN rates, whereas others do not. Also, though most FPRs are calculated by dividing the number of false positives by the number of pairs of impressions compared that did not come from the same area of skin, some are calculated by dividing the number of false positives by the number of identification conclusions[4] (Wertheim, Langenburg & Moenssens, 2006) or the total number of definitive conclusions (Koehler & Liu, 2020).

In addition to false positive or false negative rates, some studies also present error rates in the form of confidence intervals. A confidence interval is a range of values between which the true value is likely to lie. For example, in Ulery (2011), the FPR of 0.15% is calculated with a 95% confidence interval of 0.06–0.3%. What this means is that if the study was repeated, the authors expect the error rate to be between those lower and upper bounds 95% of the time. Another way of presenting this data is to say that it corresponds to a false positive error occurring every 1667 (lower bound) or 333 (upper bound) comparisons. In a review of some of the studies, PCAST (2016) chose to present the error rates in this way but have been criticised for only referencing the upper bound (Langenburg, 2019; OSAC, 2016; Pacheco, Cerchiai & Stoiloff, 2019; Wilkinson, Richard & Hockey 2018). PCAST stated that the Pacheco, Cerchiai & Stoiloff (2014) study yielded a false positive rate that 'could be as high as 1 in 18 cases'. This figure was based on the incorrect FPR rate published by that study as previously discussed, but the criticisms stemmed from the fact that presenting only the upper bound in this way could be seen as a 'worst case scenario' for fingerprint evidence rather than that the authors of the study expect the error rate to be below that 95% of the time (Wilkinson, Richard & Hockey 2018).[5]

Another factor that makes it difficult to compare studies is that they do not all define what an error is in the same way. For instance, some studies will attempt to differentiate between clerical false positives and technical ones, whereas others will count them all as false positives. In Pacheco,

---

4 This is known as a False Positive Discovery Rate (FPDR) (Koehler, 2008).
5 OSAC (2016) calculated the correct upper bound for the study was 1 in 67 or 1 in 256 (if the false positives that appeared to be clerical errors are excluded).

Cerchiai & Stoiloff (2014), although the authors thought 35 of the 42 false positives were clerical, they did not differentiate between them. Had they done so the FPR would have been 0.15% (including inconclusives) rather than 0.93%. The reason Pacheco, Cerchiai & Stoiloff (2014) suspected these errors were clerical was that the examiner had identified the correct person but not the correct area of their skin. However, the authors were unable to conclude with certainty that such errors were not technical and caused by the area of skin the examiner designated appearing similarly to the actual area that made the mark. Other studies have employed measures to assess whether a false positive is likely to be clerical or technical (Langenburg, 2009; Wertheim, Langenburg & Moenssens, 2006), but irrespective of that, many argue that false positives where the examiner identified the correct person but the wrong area of skin of that person should be treated differently from errors where they identified the wrong person (Ausdemore, Hendricks & Neumann, 2019; Koehler 2017; Koehler & Liu, 2020). Koehler & Liu (2020) explain that to the criminal justice system errors such as these are generally inconsequential as (if it is a source-level proposition that is in question) what is relevant is who made the mark not which area of skin of that person made the mark.

Though the studies have limitations and are difficult to compare because of their varying objectives, designs and ways of presenting or calculating error rates, they do support certain conclusions.

First, the rate of errors varies according to the type of error. All the studies that measured both types found that false negatives occur much more frequently than false positives (Eldridge, De Donno & Champod, 2021; Langenburg, 2009; Langenburg, Champod & Genessay, 2012; Langenburg, Champod & Wertheim, 2009; Mnookin et al., 2016; Neumann et al., 2013; Pacheco, Cerchiai & Stoiloff, 2014; Ulery et al., 2011). For instance, Ulery et al. (2011) reported 75 times as many false negatives as false positives.

The studies also show that some examiners made more errors than others (Eldridge, De Donno & Champod, 2021; Langenburg, Champod & Genessay, 2012; Langenburg, Champod & Wertheim, 2009; Pacheco, Cerchiai & Stoiloff, 2014; Ulery et al., 2011). In Wertheim, Langenburg & Moenssens (2006), over half the 92 examiners with more than a years' experience did not make any errors, but eight examiners made three each, and in Mnookin et al. (2016), over 25% of the total number of false positives were made by the same examiner.

Third, error rates were also shown to vary according to the mark (Koehler & Liu, 2020; Langenburg, 2009; Langenburg, Champod & Wertheim, 2009; Mnookin et al., 2016; Ulery et al., 2011). For instance, the FPRs for the close non-matches used in Koehler & Liu (2020) were over 100 times greater than the overall FPR produced by Ulerly et al. (2011), and in Evett & Williams (1995), 9 of 10 false negatives occurred with the same comparison, and in Neumann et al. (2013) 11 of 13 false positives occurred on one pair.

Last, all the studies that included verification (or an approximation of it) found that error rates were lower after verification (Langenburg, 2009; Pacheco, Cerchiai & Stoiloff, 2014; Ulery et al., 2011; Wertheim, Langenburg & Moenssens, 2006).[6]

A5.3.1 Conclusion

There is no field-wide error rate for fingerprint examination, but the two largest studies produced false-positive rates that both can be calculated as less than 1% before verification. As verification has been shown to be very effective at detecting false positives and almost all organisations verify all identifications, the false-positive rates that enter the criminal justice system are likely to be

---

6 In Langenburg (2009), whilst the verifiers detected all the false positives made by the original examiners, they did not detect any of the false negatives so the FNR was the same after verification as it would have been before.

considerably lower (Black, 2012; Langenburg, 2009; Pacheco, Cerchiai & Stoiloff, 2014; Ulery et al., 2011; Wertheim, Langenburg & Moenssens, 2006).

Though the studies provide objective information which can be used to evaluate the general reliability of fingerprint evidence, the error rates do not tell us much about the probability of error in a particular case the examiner has made. Because particular examiners working under particular conditions with particular marks made a particular error 0.1% of the time does not mean that there is a 0.1% chance that the conclusion an examiner has made in the case in question is erroneous. Whilst participants in that study made errors at that rate, the chance of error in any given case may be lower (or higher) than that. The factors that need to be considered to assess the chance of error in a particular case include the discriminating power of the details (their specificity) in terms of their amount, clarity and rarity, the complexity of the comparison, the examiners training, experience and proficiency and the quality control systems in place in their organisation.

References

Ausdemore, M, A, Hendricks, J, H & Neumann, C 2019, 'Review of Several False Positive Error Rate Estimates for Latent Fingerprint Examination Proposed Based on the 2014 Miami-Dade Police Department Study', *Journal of Forensic Identification*, 69 (1), pp. 59–81.

Black, J, P 2012, 'Is There a Need for 100% Verification (Review) of Latent Print Examination Conclusions?', *Journal of Forensic Identification*, 62 (1), pp. 80–100.

Campbell, A 2011, *The Fingerprint Inquiry Report*, available at: http://www. thefingerprintinquiryscotland.org.uk/inquiry/3127-2.html.

Champod, C, Lennard, C, Margot, P & Stoilovic, M 2016, *Fingerprints and Other Ridge Skin Impressions*, Second Edition, CRC Press, Boca Raton.

Cole, S. A 2005, 'More Than Zero: Accounting for Error in Latent Fingerprint Identification', *The Journal of Criminal Law & Criminology*, 95 (3), pp. 985–1078.

Dror, I, E & Charlton, D 2006, 'Why Experts Make Errors', *Journal of Forensic Identification*, 56 (4), pp. 600–616.

Dror, I, E, Charlton, D & Péron, A, E 2006, 'Contextual Information Renders Experts Vulnerable to Making Erroneous Identifications', *Forensic Science International*, 156 (1), pp. 74–78.

Dror, I, E & Scurich, N 2020, '(Mis)use of Scientific Measurements in Forensic Science', *Forensic Science International, Synergy* 2, pp.333–338.

Eldridge, H, De Donno, M & Champod, C 2021, 'Testing the Accuracy and Reliability of Palmar Friction Ridge Comparisons – A Black Box Study', *Forensic Science International*, 318, 110457.

Evett, I, W & Williams, R, L 1995, 'A Review of the Sixteen Points Fingerprint Standard in England and Wales', *Fingerprint Whorld*, 21 (82), pp. 125–141.

Garrett, B, L & Mitchell, G 2018, 'The Proficiency of Experts', *University of Pennslyvania Law Review*, 166, pp. 901–960.

Grieve, D, L 1996, 'Possession of Truth', Editorial, *Journal of Forensic Identification*, 46 (5), pp. 521–528.

Grieve, D, L 2001, 'Simon Says', Editorial, *Journal of Forensic Identification*, 51 (1), pp. 85–97.

Haber, L & Haber, R, N 2009, *Challenges to Fingerprints*, Lawyers & Judges Publishing Company, Inc., Tucson, Arizona.

Haber, R, N & Haber, L 2014, 'Experimental Results of Fingerprint Comparison Validity and Reliability: A Review and Critical Analysis', *Science and Justice*, 54, pp. 375–389.

Hall, L, J & Player, E 2008, 'Will the Introduction of an Emotional Context Affect Fingerprint Analysis and Decision-Making?', *Forensic Science International*, 181, pp. 36–39.

IEEGFI (Interpol European Expert Group on Fingerprint Identification) 2004, Method for Fingerprint Identification II, http://www.latent-prints.com/images/ieegf2.pdf.

Koehler, J, J 2008, 'Fingerprint Error Rates and Proficiency Tests: What They Are and Why They Matter', *Hastings Law Journal*, 59, pp. 1077–1098.

Koehler, J, J 2017, 'Forensics or Fauxrensics? Ascertaining Accuracy in the Forensic Sciences', *Arizona State Law Journal*, 49, pp. 1369–1416.

Koehler, J, J & Liu, S 2020, 'Fingerprint Error Rate on Close Non-Matches', *Journal of Forensic Sciences*, doi: https://doi.org/10.1111/1556-4029.14580.

Koertner, A, J & Swofford, H, J 2018, 'Comparison of Latent Print Proficiency Tests with Latent Prints Obtained in Routine Casework Using Automated and Objective Quality Metrics', *Journal of Forensic Identification*, 68 (3), pp. 379–388.

Langenburg, G, M 2006, 'Pilot Study: A Statistical Analysis of the ACE-V Methodology – Analysis Stage', *Journal of Forensic Identification*, 54(1), pp. 64–79.

Langenburg, G 2009, 'A Performance Study of the ACE-V Process: A Pilot Study to Measure the Accuracy, Precision, Reproducibility, Repeatability, and Biasability of Conclusions Resulting from the ACE-V Process', *Journal of Forensic Identification*, 59 (2), pp. 219–257.

Langenburg, G 2011, 'Scientific Research Supporting the Foundations of Friction Ridge Examinations', in National Institute of Justice, *The Fingerprint Sourcebook*, www.nij.gov.

Langenburg, G 2019, Letter Regarding: 'Review of Several False Positive Error Rate Estimates for Latent Fingerprint Examination Proposed Based on the 2014 Miami-Dade Police Department Study', *Journal of Forensic Identification*, 69 (1), pp. 82–90.

Langenburg, G, Champod C & Genessay, T 2012, 'Informing the Judgments of Fingerprint Analysts Using Quality Metric and Statistical Assessment Tools', *Forensic Science International*, 219, pp. 183–198.

Langenburg, G, Champod, C & Wertheim, P 2009, 'Testing for Potential Contextual Bias Effects During the Verification Stage of the ACE-V Methodology when Conducting Fingerprint Comparisons', *Journal of Forensic Science*, 54 (3), pp. 571–582.

Liu, S, Champod, C, Wu, J & Luo, Y 2015, 'Study on Accuracy of Judgments by Chinese Fingerprint Examiners', *Journal of Forensic Science and Medicine*, 1:33-7, pp. 33–37.

Langenburg, G, Hall, C & Rosemarie, Q 2015, 'Utilizing AFIS Searching Tools to Reduce Errors in Fingerprint Casework', *Forensic Science International*, 257, pp. 123–133.

Max, B 2019, Letter Regarding: 'Review of Several False Positive Error Rate Estimates for Latent Fingerprint Examination Proposed Based on the 2014 Miami-Dade Police Department Study', *Journal of Forensic Identification*, 69 (1), pp. 117–119.

McKie, I,& Russell, M 2007, *The Price of Innocence*, Birlinn.

Mnookin, J, Kellman, P, J, Dror, I, Erlikhman, G, Garrigan, P, Ghose, T, Metler E & Charlton, D 2016, 'Error Rates for Latent Fingerprinting as a Function of Visual Complexity and Cognitive Difficulty', U.S. Department of Justice, 249890, May, 2009-DN-BX-K225.

Montooth, M, S 2019, 'Errors in Latent Print Casework Found in Technical Reviews', *Journal of Forensic Identification*, 69 (2), pp. 125–140.

Neumann, C, Champod, C, Yoo, M, Genessay, T & Langenburg, G 2013, 'Improving the Understanding and the Reliability of the Concept of 'Sufficiency' in Friction Ridge Examination', National Institute of Justice – Office of Justice Program, Award 2010-DN-BX-K267.

NIST 2012, *Latent Print Examination and Human Factors: Improving the Practice through a Systems Approach*, Report of the Expert Working Group on Human Factors in Latent Print Analysis, US Department of Commerce.

NRC 2009, *Strengthening Forensic Science in the United States: A Path Forward*, National Academy Press, Washington, DC.

OIG, 2006, *A Review of the FBI's handling of the Brandon Mayfield Case*, U.S. Department of Justice, https://oig.justice.gov/special/s0601/final.pdf.

OSAC 2016, Friction Ridge Subcommittee's Response to the President's Council of Advisors on Science and Technology's (PCAST) Request for Additional References, Submitted December 14, https://www.nist.gov/organization-scientific-area-committees-forensic-science/friction-ridge-subcommittee.

Pacheco, I, Cerchiai, B & Stoiloff, S 2014, 'Miami-Dade Research Study for the Reliability of the ACE-V Process: Accuracy & Precision in Latent Fingerprint Examination', National Institute of Justice – Office of Justice Program, Award 2010-DN-BX-K268.

Pacheco, I, Cerchiai, B & Stoiloff, S 2019, Special Feature, Letter regarding: 'Review of Several False Positive Error Rate Estimates for Latent Fingerprint Examination Proposed Based on the 2014 Miami-Dade Police Department Study', *Journal of Forensic Identification*, 69 (1), pp. 91–103.

PCAST 2016, *Forensic Science in Criminal Courts: Ensuring Scientific Validity of Feature-Comparison Methods*, September, https://obamawhitehouse.archives.gov/sites/default/files/microsites/ostp/PCAST/pcast_forensic_science_report_final.pdf.

Rairden, A, Garrett, B, L, Kelley, S, Murrie, D & Castillo, A 2018, 'Resolving Latent Conflict: What Happens When Latent Print Examiners Enter the Cage?', *Forensic Science International*, 289, pp. 215–222.

Ray, E & Dechant, P, J 2013, 'Sufficiency and Standards for Exclusion Decisions', *Journal of Forensic Identification*, 63 (6), pp. 675–697.

Saks, M, J & Koehler, J, J 2005, 'The Coming Paradigm Shift in Forensic Identification Science', *Science*, 309, pp. 892–895.

Schiffer, B & Champod, C 2007, 'The Potential (Negative) Influence of Observational Biases at the Analysis Stage of Fingermark Individualisation', *Forensic Science International*, 167, pp. 116–120.

SWGFAST (Scientific Working Group on Friction Ridge Analysis, Study and Technology) 2012, Standard for the Definition and Measurement of Rates of Errors and Non-Consensus Decisions in Friction Ridge Examination (Latent/Tenprint), Document #15, Ver. 2.0, 11/15/12.

Tangen, J, M, Thompson, M, B & McCarthy, D, J 2011, 'Identifying Fingerprint Expertise, *Psychological Science*, 22 (8), pp. 995–997.

Triplett, M 2021, Fingerprint Dictionary, http://fprints.nwlean.net/.

Ulery, B, T, Hicklin, R, A, Buscaglia, J & Roberts, M, A 2011, 'Accuracy and Reliability of Forensic Latent Fingerprint Decisions', *PNAS*, 108 (18), pp. 7733–7738.

Ulery, B, T, Hicklin, R, A, Buscaglia J & Roberts, M, A 2012, Repeatability and Reproducibility of Decisions by Latent Fingerprint Examiners, *PLoS ONE*, 7 (3), e32800. https://doi.org/10.1371/journal.pone.0032800.

Wertheim, K 2001, *The Weekly Detail*, Monday, August 6, http://www.clpex.com/legacy/TheDetail/1-99/TheDetail01.htm.

Wertheim, K, Langenburg, G & Moenssens, A 2006, 'A Report of Latent Print Examiner Accuracy During Comparison Training Exercises', *Journal of Forensic Identification*, 56 (1), pp. 55–93.

Wilkinson, D, Richard, D & Hockey, D 2018, 'Expert Fingerprint Testimony Post – PCAST – A Canadian Case Study', *Journal of Forensic Identifictaion*, 68 (3), pp. 299–332.

Index